ACS SYMPOSIUM SERIES **580**

Carbohydrate Modifications in Antisense Research

Yogesh S. Sanghvi, EDITOR
Isis Pharmaceuticals

P. Dan Cook, EDITOR
Isis Pharmaceuticals

Developed from a symposium sponsored
by the Division of Carbohydrate Chemistry
at the 207th National Meeting
of the American Chemical Society,
San Diego, California,
March 13–17, 1994

American Chemical Society, Washington, DC 1994

Library of Congress Cataloging-in-Publication Data

Carbohydrate modifications in antisense research / Yogesh S. Sanghvi, editor, P. Dan Cook, editor.

　　p.　　cm.—(ACS symposium series, ISSN 0097–6156; 580)

"Developed from a symposium sponsored by the Division of Carbohydrate Chemistry at the 207th National Meeting of the American Chemical Society, San Diego, California, March 13–17, 1994."

Includes bibliographical references and indexes.

ISBN 0–8412–3056–0

　　1. Antisense　nucleic　acids—Congresses.　2. Oligonucleotides—Derivatives—Congresses. 3. Carbohydrates—Congresses.

　　I. Sanghvi, Yogesh S., 1956–　　. II. Cook, P. Dan, 1944–　　. III. American Chemical Society. Division of Carbohydrate Chemistry. IV. Series.

QP623.5.A58C37　1994
547.7′9—dc20　　　　　　　　　　　　　94–38930
　　　　　　　　　　　　　　　　　　　　CIP

The paper used in this publication meets the minimum requirements of American National Standard for Information Sciences—Permanence of Paper for Printed Library Materials, ANSI Z39.48–1984. ∞

PRINTED IN THE UNITED STATES OF AMERICA

Foreword

THE ACS SYMPOSIUM SERIES was first published in 1974 to provide a mechanism for publishing symposia quickly in book form. The purpose of this series is to publish comprehensive books developed from symposia, which are usually "snapshots in time" of the current research being done on a topic, plus some review material on the topic. For this reason, it is necessary that the papers be published as quickly as possible.

Before a symposium-based book is put under contract, the proposed table of contents is reviewed for appropriateness to the topic and for comprehensiveness of the collection. Some papers are excluded at this point, and others are added to round out the scope of the volume. In addition, a draft of each paper is peer-reviewed prior to final acceptance or rejection. This anonymous review process is supervised by the organizer(s) of the symposium, who become the editor(s) of the book. The authors then revise their papers according to the recommendations of both the reviewers and the editors, prepare camera-ready copy, and submit the final papers to the editors, who check that all necessary revisions have been made.

As a rule, only original research papers and original review papers are included in the volumes. Verbatim reproductions of previously published papers are not accepted.

M. Joan Comstock
Series Editor

Contents

Preface.. **vii**

1. **Carbohydrates: Synthetic Methods and Applications in Antisense Therapeutics: An Overview**.. **1**
   Yogesh S. Sanghvi and P. Dan Cook

DEPHOSPHONO LINKAGES

2. **Novel Backbone Replacements for Oligonucleotides**....................... **24**
   Alain De Mesmaeker, Adrian Waldner, Jacques Lebreton, Valérie Fritsch, and Romain M. Wolf

3. **Synthesis of Nonionic Oligonucleotide Analogues**........................... **40**
   Joseph A. Maddry, Robert C. Reynolds, John A. Montgomery, and John A. Secrist III

4. **Synthesis and Hybridization Properties of DNA Oligomers Containing Sulfide-Linked Dinucleosides**.. **52**
   George Just and Stephen H. Kawai

MODIFICATIONS OF SUGAR MOIETIES

5. **4′-Thio-RNA: A Novel Class of Sugar-Modified β-RNA**................ **68**
   Laurent Bellon, Claudine Leydier, Jean-Louis Barascut, Georges Maury, and Jean-Louis Imbach

6. **Hexopyranosyl-Like Oligonucleotides**... **80**
   Piet Herdewijn, Hans De Winter, Bogdan Doboszewski, Ilse Verheggen, Koen Augustyns, Chris Hendrix, Tula Saison-Behmoaras, Camiel De Ranter, and Arthur Van Aerschot

7. **α-Bicyclo-DNA: Synthesis, Characterization, and Pairing Properties of α-DNA-Analogues with Restricted Conformational Flexibility in the Sugar–Phosphate Backbone**................................. **100**
   M. Bolli, P. Lubini, M. Tarköy, and C. Leumann

8. 2′,5′-Oligoadenylate Antisense Chimeras for Targeted
   Ablation of RNA 118
   Paul F. Torrence, Wei Xiao, Guiying Li, Krystyna Lesiak,
   Shahrzad Khamnei, Avudaiappan Maran, Ratan Maitra,
   Beihua Dong, and Robert H. Silverman

9. Branched Nucleic Acids: Synthesis and Biological
   Applications 133
   R. H. E. Hudson, K. Ganeshan, and M. J. Damha

   MODIFICATIONS OF PHOSPHODIESTER LINKAGE

10. Anti-Human Immunodeficiency Virus Activity and Mechanisms
    of Unmodified and Modified Antisense Oligonucleotides 154
    T. Hatta, S.-G. Kim, S. Suzuki, K. Takaki, and H. Takaku

11. Carboranyl Oligonucleotides for Antisense Technology
    and Boron Neutron Capture Therapy of Cancers 169
    Raymond F. Schinazi, Zbigniew J. Lesnikowski,
    Géraldine Fulcrand-El Kattan, and David W. Wilson

    RNA AND RNA ANALOGUES

12. Advances in Automated Chemical Synthesis
    of Oligoribonucleotides 184
    Nanda D. Sinha and Stephen Fry

13. Anti-Human Immunodeficiency Virus Activity of a Novel
    Class of Thiopurine-Based Oligonucleotides 199
    Rich B. Meyer, Jr., Alexander A. Gall,
    and Vladimir V. Gorn

    STRUCTURAL STUDY

14. New Twists on Nucleic Acids: Structural Properties
    of Modified Nucleosides Incorporated
    into Oligonucleotides 212
    Richard H. Griffey, Elena Lesnik, Susan Freier,
    Yogesh S. Sanghvi, Kelly Teng, Andrew Kawasaki,
    Charles Guinosso, Patrick Wheeler, Venkatraman Mohan,
    and P. Dan Cook

Author Index 225

Affiliation Index 225

Subject Index 226

Preface

THE THERAPEUTIC UTILITY OF OLIGONUCLEOTIDES has gained increasing credibility and attention in the past decade because of in vitro and in vivo efficacy. The number of scientists committed to oligonucleotide drug discovery and development has increased from a handful to perhaps as many as several thousand. The pace of discoveries has been so rapid and global that it has become difficult for scientists intimately involved with antisense technology to remain abreast of recent advances. It is an even bigger challenge to disseminate this information to newcomers in the field of antisense technology. To our surprise, there is very little crossover of information between scientists who apply this technology to design drugs for therapeutics and those who specialize in carbohydrate chemistry. We believe that these scientists will be quite interested in the similarities between the two disciplines. This thought motivated the organization of the symposium on which this volume is based.

This book presents the latest research results on carbohydrate chemistry required to create new oligonucleotides for antisense applications. Much of the synthetic chemistry effort directed toward development of improved antisense agents has focused on modifying the carbohydrate portion of oligonucleotides to achieve stability toward nuclease degeneration and to enhance the binding affinity to a target strand of RNA. Several chapters describe phosphate modification to include sulfur (i.e., phosphorothioates) or a carboranyl moiety, as well as the complete replacement of the phosphate linkage by ethers, sulfides, amides, and amines to provide a nonionic backbone for oligonucleotides. The utility of uniquely branched or 2′,5′-linked oligonucleotides for antisense research has been included. Some drastic sugar ring alterations, such as 4′-thio sugars and hexopyranosyl- and bicyclo-DNA analogues, have been prepared, and their applications are discussed. Contributions on base-modified thiopurine analogues and the influence of sugar modifications on the conformational properties of oligonucleotides are presented. The need for pure synthetic RNA in antisense research is crucial, and this topic is also discussed.

Contributors to this volume are chemists and biologists working in university and industrial settings in many countries toward a common goal. Their research efforts should be valuable to those who conduct interdisciplinary research and to those who wish to enter this field without any prior background. Taken together, these chapters provide an appropriate historical background and summary of several recent developments in the concepts and synthetic strategies of antisense applications using carbohydrate chemistry. The overview chapter provides a more complete picture of the current status and future prospects of this emerging technology.

Acknowledgments

We acknowledge the contributing authors for the time and effort expended in preparing this volume. The credit for an up-to-date account of rapidly moving technology goes to all of the authors. We thank the many expert reviewers who insisted on inclusion of the latest results and provided their comments in a timely manner. We gratefully acknowledge the Division of Carbohydrate Chemistry for its support in organizing the symposium and publication of this volume. We appreciate the financial support provided by Isis Pharmaceuticals, Ciba-Geigy, Gilead Sciences, and Millipore, which allowed us to organize a successful symposium.

YOGESH S. SANGHVI
P. DAN COOK
Isis Pharmaceuticals
2292 Faraday Avenue
Carlsbad, CA 92008

July 19, 1994

Chapter 1

Carbohydrates: Synthetic Methods and Applications in Antisense Therapeutics

An Overview

Yogesh S. Sanghvi and P. Dan Cook

Isis Pharmaceuticals, 2292 Faraday Avenue, Carlsbad, CA 92008

This book presents a review of recent developments and progress in the carbohydrate-related chemistry of novel nucleic acid mimics and their applications in antisense therapeutics. In recent years, significant advances have been made in understanding the essential elements of designing more effective antisense molecules. One of the main focal points has been on the modifications of the carbohydrate portion of an oligonucleotide to achieve stability towards nuclease degradation and to enhance the binding affinity for a given target. The successes and failures of this research in our laboratory and others are discussed. The possible future directions of research in carbohydrate-related modifications of antisense molecules are projected. The focus on potential improvements in stability against cellular nucleases, affinity towards designated targets, and cellular uptake of these novel oligomers remains a high priority.

The antisense concept of drug discovery is based on the inhibition of gene expression at the message level. Thus, RNA is the pharmacological receptor in this discovery approach. Sequence-specific binding of a modified oligomer to an RNA target is obtained primarily by complementary hydrogen bonding of nucleobases of each macromolecule according to Watson-Crick base pairing rules. The binding affinity is obtained predominantly from consecutive stacking of the aromatic nucleobases of the modified oligomer. RNA is structured into secondary and tertiary forms, which often precludes simple complementary binding of an oligonucleotide to a single-stranded region of the RNA. Moreover, simply sequence specific binding to a targeted portion of RNA may not result in an antisense biological effect. Antisense biological activity is most often reported to result from ribonucleic acid H (RNase H) mediated cleavage of the targeted RNA involved in the oligonucleotide-RNA heteroduplex (1), although other mechanism also exist.

Several important questions can be posed when considering the process of making antisense drugs out of candidate compounds. These same questions also can be asked of antisense oligonucleotides. Thus as with all other classes of drugs, these questions include the variability of the pharmacokinetic properties that oligonucleotides may possess and the proof of the pharmacological principle of the drug candidate. In

0097–6156/94/0580–0001$08.36/0

other words, can a useful *in vivo* biological effect result?, will the specificity of drug action be as great as hoped?, what are the toxicity liabilities?, can the drug be scaled up?, and what will the scope of medicinal chemistry be on discovery and development? With respect to antisense oligonucleotides should any major problems arise in any of these areas, we believe medicinal chemistry has the tools to overcome any such problem (2).

Modified oligonucleotides as a class of materials have the most promise in the antisense drug discovery approach. These can be highly modified; indeed, a combination of modifications may be selected to produce superior antisense agents. However, modifications of oligonucleotides to enhance binding affinity, nuclease resistance, and other pharmacokinetic and pharmacodynamic properties typically preclude the resulting modified oligonucleotide-RNA heteroduplex from serving as a substrate for RNase H mediated degradation.

Phosphorothioate oligonucleotides (P=S) have received the most attention as antisense drug candidates. This class of oligonucleotides has one of the non-bonding oxygen atoms in the phosphodiester linkage replaced with a sulfur atom. The negative charge is maintained in the thiophosphate linkage. No other first generation oligonucleotides, such as methylphosphonates, phosphoramidates, or α-oligonucleotides or modified oligonucleotides of recent vintage, have exhibited the level of biological activity as the phosphorothioates. Indeed, the development of phosphorothioates has progressed to the level of human clinical trials. Four phosphorothioates are currently being studied as antiviral and antitumor agents. These are Isis's 2105 and 2922, Lynx's OL-1, and Hybridon's GEM-91 (3). This is a remarkable accomplishment considering that the antisense approach is only sixteen years old. More importantly, the use of second generation antisense oligonucleotides is rapidly growing, as witnessed by presentations at this symposium.

Sites that can be modified in oligonucleotides are limited due to the necessity to maintain Watson-Crick hydrogen bonding and base stacking required for oligonucleotide specificity and binding affinity. To target any sequence, modified oligonucleotides require, at the minimum, four different heteroaromatic structures linked together. A sugar-phosphate moiety provides the linkage between bases in natural nucleic acids. A variety of linkage changes replacing one or more of the four-atom linkers between the C-4´ and C-3´ atoms of adjacent sugars have been described, some in this book. Linkage changes typically reflect such carbohydrate modifications as the C-4´, and C-3´ atoms provide the connection points for the linker between the sugar moieties. Thus, direct modifications in the pentofuranosyl moiety (sugar), modifications in the manner the pentofuranose is connected to the heterocycles and inter-connected to other sugars (C-4´ to C-3´ linkages), and pendant moieties attached to the C-2´-position or other positions of the sugar account for the vast majority of oligonucleotide modifications that enhance their antisense properties (4). These areas of oligonucleotide modifications can be seen in **Figure 1,** which depicts a repeating dimer unit of an oligonucleotide.

The focus of second generation oligonucleotide modifications has centered on the carbohydrate moiety (pentofuranosyl group). The carbohydrate occupies a central connecting manifold that also positions the nucleobases for effective stacking. In fact, except in the case of peptide nucleic acids (PNA) (5), all modifications of oligonucleotides that enhance their pharmacokinetic and pharmacodynamic properties have retained at least in part, the carbohydrate moiety. This phenomenon clearly illustrates the connection of this symposium to its sponsor, the Division of Carbohydrate Chemistry of the American Chemical Society.

Carbohydrate Structural Modifications

In recent years, several comprehensive reviews on the synthesis and utility of modified oligonucleotides for antisense therapeutics have been published (6). Therefore, the content of this overview is restricted to publications concerning carbohydrate-based modifications in antisense research from our group and other investigators, which appeared in 1993 and the first half of 1994. This account does not include developments

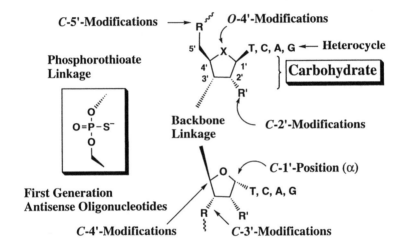

Figure 1: Types of Oligonucleotide Modifications

in modifications of the phosphorus atom of the phosphodiester linkage, such as phosphorothioates (7), phosphorodithioates (8), methylphosphonates (9), phosphoramidates (10), boranophosphates (11), phosphonofluoridates (12), and phosphate esters (13), which are adequately covered elsewhere. In each section, we have described the recent advances made in modifying the carbohydrate moiety (**Figure 1**) to enhance the nuclease stability and target affinity (hybridization) as well as other pharmacokinetic and pharmacodynamic properties of oligonucleotide analogs.

Dephosphono Linkages (Linkages without the Phosphorus Atom). In the last five years, we have witnessed a steady rise in the number of publications and review articles (14) related to the methods of synthesis and use of nonionic, achiral linkages that replaced the natural phosphate backbone (**Table I**). We believe that data concerning these phosphate backbone surrogates, as discussed in this section, begin to allow an understanding of the minimal structural features required to mimic a naturally occurring linkage. Replacement of the phosphate backbone of an oligonucleotide has several distinct advantages in terms of its antisense properties: it confers the desired stability towards cellular nucleases; modified oligomers may have increased cellular uptake due to its neutral characteristics; the chirality imposed by the phosphorothioate will be removed; major advantages may be realized in the economics of large scale synthesis in solution.

The synthetic efforts required to create these linkages is often challenging, but with recent advances in synthetic methodologies it will become more routine. The backbone modified dimers listed in **Table I** are readily converted to their 5'-O-dimethoxytrityl and 3'-O-phosphoramidite or H-phosphonate derivatives following standard protocols. These activated/protected dimers are then incorporated into oligonucleotides *via* automated DNA synthesis. The foregoing dimer incorporation strategy has been very successful in producing small quantities of oligomers containing a variety of linkages, thus allowing rapid chemical, biophysical, and biological evaluation of antisense oligomers with alternating phosphodiesters and the novel linkages. Several groups have taken a major step forward by preparing fully substituted oligonucleotides with a variety of linkages (see **Table I** for details).

In 1993, De Mesmaeker et al. (15) and Just et al. (16) independently reported a nonionic, achiral amide linkage (amide-3, **Table I**) that served as an elegant replacement for the phosphate linkage in oligonucleotides. Amide-3 (**3**) modification appears to be one of the best linkages among the family of amide-1-5 (**1-8**) (17-20) due to its favorable antisense properties. A full account of the synthesis and properties of each of these amides have been covered in this volume of the series by De Mesmaeker et al. (**Chapter 2**). In addition, De Mesmaeker et al. have described urea (**9**) (21) and carbamate (**11**) (22) linkages in detail. Herdwijn et al. have reported on the synthesis of a thiourea **10** (23) linkage in a similar manner. Recognizing the usefulness of carbamate-2 (**12**) linkage, Kutterer and Just (21) have synthesized dimeric **12**, whereas, Agrawal et al. have synthesized di-, tri-, and tetrameric oligonucleosides and successfully incorporated these into antisense oligonucleotides (24). These chimeric oligomers were found to enhance the nuclease resistance with only slight destabilization of the duplex.

Nitrogen-containing modifications, amino-1-5 (**13-17**), have been synthesized and studied for enhanced cellular absorption (25, 26). It was believed that these linkages **12-16** would be partially protonated at physiological pH and would assist in cellular uptake. The electrostatic and hydration factors between the protonated nitrogen atom of the amino linkers and the anionic phosphate backbone of the complementary target strand are expected to further increase the thermal stability of these duplexes. A method for the synthesis of a fully modified trimer (T*T*T; * = 3'-NH-CH$_2$-CH$_2$-CH$_2$-4') is reported by Saha et al. (25).

Peterson et al. reported an amino-4 (**16**) linkage along with an interesting cyclic amino-5 (**17**) linkage; both were 5-atom linkers representing a stretched backbone (27). Bergess et al. investigated a short 3-atom oxyamide linkage (**18**) as a backbone surrogate (28). A thymidine dimer **17** was synthesized containing the oxyamide linkage and was shown to hybridize to complementary DNA with slightly inferior affinity when compared to the unmodified sequence. The oxyamide chemistry is of particular interest

Table I. Nucleoside Dimers Containing Dephosphono Linkages

No.	Name	B_1^1	B_2^1	X_1	X_2	L_1-L_2-L_3-L_4	T_m^2	ND^3	Ref.
1	Amide-1	T	T	H	H	NR-CO-CH$_2$-CH$_2$	-	+	17, C[2]
2	Amide-2	T	T	H	H	CH$_2$-CH$_2$-NH-CO	-	+	18, C[2]
3	Amide-3	T	T,C	H	H	CH$_2$-CO-NH-CH$_2$	+	+	15, 16
4	Amide-3	C	T	H	H	CH$_2$-CO-NH-CH$_2$	+	+	15, C[2]
5	Amide-3	A	G	H	H	CH$_2$-CO-NH-CH$_2$	+	+	15, C[2]
6	Amide-3	G	T	H	H	CH$_2$-CO-NH-CH$_2$	+	+	15, C[2]
7	Amide-4	T	T	H	H	CH$_2$-NH-CO-CH$_2$	+	+	19, C[2]
8	Amide-5	T	T	H	H	CO-NH-CH$_2$-CH$_2$	-	+	20, C[2]
9	Urea	T	T	H	H	NR-CO-NR-CH$_2$	-	+	21, C[2]
10	Thiourea	T	T	H	H	NH-CS-NH-CH$_2$	NT	NT	23
11	Carbamate-1	T	T	H	H	O-CO-NR-CH$_2$	-	-	22
12	Carbamate-2[4]	T	T	H	H	NR-CO-O-CH$_2$	-	+	24, 21
13	Amino-1[5]	T	T	H	H	NH-CH$_2$-CH$_2$-CH$_2$	-	+	25
14	Amino-2	T	T	H	H	CH$_2$-NR-CH$_2$-CH$_2$	-	+	26
15	Amino-3	T	T	H	H	CH$_2$-CH$_2$-NR-CH$_2$	-	+	26
16	Amino-4	T	T	H	H	O-CH$_2$-CH$_2$-NH-CH$_2$	-	+	27
17	Amino-5	T	T	H	H	N⌣N-CH$_2$	NT	NT	27
18	Oxyamide	T	T	H	H	O-NH-CO	-	+	28
19	Oxime	T	T	H	H	CH=N-O-CH$_2$	- -	+	29
20	Methyleneimino	T	T	H	H	CH$_2$-NH-O-CH$_2$	-	+	29
21	MMI-1[6]	T	T,C,G	H	H	CH$_2$-NMe-O-CH$_2$	+	+	29, 33
22	MMI-2	C*	C*,T	H	H	CH$_2$-N(CH$_3$)-O-CH$_2$	+ +	+	29, 33
23	MMI-3	A	C	H	H	CH$_2$-N(CH$_3$)-O-CH$_2$	+	+	29, 33
24	MMI-4	G	C	H	H	CH$_2$-N(CH$_3$)-O-CH$_2$	+	+	29, 33
25	MDH	T	T	H	H	CH$_2$-NMe-NMe-CH$_2$	+	+ +	30, C[14]
26	MOMI	T	T	H	H	CH$_2$-O-N(CH$_3$)-CH$_2$	-	+	29

Continued on next page

Table I. Continued

27	HMIM	T	T	H	H	O-N(CH$_3$)-CH$_2$-CH$_2$	NT	NT	31
28	Ethylene Glycol	T	T	H	H	O-CH$_2$-CH$_2$-O	-	+	35, C^{14}
29	Propoxy-1	T	T	H	H	CH$_2$-CH$_2$-CH$_2$-O	-	NT	35, C^{14}
30	Propoxy-2	T	T	H	H	O-CH$_2$-CH$_2$-CH$_2$	-	+	36, 37
31	Thioether	T	T	H	H	S-CH$_2$-CH$_2$-CH$_2$	+	+	36
32	All Carbon-1	T	T	H	H	CH$_2$-CH$_2$-CH$_2$-CH$_2$	-	+	38, 39
33	All Carbon-2	T	T	H	H	CH$_2$-CO-CH$_2$-CH$_2$	-	+	39, C^2
34	All Carbon-3	A	A	OH	OH	CH$_2$-CO-CH=CH	-	+	40
35	Formacetal-1[7]	T	T,C*	H	H	O-CH$_2$-O-CH$_2$	-	+	41
36	Formacetal-2	All	All	H	H	O-CH$_2$-O-CH$_2$	-	+	41, 44
37	Formacetal-3	PU	PU	H	H	O-CH$_2$-O-CH$_2$	+ +	+	42
38	Thioformacetal-1	T	T	H	H	S-CH$_2$-O-CH$_2$	+	+	41
39	Thioformacetal-2	PU	PU	H	H	S-CH$_2$-O-CH$_2$	+ +	+	42
40	r-Thioformacetal	T	T	OH	H	S-CH$_2$-O-CH$_2$	-	+	43
41	N-Guanidine-1	T	T	H	H	NH-C(=N-CN)-NH-CH$_2$	-	+	45
42	N-Guanidine-2	T	C,A,G	H	H	NH-C(=N-CN)-NH-CH$_2$	-	+	46
43	N-Guanidine-3	T	T	H	H	NH-C(=N-R)-CH$_2$	+	+	23
44	Sulfite	A	A	OH	OH	O-SO-O-CH$_2$	-	+	47
45	Sulfonamide	T	T	H	H	NH-SO$_2$-CH$_2$-CH$_2$	-	+	48
46	Sulfamoyl	T	T	H	H	O-SO$_2$-NR-CH$_2$	-	NT	49
47	Sulfonate[8]	T	T	H	H	O-SO$_2$-CH$_2$-CH$_2$	-	+	50
48	Sulfide-1	U	U	H	H	CH$_2$-S-CH$_2$-CH$_2$	+	+	51
49	Sulfide-2	A	A	H	H	CH$_2$-S-CH$_2$-CH$_2$	+	+	51
50	Sulfide-3	T	T	OH	H	CH$_2$-S-CH$_2$-CH$_2$	+	+	51
51	Sulfide-4	T	T	H	H	CH$_2$-S-CH$_2$-CH$_2$	+	+	52
52	Sulfide-5	T	T	R*	R*	CH$_2$-CH$_2$-S-CH$_2$	+	+	53, C^4
53	Sulfide-6	T	T	R*	H	CH$_2$-CH$_2$-S-CH$_2$	+	+	53, C^4
54	Sulfone	T	T	R*	R*	CH$_2$-CH$_2$-SO$_2$-CH$_2$	+	+	53, C^4
55	Carboxyl	T	T	H	H	O-CO-CH$_2$-CH$_2$	-	NT	C^3
56	Carboxyamide	T	T	H	H	O-CH$_2$-CH$_2$-NH-CO	+	NT	54
57	Gly-amide	T	T	H	H	NH-CO-CH$_2$-NH-CO	+	NT	14d
58	Silyl	All	All	H	H	O-Si-(iPr)$_2$-O-CH$_2$	+	+	55, C^3

[1] T = Thymine; C = Cytosine; A = Adenine; G = Guanine; C* = 5-methylcytosine; U = Uracil; PU = 5-propyneuracil; R* = H, OH, OCH$_3$; All = T, C, A, G; [2,3] T_m = melting temperature, ND = nuclease data, + = good, ++ = very good, - = bad, - - - = worst; NT = Not tested; [4] 3-mer and 4-mer have been made (Ref. 24); [5] 3-mer has been made (Ref. 25); [6] (T*T)$_n$, n=2,4,6 have been made (Ref. 33); [7] 3-mer has been made (Ref. 41); [8] 8-mer has been made (Ref. 50) R' = SO$_2$CH$_3$, C^2, C^3, C^4, C^{14} = Described in the corresponding Chapters.

because it could be performed on a solid-phase peptide synthesizer, generating oxyamide-linked oligonucleotides.

Substantial progress has been made by the Isis group in utilizing nitrogen-containing backbones. These include oxime **19** (*29*), methyleneimino **20** (*29*), methylene (methylimino) **21-24** (MMI-1-4) (*29*), methylene (dimethylhydrazo) **25** (MDH) (*30*), methyleneoxy (methylimino) **26** (MOMI) (*29*), and hydroxy (methyliminomethylene), **27** (HMIM) (*31*) linkages. We have synthesized a variety of dimers (**14, 15, 19, 29**) and explored their structure-activity relationship in oligonucleotides. The effects on biochemical and biophysical properties of oligomers containing these modifications have been studied. We believe that MMI and MDH linkages are two of the more promising candidates for incorporation into antisense oligonucleotides. Our preliminary results with nuclease studies indicated that oligomers containing an alternating phosphodiester and MMI or MDH linkage were significantly stable towards degradation by endo- and exonucleases (*32*).

In order to construct antisense oligonucleotides with alternating phosphodiester and MMI linkages, a set of sixteen dimers from eight nucleoside building-blocks must be synthesized. In order to prepare mixed sequences of interest, we have prepared eight derivatives of 2´-deoxynucleosides (T, C, A, G) with an appropriate 5´-*O*-amino or 3´-*C*-formyl group. Utilizing these monomers, we have synthesized a total of six dimers **21-24** containing an MMI linkage and mixed nucleosidic bases (*33*). Incorporation of these mixed-base dimers (**21-24**) into antisense oligonucleotides is in progress. In addition, we began to explore the possibility of preparing fully substituted oligomers containing MMI linkages. Our initial attempts have resulted in the successful synthesis of $(T*T)_n$ (n = 2, 4, 6; * = MMI linkage) homopolymers (*34*). The transfer the MMI chemistry to solid-support, automated synthesis to produce antisense oligonucleotides containing MMI linkages is in progress.

Using a Vorbrüggen-type glycosylation reaction, Teng and Cook have prepared two novel thymidine dimers bridged *via* ethylene glycol **28** and propoxy **29** linkers (*35*). The results of incorporating dimers **28** and **29** into antisense oligonucleotides are summarized by Griffey *et al.* in **Chapter 14** of this series. Cao *et al.* prepared a series of oligomers containing 3´-allyl ether, 3´-allylsulfide, and their reduced derivatives **30** and **31**, respectively (*36*). These modifications when incorporated into antisense oligonucleotides exhibited a modest destabilization of the duplex compared to the unmodified oligomer. Caulfield *et al.* also reported similar results with propoxy-2 (**30**) modification (*37*).

An all carbon linkage **32** was first reported by Butterfield *et al.* (*38*). An improved synthesis of dimers **32** and **33** has been reported by Lebreton *et al.* (*39*). A detailed account of the T_m results on modifications **32** and **33** is presented in **Chapter 2** of this book. Lee and Wiemer have reported the synthesis of an adenosine-adenosine dimer linked by an enone-type (**34**) bridge (*40*).

Matteucci and co-workers have prepared a series of formacetal (**35-37**) and thioformacetal (**38, 39**) bridged dimers (*6, 41-43*). An alternative route to the formacetal dimer **36** and its trimer have been reported by Quaetflieg *et al.* (*44*). The dimer **37** bears two modifications, one in the heterocyclic base and the other in the backbone. Similarly, Cao and Matteuci (*43*) reported a ribothioformacetal (**40**) linked dimer expecting to improve the affinity. However, this modification (**40**) binds to complementary RNA with reduced affinity as compared to the unmodified oligomer. Griffey *et al.* in **Chapter 14** of this book have suggested that increased affinity of an antisense oligomer for a complementary RNA target can be achieved by decreasing the entropic motion of the sugar while maintaining a preorganized structure with an RNA-like conformation.

Herdewijn and co-workers have synthesized oligothymidylates containing thymidine dimers with a variety of *N*-substituted guanidine linkages (**41-43**). Hybridization studies revealed that the *N*-methylsulfonyl **43** (*23*) substituted guanidine linkage best mimics the natural phosphodiester bridge. They also indicated that these analogs with an appropriate *N*-substituent remain neutral (pKa ~1) at physiological p*H*. Herdewijn *et al.* have also reported the synthesis and properties of *N*-cyano-substituted guanidine-linked mixed-base dimers **42** (*46*). Oligonucleotides composed of an alternating phosphodiester and **43** linkage exhibited a ΔT_m of -2.4° C/modification.

Synthesis of sulfur-containing backbone modifications, such as **45-54,** has been accomplished by several research groups (*47-52*). Wang *et al.* have reported the synthesis of diadenylyl sulfite **45** as an analog of ribonucleoside (*47*). A sulfonamide-linked thymidine dimer **45** has been prepared and incorporated into oligonucleotides by Widlanski and co-workers (*48*). Their studies with **45** indicated that sulfonamide-linked oligomers do not form stable duplexes with complementary DNA. Binding affinity studies of the corresponding RNA:DNA hybrid are in progress. Maddry *et al.* have also described the synthesis of **45** in **Chapter 3** of this volume. An efficient synthesis of a thymidine dimer containing a sulfamoyl bridge **46** has been reported by Dewynter and Montero (*49*). Recently, Widlanski *et al.* have developed a simplified and straightforward synthesis of an octomer-containing sulfonate **48** (*50*). This was accomplished with greater than 90% coupling efficiency. In addition to the thymidine dimer **48,** they have also prepared C*T, G*T and A*T mixed-base dimers. Oligomers containing a single sulfonate linkage exhibited a ΔT_m of -1.0-3.0° C, depending on the position of the modification.

Benner *et al.* have actively pursued the synthesis of oligonucleotide analogs containing dimethylene sulfide-1-3 (**48-50**) linkages. In a short communication (*51*), they described the methodology used to prepare a 2´-deoxyuridine dimer bridged *via* a sulfide linkage. Just and his group have been involved with the synthesis and studies of sulfide-linked oligonucleotide analogs (sulfide-5,6, **52-54**) for a number of years (*52, 53*). Their efforts have produced interesting backbone chemistry described in **Chapter 4** of this series. They have also described the synthesis and properties of sulfone-linked (**54**) oligomers in the same chapter.

Maddry *et al.* have briefly discussed the synthesis of a thymidine dimer containing a carboxyl (**55**) bridge in **Chapter 3**. Even though a four atom backbone modification appears to be the choice of the majority of researchers, some have prepared five atom linkers such as carboxamide **56** (*54*). The T_m studies of oligomers containing carboxamide linkage **56** concluded that the modifications destabilized the duplex by 2-4° C depending on the extent of incorporations. Varma *et al.* have described a 5-atom glycine-amide linker **57**, based on peptide chemistry (*14d*). They have assembled a fully modified soluble polymer of thymidine containing several glycine-amide linkages and stated that it forms a stable duplex with its complement. A full report on this work is not published as yet.

Saha *et al.* have developed an efficient synthetic method for solution and automated solid-phase synthesis of dialkylsilyl-linked (**58**) oligonucleotide analogs (*55*). Their solid-phase automated synthesis appears to be versatile and could be utilized for the preparation of uniformly dialkylsilyl-linked DNA analogs. This method is further elaborated in **Chapter 3** by Maddry *et al.* These oligomers demonstrated good stability towards cleavage by 3´-exonucleases but considerably reduced binding affinity (ΔT_m ~ -2.0-3.0° C/modification) with complementary DNA.

In summary, considerable advances in the synthesis and applications of dephosphono backbones have been reported in the past two years. Backbone modifications alone or in combination with base or sugar modifications have been reported and a number of interesting analogs have been identified. One of the key features of this strategy is that the monomers are very stable, unlike standard phosphoramidites, and polymerization using solid-support methodologies in an efficient manner is reasonable. It is clear that a systematic evaluation of these modifications would provide a valuable insight into the design of superior backbone linkages for antisense oliognucleotides.

C-1´-**Modifications** (**Figure 2**). Among the C-1´-modifications, the α-oligonucleotide is an important and well studied class of oligonucleotides (*56*). These oligomers are prepared from *C*-1´-α- 2´-deoxy or -ribonucleosides and have been shown to be resistant to nuclease degradation. An extensive study of α-oligomers has been published elsewhere (*56*), and it is beyond the scope of this overview to summarize it here.

Dan *et al.* have reported the synthesis of a 2′-deoxyuridine analog bearing an aminoalkyl tether at the *C*-1′-position of the sugar moiety (**2**) (*57*). This building block was incorporated into oligonucleotides, and an intercalating anthraquinone group was attached to its amino functionality. The oligomers were shown to form more stable duplexes (ΔT$_m$ ~4° C/modification) with their complementary oligomers compared to unmodified oligomers. The intercalator is believed to be positioned in the minor groove of the duplexes. In a similar manner, Azhayev *et al.* prepared two building blocks (**3**) armed with dissimilar protecting groups and utilized it efficiently for the synthesis of branched oligonucleotides (*58*). This topic has been discussed in more detail by Damha *et al.* (**Chapter 9**).

The incorporation of abasic sites or their generation *via* a selectively cleavable group in an oligonucleotide is of considerable interest. Thus, phosphoramidites **4** have been prepared and utilized for monitoring acid-catalyzed depurination resulting in apurinic sites during oligonucleotide synthesis (*59*). In addition, Laayoun *et al.* have recently described a novel approach of preparing abasic oligomers (*60*). They have synthesized a modified 2′-deoxyadenosine analog **5** bearing a *C*-8-propylthio-functionality. Incorporation of **5** into oligonucleotides was accomplished by automated synthesis. Selective oxidation of the thiol function in **5** created a labile glycosidic link, which on hydrolytic treatment furnished an abasic site in the oligomer.

C-2′-Modifications (RNA Mimics). A growing number of oligonucleotides in which the *C*-2′-position of the sugar ring is modified have been reported (**Table II**). These modifications include lipophilic alkyl groups, intercalators, amphipathic amino-alkyl tethers, positively charged polyamines, highly electronegative fluoro or fluoro alkyl moities, and sterically bulky methylthio derivatives. The beneficial effects of a *C*-2′-substitution on the antisense oligonucleotide cellular uptake, nuclease resistance, and binding affinity have been well documented in the literature (*2*). In addition, excellent review articles have appeared in the last two years on the synthesis and properties of *C*-2′-modified oligonucleotides (*61*). Therefore, we will discuss only those modifications that are of significant interest and appeared in the current literature (1993-94).

The following is a summary of important findings from our recent studies on *C*-2′-substituted oligonucleotides with a wide range of functional groups, such as aliphatic (**1-7**) (*62*), aromatic (**8**) (*62*), fluoro (**9**) (*63*), and thiols (**18, 19**) (*64,65*). We have established a clear correlation between the *C*-2′-substituent size (*i.e.*, steric bulk) and the stability of a DNA:RNA duplex. Among the 2′-*O*-alkyl series, a short substituent (no larger than propyl) stabilizes the duplex, whereas a longer substituent (above 4-carbon chain) destabilizes the duplex. We reasoned that a longer alkyl-chain may not be accommodated easily in the limited space of the minor groove of a hybrid duplex. The stabilizing effects of the 2′-fluoro-modification (**9**) on RNA:DNA duplexes were shown to be superior to those of the 2′-*O*-alkyl substituents (*63*). We believe that the high electronegativity of the fluorine atom in combination with the gauche effect (see **Chapter 14** for details) causes duplex stabilization. However, 2′-fluoro-arabino-modification (**10**) also stabilizes the duplex formation, and the reason for duplex stability is not understood well (*66*).

Miller *et al.* have reported inferior duplex stability utilizing a *C*-2′-amino-modification **11** compared to an unmodified oligomer (*67*). Oligonucleotides carrying lipophilic *N*-octyl groups **12**, intercalating 2-*N*-anthraquinonyl groups **13**, and protonated *N*-dimethylamino groups **14** have been prepared (*68*). Modifications with **12** had no effects on the duplex stability, whereas modifications with **13** substantially increased the duplex stability (ΔT$_m$ ~5° C/modification). An oligomer tethered with **14** increased the T$_m$ by ~1.5° C per modification. An elaborate synthesis of *C*-2′-carbon substituted **15-16** oligonucleotide building blocks has been reported by Schmit (*69*). The oligomerization and stability studies on these compounds were not reported. Giannaris and Damha have reported the synthesis and hybridization properties of oligoarabinonucleotides **17** (*70*). Their studies indicated that the inversion of

Table II.　C-2´-Modifications

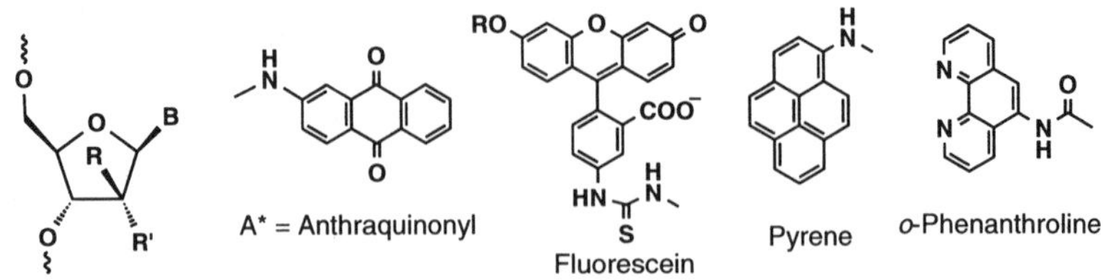

A* = Anthraquinonyl　　　Fluorescein　　　Pyrene　　　o-Phenanthroline

No.	Name	B[1]	R	R´[2]	T_m[3]	ND[4]	Ref.
1	2´-O-Methyl	All	H	OCH$_3$	+	+	61, C[14]
2	2´-O-Ethyl	A	H	OCH$_2$CH$_3$	+	+	62
3	2´-O-Propyl	All	H	O(CH$_2$)$_2$CH$_3$	++	++	62
4	2´-O-Butyl	A	H	O(CH$_2$)$_3$CH$_3$	-	+	62
5	2´-O-Pentyl	A	H	O(CH$_2$)$_4$CH$_3$	-	+	62
6	2´-O-Nonyl	A	H	O(CH$_2$)$_8$CH$_3$	- -	+	62
7	2´-O-Allyl	All	H	OCH$_2$CH=CH$_2$	+	+	62
8	2´-O-Benzyl	A	H	OCH$_2$Ph	- -	-	62
9	2´-Fluoro	All	H	F	++	- -	63, C[14]
10	2´-Fluoro-arabino	T	F	H	+	+	66
11	2´-Amino	U	H	NH$_2$	- -	NT	67
12	2´-N-Octyl	T, A, C*	H	X-(CH$_2$)$_7$CH$_3$	-	NT	68
13	2´-N-Anthraquinonyl	T	H	X-(CH$_2$)$_2$-A*	++	NT	68
14	2´-N-Dimethylamino	T, A	H	X-(CH$_2$)$_6$N(CH$_3$)$_2$	+	+	68
15	2´-Fluoromethyl	T	H	CH$_2$F, CF$_3$	NT	NT	69
16	2´-C-Substituted	T, C	H	CH$_3$, Ph, CH$_2$OH, CH$_2$NH$_2$	NT	NT	69
17	2´-Arabino	U, C, A	OH	H	+	-	70
18	2´-Methylthio	All	H	SCH$_3$	-	-	64, C[14]
19	2´-S-Tether	A	H	O(CH$_2$)$_6$S-T*	-	-	65

[1] All = T, U, C, A, G; C* = 5-MeC; [2] X = OCH$_2$CONH; A*, T* = see structures above;
[3,4] T_m = Melting temperature; + = good, ++ = very good, - = bad, - - = worst, NT = Not tested;
[4] ND = Nuclease Data; C[14] = Described in Chapter 14.

stereochemistry at the *C-2'* position of ribonucleotides had no effect on its hybridization properties. Manoharan *et al.* have introduced a lipophilic thioether-tether **18** attached to the *C-2'*-hydroxyl group for antisense applications (*65*). The thioether **18** provided a convenient handle to place a conjugate (*e.g.*, *O*-phenanthroline, fluorescein, pyrene, etc.) in the minor groove of the nucleic acid structures. We have synthesized and studied the conformational properties of 2'-methylthio-nucleosides **19** for their incorporation into antisense oligonucleotides (*64*). We have discussed the properties of this modification in greater depth in **Chapter 14**.

An oligonucleotide analog containing both sugar-- and base-modified residues has been prepared as an antiviral agent by Meyer and his colleagues. These oligonucleotides contain 1-methyl-6-thiopurine heterocyclic bases and 2'-*O*-methyl sugar modifications assembled *via* normal phosphodiester linkages. Meyer *et al.* have described the synthesis and broad spectrum antiviral activity of these oligonucleotides in **Chapter 13**. Of interest, the phosphodiester oligonucleotides described by Meyer *et al.* are among the most potent phosphodiester oligonucleotides for which cell culture activity has been reported.

Synthesis of pure oligoribonucleotides (RNAs) is essential for the hybridization and duplex structural studies of antisense oligonucleotides. In addition, larger quantities of short synthetic RNAs are required to conduct NMR studies of modified DNA:RNA duplexes. In this regard, we have included a chapter by Sinha and Fry describing the recent advances in automated synthesis of RNAs in large quantities. The methodology described in **Chapter 12** of this volume should enable production of not only RNA but also analogs of RNA on a large scale.

C-2'-5'-**Modifications** (**Analogs of 2-5 A**) (**Figure 3**). Torrence *et al.* have developed a novel strategy for cleaving unique RNA sequences with 2-5 A-dependent RNase (*71*). This RNase is a latent endonuclease that requires the unusual 2'-5'-phosphodiester-linked trimeric oligonucleotide ppp5'A2'p5'A2'p5'A for its activation and cleavage of RNA. In order to activate 2-5 A-dependent RNase to cleave RNA sequences, the trimeric 2'-5 oligomer was covalently linked to an antisense oligonucleotide providing a chimeric molecule (2-5A:oligomer). The synthesis of 2-5 A chimeras and their applications in antisense therapeutics has been described by Torrence *et al.* in **Chapter 8** of this issue. Giannaris and Damha (*72*) have synthesized oligoribonucleotides containing "all" 2'-5'-phosphodiester linkages as well as chimeric 2'-5'/3'-5' backbones (**Figure 3**). The 2'-5'-linked oligomers prepared in this study exhibited unusually high selectivity for complementary single stranded RNA over DNA (T_m 54 vs -1° C, respectively).

A further application of the 2'-5' chemistry has been extended to incorporate dephosphono linkages (**Table I**) into oligomers. Matteucci *et al.* have studied the binding properties of 2'-5' thioformacetal (**38** in **Table I**) and formacetal (**35** in **Table I**) linked oligomers with DNA and RNA complements (*73*). Their results indicate no change in the binding affinity of 2'-5'-modified oligomers compared to unmodified oligomers. In brief, dephosphono oligomers linked *via* the 2'-5' direction should have similar properties as reported for the 3'-5'-linked dephosphono oligomers. However, the synthetic efforts in preparing the 2'-5'-linked oligomers will be more demanding compared to the 3'-5'-linked oligomers.

C-3'-**Modifications** (**Figure 4**). The synthesis of oligonucleotides containing *C-3'*-modifications such as replacement of the bridging *C-3'*-oxygen atom of the phosphodiester linkage with a methylene group [3'-CH_2-P(O)_2-5'], nitrogen functionality [3'-NR-P(O)_2-5'], or sulfur atom [3'-S-P(O)_2-5'] has received considerable attention. These and other phosphate analogs have recently been thoroughly reviewed (*74*). A summary of the more interesting *C-3'*-modifications is incorporated here. A conformationally restricted nucleoside analog **1** (*C-3'*-6-ethano-bridged uridine) has been synthesized and incorporated into different oligonucleotides (*75*). The insertion of **1** in an oligonucleotide was anticipated to preorganize the sugar pucker to a 3'-endo conformation, characteristic of an A-type oligonucleotide. It has been demonstrated that

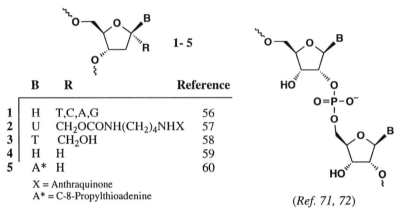

	B	R	Reference
1	H	T,C,A,G	56
2	U	$CH_2OCONH(CH_2)_4NHX$	57
3	T	CH_2OH	58
4	H	H	59
5	A*	H	60

X = Anthraquinone
A* = C-8-Propylthioadenine

Figure 2: *C*-1'-Modifications

Figure 3: *C*-2'-5'-Modifications

(*Ref. 71, 72*)

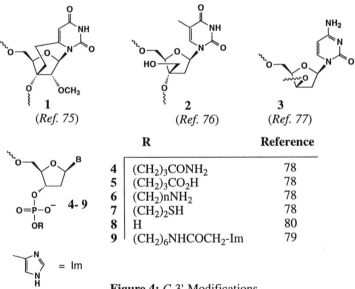

1	**2**	**3**
(*Ref. 75*)	(*Ref. 76*)	(*Ref. 77*)

	R	Reference
4	$(CH_2)_3CONH_2$	78
5	$(CH_2)_3CO_2H$	78
6	$(CH_2)nNH_2$	78
7	$(CH_2)_2SH$	78
8	H	80
9	$(CH_2)_6NHCOCH_2$-Im	79

= Im

Figure 4: *C*-3'-Modifications

an A-type oligonucleotide conformation enhances the binding affinity to RNA complement due to entropic advantages (Griffey *et al.* in **Chapter 14**). Contrary to the anticipation, these oligonucleotides displayed a lower affinity (ΔT_m ~4° C/modification) towards the complementary RNA strand. This work indicates that too much rigidity in an oligonucleotide is not desirable for enhancing the affinity.

Interestingly, Jorgensen *et al.* have described the synthesis of *C*-3′-xylo nucleoside analog **2**, which provides the desired 3′-endo conformation with reduced rigidity compared to bridged analog (*76*). Incorporation of nucleoside **2** into oligonucleotides causes no change in the affinity when placed in the middle of the sequence and hybridized to complement DNA. The 3′-end-capped sequences containing this modification **2** exhibit enhanced stability towards snake venom phosphodiester (3′-exonuclease). Given the affinity and stability of **2** to exonuclease, this *C*-3′-modification should be investigated further as antisense oligonucleotides.

Seela *et al.* recently reported (*77*) on the synthesis and properties of oligonucleotides containing 2′-deoxy-β-*D*-xylocytidine **3** as a building block. This modification did not exhibit favorable hybridization properties for antisense applications. Several studies of the synthesis of oligonucleotides tethered with an amino-, carboxy-, thio-, or phosphate function have been reported for their use in preparation of labeled oligomers. Improved procedures for the synthesis of 3′-functionalized oligonucleotides on solid-support have recently been published by Hovinen *et al.* (*78*). The *C*-3′-modifications of oligonucleotides such as **3** may serve a use in post conjugation chemistry of oligonucleotides required for diagnostic and therapeutic applications. An account of the recent developments in the *C*-3′-conjugation chemistry of oligonucleotides has been covered by Manoharan in an antisense book (1a).

C-5′-Modifications (Figure 5). The introduction of a moiety at the *C*-5′ site of an oligonucleotide in order to modulate its antisense properties has been reported. A survey of the *C*-5′-conjugation literature has recently been published (*82*). We have summarized only a selected number of papers that appeared in the current literature (1993-94).

Efficient methods for the introduction of a pyrene group at the *C*-5′-position of oligonucleotides as a fluorescent tag have been reported (*83*). Kierzek *et al.* demonstrated that *C*-5′-pyrene RNA conjugates are stable and very sensitive to their environment while having minimum perturbations on the thermodynamics of secondary and tertiary structure formation. In particular, a short linker such as **1** (**Figure 5**) appears to provide an ideal probe for binding and kinetic studies at nanomolar concentrations.

Recently, Yamana *et al.* introduced a shorter *C*-5′-linked pyrene **2** attached to a DNA (*84*). The fluorescence of these oligonucleotides was not quenched on hybridization to complementary DNA. It has been documented that a triplex can be further stabilized by addition of low levels (mM) of spermine to the complex. Therefore, Tung *et al.* (*85*) covalently linked spermine and other polyamines (**3**) to the *C*-5′-terminus of oligonucleotides in order to study the possibility of a third stand conjugate with superior triplex-forming properties. They found that appended polyamine enhanced the T_m of the resulting triplex. Potentially, conjugation of a polycationic group to an oligonucleotide may also improve its nuclease resistance and cellular uptake.

A convenient synthesis of a base-labile protected derivative of 6-mercaptohexanol **4** for the preparation of oligonucleotides containing a thiol group at the *C*-5′-end is reported by Torre *et al.* (*86*). The methodology should facilitate the labeling of oligonucleotides with reporter groups, such as biotin, fluorescent groups, and enzymes. Ramage and Wahl reported the utility of 4-(17-tetrabenzo[a, c, g, i]fluorenylmethyl)-4′,4"-dimethoxytrityl chloride (tbf-DMT group **5**) as a new hydrophobic protecting group on the *C*-5′-end of long oligonucleotides and demonstrated its use as a purification handle during HPLC separations (*87*).

The methodologies for incorporation of reporter molecules at the *C*-5′-end of antisense oligonucleotides with non-nucleosidic phosphoroamidite units have recently been investigated. Among these, phosphoramidite units were prepared from ethane-1,2-diol and propane-1,3-diol backbones (**6**). Fontanel *et al.* have reported the stereoselective phosphorylation (~90%) of the L-form of the C-5′-end attached non-nucleosidic chiral moieties by T4 polynucleotide kinase (PNK) (*88*).

Branched and Dendriatic Oligo -RNA and -DNA. The processing of heterogeneous nuclear RNA (hnRNA) in eukaryotic genes involves the accurate removal of introns and ligation of exons *via* two sequential transesterification reactions, a process known as 'splicing'. The splicing of a transcript may occur in two ways. First, in *cis*-splicing the introns are transformed into lariats, a single-stranded circular RNA with a 'tail' originating from a branch point. Second, in *trans*-splicing the introns are transformed into 'Y'-shaped or forked RNA structures. Unlike normal RNA, these so-called branched RNAs contain a branch-point adenosine, which bears vicinal 2′-5′ and 3′-5′ phosphodiester linkages. Because of the unique and precise splicing process and better understanding of the biological functions of these branched RNAs, their syntheses have become an important area of research (*89*). In this regard, Damha and his colleagues have carried out a systematic search toward the synthesis and biological applications of branched-RNA (*90*). Their success in making 'forked' shaped RNA and DNA and demonstrating their ability to hybridize to single stranded targets via the 'duplex-triplex' motif is discussed in **Chapter 9** of this volume.

Alternate routes to solid-support synthesis of small branched 'Y'-shaped RNA, including a substrate for the mammalian and yeast RNA debranching enzyme, have been conducted recently by Sproat *et al.* (*91*) and independently by Damha and Boeke (*90 c-e*) and coworkers. Solution phase synthesis of small lariat-RNA has been a forte of Chattopadhyaya and his colleagues (*92*). Recently, they reported synthesis and the solution conformation of hexameric and heptameric lariat-RNAs and their self-cleavage reactions. These results taken together should prove valuable in designing therapeutically useful catalytic RNA (ribozymes) and antisense agents.

Hexose-modified Oligonucleotides. Why pentose and not hexose-nucleic acids? This was the question that began an extensive study in Eschenmoser's laboratory on alternative nucleic acid structures (*93*). The synthesis of alternative structures and comparison of their chemical and biological properties with those of actual biomolecules (*i.e.*, DNA and RNA) has uncovered divergences in structural properties and Watson-Crick base pairing behavior of hexose-containing oligonucleotides. Above all, it revealed the importance of the sugar ring in natural DNA and RNA as being five-membered. In this context, Herdewijn and his co-workers (*94*) have synthesized and studied the properties of hexopyranosyl-like oligonucleotides. A comprehensive overview of their efforts is described in **Chapter 6** of this series. Some of the recently published work in this area by Eschenmoser *et al.* and Herdewijn *et al.* should be useful reading material for the experimental details. In conclusion, Eschenmoser's short answer to the question, why pentose- and not hexose-nucleic acids?, simply turns out to be: too many atoms!

L-and L/D-modifications (Figure 6). Recently, efforts have been devoted towards the synthesis of unnatural L-2′-deoxyribonucleosides (T, dC, dG, dA) from L-arabinose and assembled to give L-oligonucleotides (L-DNAs, **1**). In this respect, Garbesi *et al.* (*95*) reported the synthesis of fully modified L-DNAs (containing all four bases), Hashimoto *et al.* (*96*) reported on the synthesis of oligonucleotides containing L-dA and D-dA residues in alternating manner ('Meso'-DNA), and Damha *et al.* (*97*) reported on the applications of L/D-DNA chimeras ('heterochiral'-DNA). All of them reported the high stability of L-and L/D-modified oligonucleotides against degradation by various commercial phosphodiesterases and stability against digestion in human serum.

However, the debate on the hybridizatioň (Watson-Crick base pairing) properties of L-, Meso-, and heterochiral DNAs with natural oligonucleotides is ongoing. Garbesi *et al.* concluded that fully modified L-DNA, which contains all four natural bases in a random manner, does not bind single-stranded DNA or RNA in either parallel or antiparallel orientations (*95*). Thus, **1** may have limited applications for the design of antisense oligonucleotides. Similar conclusions were made by Hashimoto *et al.*, indicating that the affinity for poly (U) of dodecadeoxyadenylic acids decreased in the order of D-dA$_{12}$>L/D-dA$_{12}$>L-dA$_{12}$ (*96*).

The solution structure of the heterochiral-DNA duplex has been studied by NMR spectroscopy (*98*). The structure of the duplex is close to that of B-type DNA (sugars with S-type pucker), and the base-pair is of the Watson-Crick type. The recent work of Damha *et al.* on L/D-oligonucleotide chimeras indicated that oligonucleotides containing terminal as well as internal L-dC residue may serve as an acceptable modification for antisense oligonucleotides that have enhanced resistance to the 3′-exo nucleases, bind to complementary RNA (ΔT_m ~2-4° C/modification), and elicit RNase H activity.

***O*-4′-sugar Modifications (Figure 7).** The search for novel carbohydrate-based modifications took a drastic turn when the *O*-4′-atom (i.e. oxygen) of the sugar moiety was replaced by a sulfur atom. The synthesis of novel 2′-deoxy-4′-thio-modified oligonucleotides from 2′-deoxy-4′-thiothymidine (**1**) has been reported by Walker and his colleagues (*99*). Their studies demonstrated that self-complementary dodecanucleotides containing a single 2′-deoxy-4′-thio modification destabilized the duplex by ~ 3-5° C, compared to unmodified DNA. However, incorporation of two residues of **1** in the same sequence increased the T_m by 3° C. The CD spectrum of this duplex was very similar to a B-DNA structure indicating that the 4′-thio-sugar adopts an S-pucker (73%).

Bellon *et al.* reported the first synthesis of the ribo-analog **2** containing the 4-thio-moiety and its incorporation into oligonucleotides (*100*). A detailed account of their endeavor is covered in **Chapter 5**. In brief, their nuclease studies of oligomers containing the 4-thio-RNA residue indicated that the modified oligomers were more stable (~5 times) compared to the unmodified ribo-oligonucleotides against various nucleases and in cell culture medium. They have studied the binding affinity of 4′-thio-ribo-oligonucleotides with complementary RNAs. The studies indicated that 4′-thio-ribo-oligomers formed a more stable duplex (4′-SU$_{12}$/poly-A; T_m 46° C) under physiological conditions, compared to the unmodified duplex (U$_{12}$/A$_{12}$; T_m 32° C). These data taken together suggest that in spite of their tedious synthesis this modification should be further investigated for its utility in the antisense area.

Some efforts have been directed towards the incorporation of a 5-membered carbocyclic ring replacing the 5-membered sugar ring to prepare carbocyclic oligomers as antisense molecules. Beaucage and Iyer have recently reviewed this modification in detail (*14a*).

Bicyclic-sugar Modifications (Figure 8). In principle, it is possible to modulate the thermodynamic stability of DNA:RNA heteroduplexes by the chemical alterations of the DNA molecule. Of particular importance, Leumann and his colleagues investigated the potential of enhancing the duplex stability by incorporating a preorganized-structure, such as bicyclic-nucleosides **1,** into oligonucleotides. This conformational rigidity should lead to an entropic advantage for the Watson-Crick base-pairing process and in turn provide more stable duplexes. An excellent account of their efforts has been discussed in **Chapter 7** of this series. However, a short discussion on some of their important conclusions is included herein. The α-and β-anomeric forms of these bicyclic nucleosides have been synthesized and assembled into oligonucleotides *via* automated synthesis (*101*). The ethylene bridge of bicyclo **1** forces the sugar to adopt a 2′-endo (S-pucker) conformation resulting in an overall B-type of DNA structure. Both α-and β-bicyclo-oligonucleotides were more stable to cleavage by nucleases compared to unmodified DNA. The hybridization studies of these oligomers indicated effective base-pairing to complementary sequences. Given these properties of bicyclo **1,** the

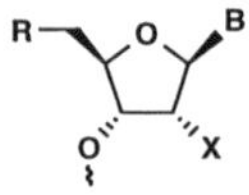

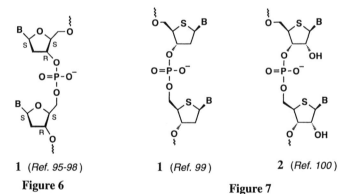

	R	B	X	Reference
1	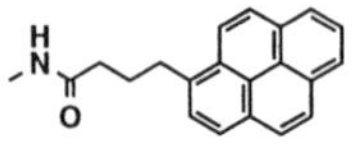 (N-pyrene amide)	C	OH	83
2	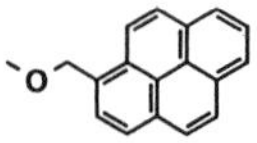 (O-pyrenyl)	T	H	84
3	-NHCOCH$_2$S(CH$_2$)$_3$-polyamine	T	H	85
4	-O-P(O)$_2$-O-(CH$_2$)$_6$SH	A, G	H	86
5	-O-DMTr-Tbf (hydrophobic group)	T	H	87
6	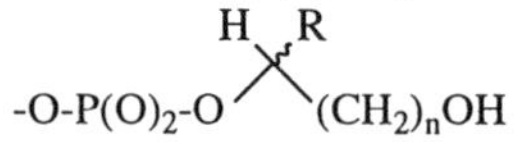 -O-P(O)$_2$-O-CH(NHR)-(CH$_2$)$_n$OH	T	H	88

Figure 5: *C*-5'-Modifications

1 (*Ref. 95-98*)

Figure 6

L-Oligomer Indicating the
Stereochemistry at *C*-1', *C*-3'
and *C*-4' Carbon Atoms.

1 (*Ref. 99*)   2 (*Ref. 100*)

Figure 7

Structure of 4'-Thiosugar Oligonucleotides
1 and 2.

applications of α- and β-anomeric oligonucleotides as potential antisense agents should be further investigated.

Altmann *et al.* recently reported the synthesis of a bicyclo-carbocyclic thymidine analog **2** and its incorporation into oligonucleotides (*102*). Their preliminary hybridization data indicate an increase in thermodynamic stability. Jones *et al.* have reported on oligonucleotides containing a conformationally restricted riboacetal **3** internucleosidic linkage (*103*). The data showed that the T_m of oligonucleotides containing **3** were lower (ΔT_m ~1° C/modification) than the T_m of the unmodified oligomer when hybridized to the same RNA complement. These results are in harmony with the concept of entropic duplex stabilization by reducing the conformational flexibility of the sugar moiety in the carbohydrate-based modified oligonucleotides.

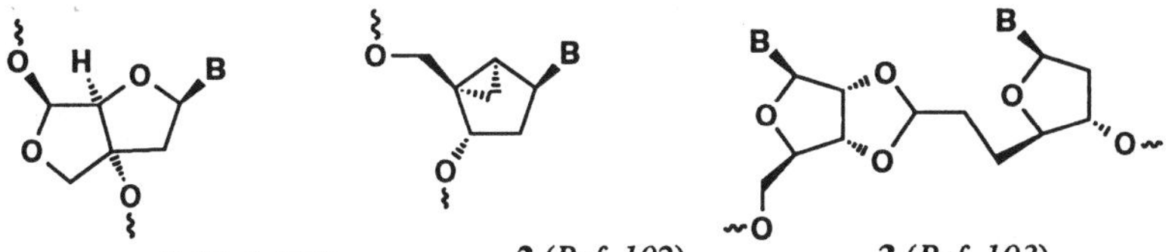

1 (*Ref. 101*)   **2** (*Ref. 102*)   **3** (*Ref. 103*)

Figure 8: Bicyclic-sugar Modification

Future Applications

Four phosphorothiate oligonucleotides are currently undergoing human clinical trials. This should certainly strengthen one's confidence in the antisense paradigm. However, although first generation phosphorothioates have performed much better than was generally anticipated, we believe that the intense medicinal chemistry of second generation analogs, some being discussed in this book, will soon provide superior anitsense agents. Clearly, the preliminary data on the effects of sugar modifications on nuclease stability, hybridization properties, RNase H activity, and for certain modifications, activity *in vitro* suggest that second generation oligonucleotides may exhibit substantially improved pharmacokinetic and pharmacodynamic properties.

Chemical modifications of oligonucleotides at nearly every carbohydrate position have been considered, and numerous potentially interesting analogs have been identified. Advances in the synthesis of uniformly modified oligomers *via* automation should reduce the turnaround time and improve the quality of these synthetic molecules. The availability of novel oligonucleotides coupled with fast-screening programs will lead to a more rapid discovery of the biologically active molecules. We believe that this American Chemical Society symposium series offers a wealth of new chemical modifications ready for a systematic evaluation. These modifications should provide important insights into structure-activity relationships of oligonucleotide analogs designed for antisense therapeutics.

Literature Cited

1. Selected books and reviews published in 1993-94 (see reference 2 for a complete list): (a) *Antisense Research and Applications*; Crooke, S. T.; Lebleu, B., Eds.; CRC Press: Boca Raton, Florida, **1993**; (b) *Protocols for Oligonucleotides and Analogs*; Agrawal, S. Ed.; Humana Press: Totowa, New Jersey, **1993**, Vol. 20; (c) *Protocols for Oligonucleotide Conjugates*, Agrawal, S., Ed.; Humana Press: Totowa, New Jersey, **1994**, vol. 26; (d) *Design and Targeted Reactions of Oligonucleotide Derivatives*; Knorre, D. G.; Vlassov, V. V.; Zarytova, V. F.; Lebedev, A. V.; Fedorova, O. S. Eds.; CRC Press: Boca Raton, Florida, **1994** ; (e) Stein, C. A.; Cheng, Y.-C.*Science* **1993**, *261*, 1004-1012; (f) Crooke, S. T. *FASEB J.* **1993**, *7*, 533-539; (g) Neckers, L.; Whitesell, L. *Am. J. Physiol.* 265 (*Lung Cell. Mol. Physiol.* 9) **1993**, L1-L12; (h) Kiely, J. S. *Annu. Rep. Med Chem.* **1994**, Vol. 29 in press.
2. Cook, P. D.: *Medicinal Chemistry Strategies for Antisense Research* in reference 1 (a), p 149-187.
3. Mehta and Isaly: A special report (55 Pages): *Antisense Technology Primer-Rational Drug Design, Oligonucleotides and Profits* **1994**, New York (Phone: 212-758-2662; Fax: 212-758-2764).
4. Crooke, S. T. *Oligonucleotide Therapeutics* in *Burger's Medicinal Chemistry and Drug Discovery*, Wolff, M. E., Ed., Vol. 5, **1994**, in press.
5. Nielsen, P. E.; Egholm, M.; Berg, R. M.; Buchardt, O. *Anti-Cancer Drug Design* **1993**, *8*, 53-63, and references cited therein.
6. (a) Milligan, J. F.; Matteucci, M. D.; Martin, J. C. *J. Med. Chem.* **1993**, *36*, 1923-1936; (b) Moser, H. E. In *Perspectives in Medicinal Chemistry*, Testa, B. et al., Eds., Verlag Helv. Chim. Acta, Basel: 1993, p279-297; (c) Persaud, S.; Jones, P. M. *J. Mol. Endocrinology* **1994**, *12*, 127-130.
7. Cohen, J. S. *Phosphorothioate Oligodeoxynucleotides* in reference 1 (a), p205-221, and references cited therein.
8. Marshall, W. S.; Caruthers, M. H. *Science* **1993**, *259*, 1564-1570.
9. Miller, P. S.; Ts'o, P.O.P.; Hogrefe, R. I.; Reynolds, M. A.; Arnold, L. J. *Anticode Oligonucleoside Methylphosphonate and Their Psoralen Derivatives* in reference 1 (a) p189-204.
10. Gryaznov, S.; Chen, J-K. *J. Am. Chem. Soc.* **1994**, *116*, 3143-3144; Bjergarde, K.; Dahl, O.; Caruthers, M. H. *Tetrahedron Lett.* **1994**, *35*, 2941-2944, and references cited therein.
11. Goudgaon, N. M.; El-Kattan, G. F.; Schinazi, R. F. *Nucleosides Nucleotides* **1994**, *13*, 849-880, and references cited therein.
12. Dabkowski, W.; Michalski, J.; Wasiak, J.; Cramer, F. *J. Chem. Soc. Perkin Trans. 1* **1994**, 817-820, and references cited therein.
13. Lesnikowski, Z. J. *Bioorg. Chem.* **1994**, *22*, 128-139; For a comprehensive list of references see 14 (a).
14. (a) Beaucage, S. L.; Iyer, R. P. *Tetrahedron.* **1993**, *49*, 6123-6194; (b) Milligan, J. F.; Matteucci, M. D.; Martin, J. C. *J. Med. Chem.* **1993**, *36*, 1923; (c) Uhlmann, E.; Peyman, A. in reference 1 (b) p355-390; (d) Varma, R. S. *Synlett* **1993**, 621.
15. (a) De Mesmaeker, A.; Waldner, A.; Lebreton, J.; Hoffmann, P.; Fritsch, V.; Wolf, R. M.; Freier, S. M. *Angew. Chem. Int. Ed. Engl.* **1994**, *33*, 226-229; (b) Lebreton, J.; Waldner, A; Lesueur, C.; De Mesmaeker, A. *Synlett* **1994**, 137-139.
16. Idziak, I.; Just, G.; Damha, M. J.; Giannari, P. A. *Tetrahedron Lett.* **1993**, *34*, 5417-5420.
17. Lebreton, J.; De Mesmaeker, A.; Waldner, A; Fritsch, V.; Wolf, R. M.; Freier, S. M. *Tetrahedron Lett.* **1993**, *34*, 6383.
18. De Mesmaeker, A.; Lebreton, J.; Waldner, A; Fritsch, V.; Wolf, R. M.; Freier, S. M. *Synlett* **1993**, 733.

19. Lebreton, J.; Waldner, A.; Fritsch, V.; Wolf, R. M.; De Mesmaeker, A. *Tetrahedron Lett.* **1994**, *35*, 5225-5228.
20. De Mesmaeker, A.; Lebreton, J.; Waldner, A.; Fritsch, V.; Wolf, R. M. *Bioorg. Med. Chem. Lett.* **1994**, *4*, 873-878.
21. Kutterer, K. M. K.; Just, G. *Bioorg. Med. Chem. Lett.* **1994**, *3*, 435-438.
22. Waldner, A.; De Mesmaeker, A.; Lebreton, J. *Bioorg. Med. Chem. Lett.* **1994**, *4*, 405.
23. Vandendrissche, F.; Van Aerschot, A.; Voortmans, M.; Janssen, G.; Busson, R.; Van Overbeke, A.; Van Den Bossche, W.; Hoogmartens, J.; Herdewijn, P. *J. Chem. Soc.* **1993**. 1567-1575.
24. Habus, I.; Temsamani, J.; Agrawal, S. *Bioorg. Med. Chem. Lett.* **1994**, *4*, 1065-1070.
25. Saha, A. K.; Schairer, W.; Waychunas, C.; Prasad, C. V. C.; Sardaro, M.; Upson, D. A.; Kruse, L. I. *Tetrahedron Lett.* **1993**, *34*, 6017-6020.
26. (a) De Mesmaeker, A.; Waldner, A.; Sanghvi, Y. S.; Lebreton, J. *Bioorg. Med. Chem. Lett.* **1994**, *4*, 395-398; (b) Caulfield, T. J.; Prasad, C. V. C.; Prouty, C. P.; Saha, A. K.; Sardaro, M. P.; Schairer, D. W. C. *Bioorg. Med. Chem. Lett.* **1993**, *3*, 2771-2776.
27. Petersen, G. V.; Walsted, M.; Wengel, J. 207th ACS National meeting, San Diego, CA, March 13-17, 1994, CARB 61.
28. Burgess, K.; Gibbs, R. A.; Metzker, M. L.; Raghavachari, R. *J. Chem. Soc. Chem. Commun.* **1994**, 915-916.
29. Sanghvi, Y. S.; Cook, P. D. In *Nucleosides and Nucleotides as Antitumor and Antiviral Agents*; Chu, C. K.; Baker, D. C., Eds.; Plenum Press: New York, **1993**; p311-324.
30. Sanghvi, Y. S.; Vasseur, J.-J.; Debart, F.; Cook, P. D. *Collect. Czech. Chem. Commun.* Special Issue **1993**, *58*, 158-162.
31. Sanghvi, Y. S.; Cook, P. D. 11th IRT Nucleosides & Nucleotides, Leuven, Belgium, Sept. 7-11, **1994** (poster presentation).
32. Cummins, L.; Owen, S.; Sanghvi, Y. S.; Cook, P. D. *Nucleic Acids Research*. Submitted.
33. Hoshiko, T.; Fraser, A.; Perbost, M.; Dimock, S.; Cook, P. D.; Sanghvi, Y. S. 207th ACS National Meeting, San Diego, CA, March 13-17, **1994**, CARB 35 (poster presentation).
34. Cook, P. D.; Sanghvi, Y. S.; Vasseur, J.-J.; Debart, F. International Patent WO 92/20822, **1992**.
35. Teng, K.; Cook, P. D. *J. Org. Chem.* **1994**, *59*, 278-280.
36. Cao, X.; Matteucci, M. D. *Tetrahedron Lett.* **1994**, *35*, 2325-2328.
37. Caulfield, T. J.; Prasad, C. V. C.; Delecki, D. J.; Prouty, C. P.; Saha, A. K.; Upson, D. A.; Kruse, L. I. *Bioorg. Med. Chem. Lett.* **1994**, *4*, 1497-1500.
38. Butterfield, K.; Thomas, E. J. *Synlett*, **1993**, 411-412.
39. Lebreton, J.; De Mesmaeker, A.; Waldner, A. *Synlett* **1994**, 54.
40. Lee, K.; Wiemer, D. F. *J. Org. Chem.* **1993**, *58*, 7808-7812.
41. (a) Jones, R. J.; Lin, K-Y.; Milligan, J. F.; Wadwani, S.; Matteucci, M. D. *J. Org. Chem.* **1993**, *58*, 2983-2991; (b) Veal, J. M.; Gao, X.; Brown, F. K. *J. Am. Chem. Soc.* **1993**, *115*, 7139-7145.
42. Lin, K-Y.; Pudlo, J. S.; Jones, R. J.; Bischofberger, N.; Matteucci, M.; Froehler, B. C. *Bioorg. Med. Chem. Lett.* **1994**, *4*, 1061-1064.
43. Cao, X.; Matteucci, M. D. *Bioorg. Med. Chem. Lett.* **1994**, *4*, 807-810.
44. Quaedflieg, P. J. L. M.; Timmers, C. M.; Van der Marel, G. A.; Kuyl-Yeheskiely, E.; Van Boom, J. H. *Synthesis* **1993**, 627-633.
45. (a) Vandendriessche, F.; Voortmans, M.; Hoogmartens, J.; Van Aerschot, A.; Herdewijn, P. *Bioorg. Med. Chem. Lett.* **1993**, *3*, 193-198; (b) Pannecouque, C.; Schepers, G.; Rozenski, J.; Van Aerschot, A.; Claes, P.; Herdewijn, P. *Bioorg. Med. Chem. Lett.* **1994**, *4*, 1203-1206.

46. Pannecouque, C.; Vandendriessche, F.; Rozenski, J.; Janssen, G.; Busson, R.; Van Aerschot, A.; Claes, P.; Herdewijn, P. *Tetrahedron* **1994**, *50*, 7231-7246.
47. Wang, L.; Zhang, L. *Chin. Chem. Lett.* **1993**, *4*, 101-104.
48. McElroy, E. B.; Bandaru, R.; Huand, J.; Widlanski, T. S. *Bioorg. Med. Chem. Lett.* **1994**, *4*, 1071-1076.
49. Dewynter, G.; Montero, J-L. *Acad. Sci.* **1992**, *315*, 1675-1682.
50. Huang, J.; McElroy, E. B.; Widlanski, T. S. *J. Org. Chem.* **1994**, *59*, 3520-3521.
51. Huang, Z.; Benner, S. A. *Synlett* **1993**, 83-84.
52. Kawai, D. H.; Wang, D.; Giannaris, P. A.; Damha, M. J.; Just, G. *Nucleic Acids Res.* **1993**, *21*, 1473-1479.
53. Meng, B.; Kawai, S. H.; Wang, D.; Just, G.; Giannaris, P. A.; Damha, M. J. *Angew. Chem. Int. Ed. Engl.* **1993**, *32*, 729-731.
54. Chur, A.; Holst, B.; Dahl, O.; Valentin-Hansen, P.; Pedersen, E. B. *Nucleic Acids Res.* **1993**, *21*, 5179-5183.
55. Saha, A. K.; Sardaro, M.; Waychunas, C.; Delecki, D.; Kutny, R.; Cavanaugh, P.; Yawman, A.; Upson, D. A.; Kruse, L. I. *J. Org. Chem.* **1993**, *58*, 7827-7831.
56. Chaix, C.; Toulmé, J.-J.; Morvan, F.; Rayner, B.; Imbach, J.-L. in reference 1 (a) p223-234.
57. Dan, A.; Yoshimura, Y.; Ono, A.; Matsuda, A. *Bioorg. Med. Chem. Lett.* **1993**, *3(4)*, 615-618; (b) Ono, A.; Dan, A.; Matsuda, A. *Nucleic Acids Symposium Series No. 29* **1993**, 13-14.
58. Azhayev, A.; Gouzaev, A.; Hovinen, J.; Azhyeva, E.; Lönnberg, H. *Tetrahedron Lett.* **1993**, *34*, 6435-6438.
59. (a) Ravikumar, V. T.; Wyrzykiewicz, T. K.; Moham, V.; Cole, D. L. *Nucleosides Nucleotides.* **1994**, (in press); (b) Morgan, R. L.; Celebuski, J. E.; Fino, J. R. *Nucleic Acids Res.* **1993**, *21*, 4574-4576.
60. (a) Laayoun, A.; Décout, J-L.; Defrancq, E.; Lhomme, J. *Tetreahedron Lett.* **1994**, *35*, 4991-4994; (b) Laayoun, A.; Décout, J-L.; Lhomme, J. *Tetrahedron Lett.* **1994**, *35*, 4989-4990.
61. (a) Lamond, A. I.; Sproat, B. S. *FEBS* **1993**, *325*, 123-127; (b) Sproat, B. S.; Lamond, A. I. in reference 1 (a) p351-362; (c) Parmentier, G.; Schmitt, G.; Dolle, F.; Luu, B. *Tetrahedron* **1994**, *50*, 5361-5368.
62. (a) Lesnik, E. A.; Guinosso, C. J.; Kawasaki, A. M.; Sasmor, H.; Zounes, M. C.; Cummins, L. L.; Ecker, D. E.; Cook, P. D.; Freier, S. M. *Biochemistry* **1993**, *32*, 7832-7838; (b) Hughes, J. A.; Bennett, C. F.; Cook, P. D.; Guinosso, C. J.; Mirabelli, C. K.; Juliano, R. L. *J. Pharm. Sci.* **1994**, *83*, 597-600.
63. (a) Kawasaki, A. M.; Casper, M. D.; Freier, S. M.; Lesnik, E. A.; Zounes, M. C.; Cummins, L. L.; Gonzalez, C.; Cook, P. D. *J. Med. Chem.* **1993**, *36*, 831-841; (b) Monia, B. P.; Lesnik, E. A.; Gonzalez, C.; Lima, W. F.; McGee, D.; Guinosso, C. J.; Kawasaki, A. M.; Cook, P. D.; Freier, S. M. *J. Biol. Chem.* **1993**, *268*, 14514-14522; (c) Sato, Y.; Utsumi, K.; Maruyama, T.; Kimura, T.; Yamamoto, I.; Richman, D. D. *Chem. Pharm. Bull.* **1994**, *42*, 595-598.
64. Fraser, A.; Wheeler, P.; Cook, P. D.; Sanghvi, Y. S. *J. Heterocyclic. Chem.* **1993**, *30*, 1277-1287.
65. (a) Manoharan, M.; Johnson, L.K.; Tivel, K. L.; Springer, R. H.; Cook, P. D. *Bioorg. Med. Chem. Lett.* **1993**, *13*, 2765-2770; (b) Manoharan, M.; Johnson, L. K.; Bennett, C. F.; Vickers, T. A.; Ecker, D. J.; Cowsert, L. M.; Freier, S. M.; Cook, P. D. *Bioorg. Med. Chem. Lett.* **1994**, *4*, 1053-1060.
66. Kois, P.; Watanabe, K. A. *Nucleic Acids Symposium No. 29* **1993**, p215-216.

67. (a) Miller, P. S.; Bhan, P.; Kan, L-S. *Nucleosides Nucleotides* **1993**, *12,* 785-792; (b) Heidenreich, O.; Pieken, W.; Eckstein, F. *FASEB J.* **1993**, *7,* 90-96.
68. (a) Keller, T. H.; Haner, R. *Helv. Chim. Acta* **1993**, *76,* 884-892; (b) Keller, T. H.; Haner, R. *Nucleic Acids Res.* **1993**, *21,* 4499-4505.
69. (a) Schmit, C. *Synlett* **1994**, 238-240; (b) Schmit, C. *Synlett* **1994**, 241-242.
70. Giannaris, P. A.; Damha, M. J. *Can. J. Chem.* **1994**, *72,* 909-918.
71. Torrence, P. F.; Maitra, R. K.; Lesiak, K.; Khamnei, S.; Zhou, A.; Silverman, R. H. *Proc. Natl. Acad. Sci.* **1993**, *90,* 1300-1304.
72. Giannaris, P. A.; Damha, M. J. *Nucleic Acids Res.* **1993**, *21,* 4742-4749.
73. Swaminathan, S.; Matteucci, M.; Pudlo, J.; Jones, R. J. PCT International Application No. WO 93/24508, December **1993.**
74. Beaucage, S. L.; Iyer, R. P. *Tetrahedron* **1993**, *49,* 6123-6194.
75. Bévierre, M-O.; De Mesmaeker, A.; Wolf, R. M.; and Freier, S. M. *Bioorg. Med. Chem. Lett.* **1994**, *4,* 237-240.
76. Jorgensen, P. N.; Stein, P. C.; Wengel. *J. Am. Chem. Soc.* **1994,** *116,* 2231-2232.
77. Seela, F.; Wörner, K.; Rosemeyer, H. *Helv. Chim. Acta* **1994**, *77,* 883-896.
78. (a) Hovinen, J.; Guzaev A.; Azhayev, A.; Lönnberg, H. *Tetrahedron Lett.* **1993**, *34,* 5163-5166; (b) Tung, C-H.; Breslauer, K.J.; Stein, S.; *Nucleic Acids Res.* **1993**, *21,* 5489-5494.
79. Plushin, N. N.; Chen, B-C.; Anderson, L. W.; Cohen, J. S. *J. Org. Chem.* **1993**, *58,* 4606-4613.
80. (a) Kumar, A. *Nucleosides Nucleotides* **1993**, *12,* 441-447; (b) Kumar, A.; Ghosh, N. N.; Sadana, K. L.; Garg, B. S.; Gupta, K. C. *Nucleosides Nucleotides.* **1993**, *12,* 565-584.
81. Nakamura, Y.; Akiyama, T.; Bessho, K.; Yoneda, F. *Chem. Pharm. Bull.* **1993**, *41,* 1315-1317.
82. Manoharan, M. in reference 1(a) p303-349; See reference 1(c) for a comprehensive list.
83. Kierzek, R.; Li, Yi.; Turner, D. H.; Bevilacqua, P. C. *J. Am. Chem. Soc.* **1993**, *115,* 4985-4992.
84. Yamana, K.; Nunota, K.; Nakano, H.; Sangen, O. *Tetrahedron Lett.* **1994**, *35,* 2555-2558.
85. Tung, C-H.; Bresalauer, K. J.; Stein, S. *Nucleic Acids Res.* **1993**, *21,* 5489-5494.
86. De la Torre, B. G.; Avino, A. M.; Escarceller, M.; Royo, M.; Albericio, F.; Eritja, R. *Nucleosides Nucleotides* **1993**, *12,* 993-1005.
87. Ramage, R.; Wahl, F.O. *Tetreahedron Lett.* **1993**, *34,* 7133-7136.
88. Fontanel, M-L.; Bazin, H.; Téoule, R. *Nucleic Acids Res.* **1994**, *22,* 2022-2027.
89. Beaucage, S. L.; Lyer, R. P. *Tetrahedron* **1993**, *49,* 10441-10488.
90. (a) Hudson, R. H. E.; Damha, M. J. *J. Am. Chem. Soc.* **1993**, *115,* 2119-2124; (b) Amato, I. *Science.* **1993**, *260,* 491; (c) Nam, K.; Hudson, R.H.E.; Chapman, K.B.; Ganeshan, K.; Damha, M.J.; Boeke, J. D. *J. Biol. Chem.* **1994**, in press; (d) Ganeshan, K.; Nam, K.; Hudson, R.E.H.; Tadey, T.; Braich, R.; Purdy, W.C.; Boeke, J.D.; Damha, M.J. 11th IRT Nucleosides & Nucleotides, Leuven, Belgium, Sept. 7-11, 1994 (poster presentation); (e) Damha, M.J.; Ganeshan, K. 10th IRT Nucleosides & Nucleotides, Park City, Utah, Sept. 16-20, 1992 (poster presentation # 58).
91. Sproat, B. S.; Beijer, B.; Grotli, M.; Ryder, U.; Morand, K. L.; Lamond, A. I. *J. Chem. Soc. Perkin. Trans.* 1 **1994**, 419-431.
92. Rousse, B.; Puri, N.; Viswanadham, G.; Agback, P.; Glemarec, C.; Sandström, A.; Sund, C.; Chattopadhyayta, J. *Tetrahedron* **1994**, *50,* 1777-1810.

93. (a) Eschenmoser, A. *Chemistry & Biology.* 15 April **1994**, Introductory issue; (b) Pitsch, S.; Wendeborn, S.; Jaun, B.; Eschenmoser, A. *Helv. Chim. Acta* **1993**, *76*, 2161-2183; (c) Hunziker, J.; Roth, H-J.; Böhringer, M.; Giger, A.; Diederichsen, U.; Göbel, M.; Krishnan, R.; Jaun, B.; Leumann, C.; Eschenmoser, A. *Helv. Chim. Acta* **1993**, *76*, 259-352.

94. (a) Augustyns, K.; Rozenski, J.; Van Aerschot, A.; Janssen, G.; Herdewijn, P. *J. Org. Chem.* **1993**, *58*, 2977-2982; (b) Augustyns, K.; Godard, G.; Hendrix, C.; Van Aerschot, A.; Rozenski, J.; Saison-Behmoaras, T.; Herdewijn, P. *Nucleic Acids Res.* **1993**, *21,* 4670-4676.

95. Garbesi, A.; Capobianco, M. L.; Colonna, F. P.; Tondelli, L.; Arcamone, F.; Manzini, G.; Hilbers, C. W.; Aelen, J. M. E.; Blommers, M. J. J. *Nucleic Acids Res.* **1993**, *21*, 4159-4165.

96. Hashimoto, Y.; Iwanami, N.; Fujimori, S.; Shudo, K. *J. Am. Chem. Soc.* **1993**, *115*, 9883-9887.

97. Damha, M. J.; Giannaris, P. A.; Marfey, P. *Biochemistry* **1994**, *33*, 7877-7885.

98. (a) Blommers, M. J. J.; Tondelli, L.; Garbesi, A. *Biochemistry* **1994**, *33*, 7886-7896; (b) Urata, H.; Ueda, Y.; Suhara, H.; Nishioka, E.; Akagi, M. *J. Am. Chem. Soc.* **1993**, *115*, 9852-9853; (c) Urata, H.; Ueda, Y.; Suhara, H.; Nishioka, E.; Akagi, M. *Nucleic Acids Symposium Series No. 29*, **1993**, 69-70.

99. Hancox, E. L.; Connolly, B. A.; Walker, R. T. *Nucleic Acids Res.* **1993**, *21*, 3485-3491.

100. Bellon, L.; Barascut, J.-L.; Maury, G.; Divita, G.; Goody, R.; Imbach, J.-L. *Nucleic Acids Res.* **1993**, *21*, 1587-1593.

101. Tarköy, M.; Bolli, M.; Leumann, C. *Helv. Chim. Acta* **1994**, *77*, 716-744.

102. Altmann, K.-H.; Kesselring, R.; Francotte, E.; Rihs, G. *Tetrahedron Lett.* **1994**, *35*, 2331-2334.

103. Jones, R. J.; Swaminathan, S.; Milligan, J. F.; Wadwani, S.; Froehler, B. C.; Matteucci, M. D. *J. Am. Chem. Soc.* **1993**, *115*, 9816-9817.

RECEIVED August 23, 1994

DEPHOSPHONO LINKAGES

Chapter 2

Novel Backbone Replacements for Oligonucleotides

Alain De Mesmaeker, Adrian Waldner, Jacques Lebreton, Valérie Fritsch, and Romain M. Wolf

Central Research Laboratories, Ciba-Geigy Ltd., CH-4002 Basel, Switzerland

A new type of backbone replacement in antisense oligodeoxynucleotides (ODN) containing an amide group is described. The syntheses of five isomeric amide modifications are described together with their binding affinities to complementary RNA and DNA strands. Among these backbone replacements, the amides **3** and **4** have a similar affinity for an RNA strand as the wild type. The properties of those amide modifications are compared to more flexible (carbon chain internucleosidic linkage) and to more rigid (ureas, carbamates) backbones. The modified ODNs display high resistance towards nucleases.

Antisense oligodeoxynucleotides (ODN) constitute a promising class of antiviral drugs that offer a new and highly selective chemotherapeutic strategy to treat human diseases (1). The ability of the antisense oligonucleotides to interfere specifically with an mRNA target provides a powerful tool for the control of cellular and viral gene expression. However, this strategy is restricted by the poor resistance towards cellular nucleases and by the low cellular uptake of the natural phosphodiester linked antisense ODNs (wild type). Therefore, chemical modifications of ODNs are required for their effective use as potential antisense drugs (2,3). So far, most of the replacements of the phosphodiester linkage leading to an increased resistance towards nucleases cause a decrease in the affinity for the complementary RNA strand (1,2). Two exceptions to this were recently reported dealing either with an N-methylhydroxylamine or with a thioformacetal as backbone replacement (2a,2c).

We proposed to replace the natural phosphodiester group by an amide function, 1-5 (3,4) (Figure 1). The amide bond is compatible with the conditions required for the solid phase synthesis and is also more stable under physiological conditions than the phosphodiester bond. Furthermore, the overall charge reduction of the ODNs containing neutral amide bonds should favor the penetration of the

0097–6156/94/0580–0024$08.00/0

ODNs through the negatively charged cellular membranes (*5*). The amide moiety is readily accessible by simple synthetic methods and is achiral, thereby avoiding diastereomeric mixtures in ODNs. Amides **1** and **5**, as well as amides **3** and **4**, are two structural isomers that display the same geometry containing neutral amide bonds, which should favor the penetration of the ODNs through the negatively charged cellular membranes (*5*). The amide moiety is readily accessible by simple synthetic methods and is achiral, thereby avoiding diastereomeric mixtures in ODNs. Amides **1** and **5**, as well as amides **3** and **4**, are two structural isomers that display the same geometry.

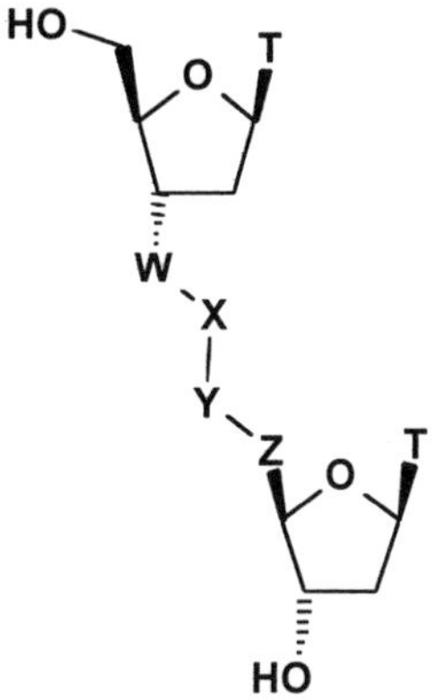

	W	X	Y	Z
1	NH	CO	CH$_2$	CH$_2$
2	CH$_2$	CH$_2$	NH	CO
3	CH$_2$	CO	NH	CH$_2$
4	CH$_2$	NH	CO	CH$_2$
5	CO	NH	CH$_2$	CH$_2$
6	NR^2	CO	NR^1	CH$_2$
7	O	CO	NR	CH$_2$
8	NR	CO	O	CH$_2$
9	CH$_2$	CO	CH$_2$	CH$_2$
10	CH$_2$	CH$_2$	CH$_2$	CH$_2$

Fig. 1 **Novel backbone replacements for oligonucleotides**

According to molecular models, these five isomers can be incorporated in an A-type duplex structure. Our detailed molecular modeling study (V. Fritsch et al. submitted) confirmed this as far as energy minimization was concerned. In addition, we decided to investigate the role and the importance of the rigidity of the backbone. Therefore, we studied the ureas **6** and the carbamates **7** and **8**, where the additional heteroatom conjugated to the carbonyl group increases the rigidity (Figure 1). On the other hand, we also studied the much more flexible analogs **9** and **10** (Figure 1). In particular, the backbone replacement **10** would be the most flexible analog in this series of compounds. We describe here the syntheses of these derivatives, their incorporation into ODNs, and some melting temperature data of the duplexes formed with complementary RNA strands. The resistance of the ODNs containing these modifications towards nucleases is also mentioned.

Synthesis of the Dimers 1-10

Synthesis of the Amide 1 Dimers (3'-NR-CO-CH$_2$-5') (*3,6*). The amine **11** (*7*) and the alcohol **12** (*8*) were both prepared from thymidine (Scheme 1). The alcohol **12** was oxidized to the aldehyde followed by *Wittig* reaction. Hydrogenation of the C=C bond and saponification of the methyl ester afforded **13** (*9*). Compond **13** was coupled with the amine 11 (*10*) to dimer 14. Deprotection in 14 gave 1. The 4,4'-dimethoxytrityl and the phosphoramidite groups were introduced at the 5'- and 3'-ends, respectively, by standard methods (*11*). NOE experiments indicated that the dimer 1 is present in solution exclusively as the more stable s-trans rotamer. We also prepared two N-alkylated amide derivatives (Scheme 2). Dimer 14 was fully protected (*12*) followed by methylation to 16. N-propylation was accomplished via the allyl derivative followed by hydrogenation of the C=C bond. Silyl and BOM protective groups in 16 and 17 were removed to give the fully deprotected dimers 18 and 19.

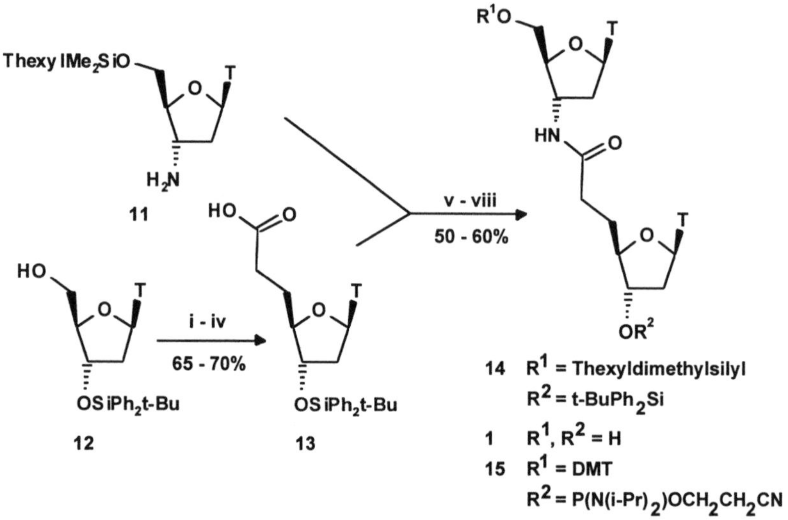

Scheme 1: i) DCC, Pyridinium trifluoroacetate, DMSO. ii) Ph_3P=CHCOOMe, CH_2Cl_2. iii) H_2, Pd/C 10%, MeOH. iv) KOH (2M), MeOH/H_2O (7:3). v) Et_3N, O(1H-benzotriazolyl)-N,N,N,N'-tetramethyluronium tetrafluoroborate (TBTU), N-hydroxybenzo-triazole (NHBT), CH_3CN, **11**, Et_3N. vi) Tetra-n-butylammonium fluoride (TBAF), THF. vii) 4,4'-Dimethoxytritylchloride (DMTCl), pyr. viii) $(i-Pr_2N)_2PO-CH_2CH_2CN$, $(i-Pr)_2NH_2^+$ tetrazole⁻, CH_2Cl_2.

Synthesis of the Amide 2 Dimer (3'-CH₂-CH₂-HN-CO-4') (*3,13*). We synthesized the amine **26** (Scheme 3) and the acid **27** (Scheme 4) from thymidine **22**. Thiocarbonate **23** was treated with allyltributyl tin to give **24** (*14*) as the sole isomer. The cleavage of the C=C bond in **24** was performed via dihydroxylation and diol cleavage. After the reduction of aldehyde **25**, tosylation, and substitution with LiN_3, the azido nucleoside was converted to the amine **26** (*15*). Acid **27** (Scheme 4) was prepared from thymidine **22** (*8*). Dimer **28** was obtained by coupling of **27** with the amine **26**. After deprotection, dimer **2** was transformed to **29**.

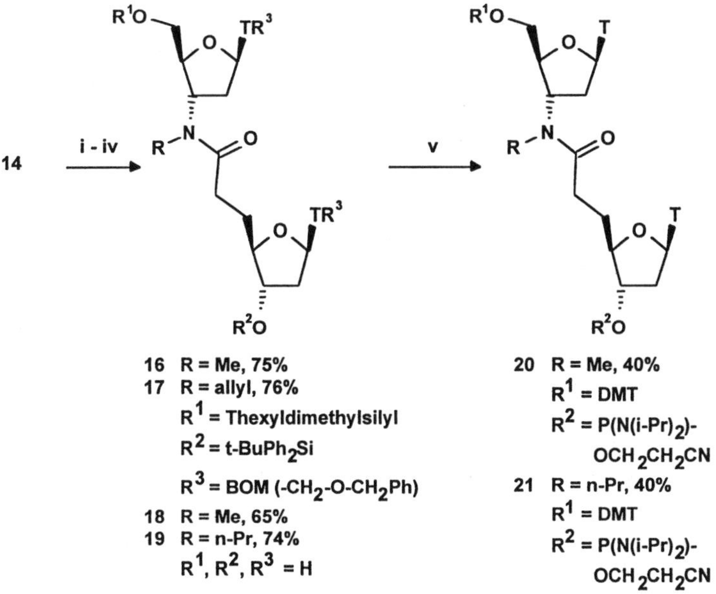
R¹O
TR³
O
14
i - iv
R
N
O
TR³
O
R²O
v
R¹O
T
O
R
N
O
T
O
R²O
16 R = Me, 75%
17 R = allyl, 76%
R¹ = Thexyldimethylsilyl
R² = t-BuPh2Si
R³ = BOM (-CH2-O-CH2Ph)
18 R = Me, 65%
19 R = n-Pr, 74%
R¹, R², R³ = H
20 R = Me, 40%
R¹ = DMT
R² = P(N(i-Pr)2)-
OCH2CH2CN
21 R = n-Pr, 40%
R¹ = DMT
R² = P(N(i-Pr)2)-
OCH2CH2CN

Scheme 2: i) BOMCl, DBU, CH_3CN, 0° C. ii) NaH, MeI for **16**, allyliodide for **17**, THF, 55-60° C. iii) TBAF, THF, 0° C. iv) H_2, Pd/C 10%, MeOH. v) see vii, viii) (Scheme 1).

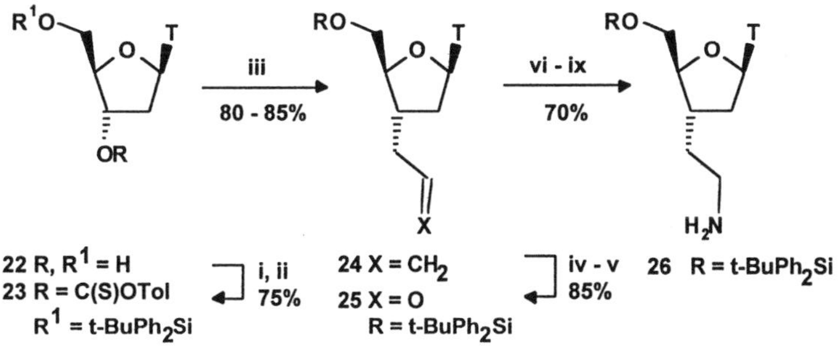
R¹O
T
O
OR
iii
80 - 85%
RO
T
O
X
vi - ix
70%
RO
T
O
H2N
22 R, R¹ = H
23 R = C(S)OTol
R¹ = t-BuPh2Si
i, ii
75%
24 X = CH2
25 X = O
R = t-BuPh2Si
iv - v
85%
26 R = t-BuPh2Si

Scheme 3: i) t-$BuPh_2SiCl$, imidazole, DMF. ii) Tol-OCS-Cl, DMAP, Et_3N, CH_2Cl_2. iii) allyltributyltin, AIBN, PhH (0.1 M), 80° C. iv) N-methylmorpholine oxide (NMMO), OsO_4, acetone/H_2O (4:1). v) $NaIO_4$, dioxane/H_2O (3:1). vi) $NaBH_4$, MeOH. vii) TsCl, DMAP, pyr, $CHCl_3$. viii) LiN_3, NaI, DMF, 100° C. ix) n-Bu_3SnH, AIBN, PhH, 80° C.

Synthesis of the Amide 3 Dimers (3′-CH$_2$-CO-NR-CH$_2$) (*3,16*). Oxidation of aldehyde **25** (Scheme 5) to **30** was efficiently realized using sodium chlorite (*17*). Amines **32a-32c**, which were coupled with acid **30**, were obtained by substitution of the tosyl group in **31** with methyl- or isopropylamine (for **32b**, **32c**) or with LiN$_3$ (*18*) followed by reduction (for **32a** (*19*)). Only the depicted rotamer was detected by ^{1}H-NMR in the case of the amide dimer **33a**, whereas in amides **33b**, **33c** both rotamers were populated (^{1}H-NMR NOE experiments; ratio $\cong$ 4:1 for **33b**, **33c** in CDCl$_3$). We synthesized the eight building blocks **30** and **32** having the four different bases as well as five dimers (*16b*).

Synthesis of the Amide 4 Dimer (3′-CH$_2$-NH-CO-5′) (*3,20*). Aldehyde **36** (Scheme 6) is the key intermediate for the synthesis of amide **4** and **5**. The known procedures for **36** (*2a,21*) are very unsatisfactory. Therefore, we developed an efficient synthesis using a stereoselective radical addition with -tri-n-butyl-tinstyrene (*22*). A single isomer, **35**, was obtained. The C=C bond was transformed via **36** to the amine **37**. Acid **40** was obtained from aldehyde **38** (*23*) by *Wittig* reaction to **39** (*24*), followed by methanolysis. After the hydrolysis of the methylester to acid **40**, coupling with amine **37** and the further elaboration of the resulting dimer **41** to **42** was performed as described for the previous cases.

Synthesis of the Amide 5 Dimer (3′-CO-NH-CH$_2$-5′) (*3,25*). Aldehyde **36** was oxidized to the methyl ester (Scheme 7) followed by base protection. The methyl ester was saponified to acid **43**. The coupling of amine **44** (*26*) with acid **43** was achieved by using the *Ghosez* reagent (*27*), followed by N(3)-deprotection using DDQ.

Synthesis of the Urea Dimers 6 (3′-NR^2-CO-NR^1-5′) (*28*). Amine **47** (Scheme 8), obtained from the known azide (*7*) by N(3)-BOM-protection and reduction, was trifluoroacetylated. Methylation to **48** was followed by cleavage of BOM, accompanied by deprotection of the 5′-silicon group. Deacetylation of the amide and 5′-protection led to **49**. Amine **32a** was coupled to the p-nitrophenyl carbamate obtained in situ from **49**. Amines **32b, c** were coupled with the 3′-phenyl carbamate **52** (Scheme 9).

Synthesis of the Carbamate Dimers 7 (3′-O-CO-NR-5′) and 8 (3′-NR-CO-O-5′) (*29*). The incorporation of the carbamate **56** in ODNs was already reported, but the results on the duplex formation between these ODNs and their complements are contradictory (*30*). **55** (Scheme 10) was coupled with the amine **32**. For the synthesis of **60**, amines **58** were coupled with **59** (Scheme 11). The trityl group in **60** (R, R^1 = H, R^2 = trityl) was replaced by the DMT group, and the phosphoramidite **61** was obtained as described before.

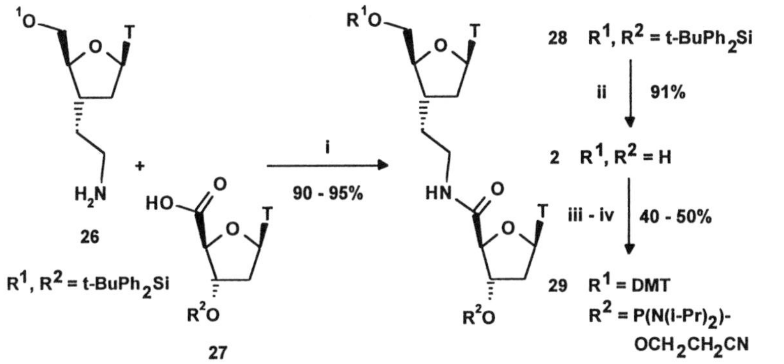

Scheme 4: i) N-methylmorpholine (NMM), TBTU, NHBT, CH_3CN, then **26**, NMM. ii) TBAF, AcOH, THF. iii, iv) see vii, viii (Scheme 1).

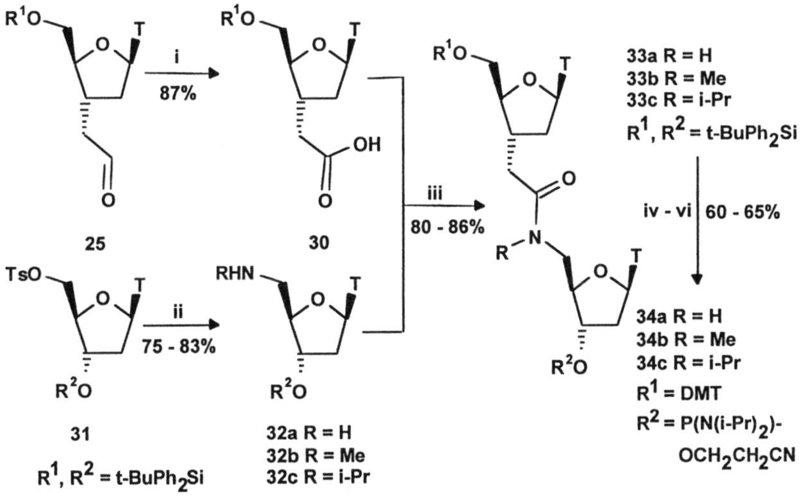

Scheme 5: i) $NaClO_2$, 2-methyl-2-butene, NaH_2PO_4, t-BuOH, H_2O. ii) RNH_2, THF (R = H, see (19), R = Me (83%), R = i-Pr (75%)). iii) **30**, NMM, TBTU, NHBT, CH_3CN, then **32**, NMM (R = H (86%), R = Me (80%), R = i-Pr (85%)). iv) TBAF, AcOH, THF. v, vi) see vii, viii (Scheme 1).

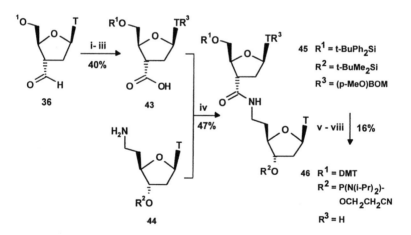

Scheme 6: i) Ph-CH=CH-SnBu₃, AIBN, PhH, 80° C. ii) NMMO, OsO₄, NaIO₄, dioxane/H₂O (3:1). iii) NaBH₄, MeOH. iv) TsCl, DMAP, CHCl₃. v) LiN₃, NaI, DMF, 100° C. vi) n-Bu₃SnH, AIBN, PhH, 80° C. vii) 1,3-dithia-2-cyclohexylidene triphenylphosphorane, THF, -78° C → RT. viii) HgCl₂, MeOH/H₂O (9:1), reflux. ix) NaOH (0.5 M), H₂O, THF. x) NMM, TBTU, NHBT, CH₃CN, **37**, NMM. xi) TBAF, THF, AcOH. xii, xiii) see vii, viii (Scheme 1).

Scheme 7: i) PDC, MeOH, DMF. ii) p-MeOBOMCl, DBU, CH₂Cl₂/ CH₃CN (1:1). iii) NaOH (0.1 M), THF/H₂O (2:1). iv) 1-Chloro-N,N,2-trimethylpropenylamine, **43**, CH₂Cl₂, then **44**, Et₃N. v) DDQ, CH₂Cl₂/H₂O (18:1). vi) TBAF, AcOH. vii, viii) see vii, viii (Scheme 1).

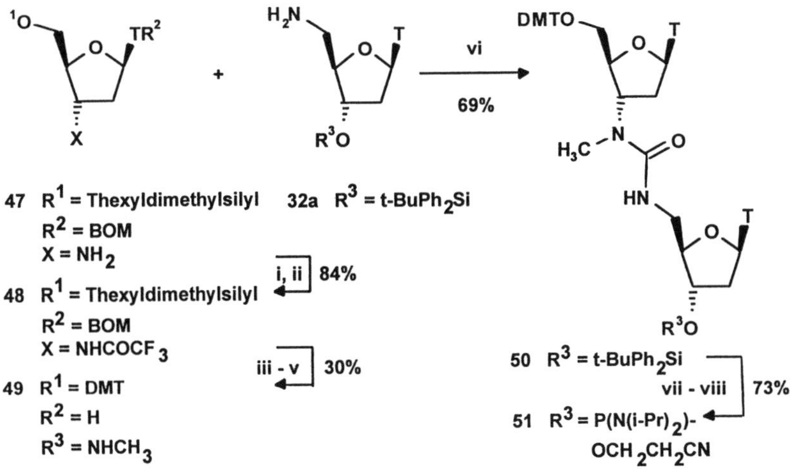

Scheme 8: i) $(CF_3CO)_2O$, pyr, CH_2Cl_2, 0° C. ii) MeI, NaH, THF. iii) H_2, Pd/C 10%, MeOH. iv) KOH, MeOH, H_2O. v) DMTCl, pyr. vi) p-$NO_2C_6H_4OCOCl$, **49**, pyr, then **32a**. vii) TBAF, THF. viii) see viii (Scheme 1).

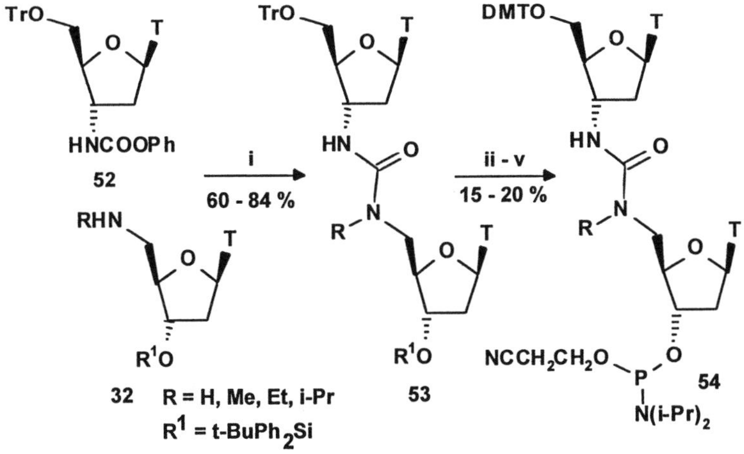

Scheme 9: i) DMF, 80° C. ii) AcOH (80%). iii) DMTCl, pyr. iv) TBAF, THF. v) see viii (Scheme 1).

Scheme 10: For R = alkyl, R^1 = diphenyl-t-butylsilyl, R^2 = OC_6H_4-p-NO_2 i) pyr, 80° C. ii) TBAF, THF, AcOH. iii) see viii) (Scheme 1). For R, R^1 = H, R^2 = imidazole i') pyr, 16 h; ii') as iii).

Scheme 11: For R, R^1 = H, R^2 = Trityl i) pyr, 80° C. ii) AcOH (80%). iii, iv) see vii, viii (Scheme 1). For R = Me, R^1 = BOM, R^2 = thexyl-dimethylsilyl i') DMF, DMAP, 80° C. ii') TBAF, THF, AcOH. iii') H_2, Pd/C (10%), MeOH. iv'-v') as iii'-iv).

**Synthesis of the Carbon Chain Internucleoside Dimers 9 (3'-CH$_2$-CO-CH$_2$-5')
and 10 (3'-CH$_2$-CH$_2$-CH$_2$-5')** *(31)*. The key step is the coupling of 64 with the
bromoacetylene 63 *(32)*. Compond 62 was obtained from the BOM protected
thymidine using *Swern* oxidation followed by the *Wittig* reaction *(33)* (Scheme 12).
Compound 62 was transformed to the bromo acetylene 63. The coupling *(34)* was
performed by adding a premixed solution of 64 and 63 in THF to CrCl$_2$
contaminated with 0.01% of NiCl$_2$ (Scheme 13). The triple bond in 65 was reduced,
and after deoxygenation *(35)*, the BOM protective groups were removed. For the
ketone 68 (Scheme 14), dimer 65 was hydrogenated. The resulting dimer was
oxidized to 68 with the *Dess-Martin* periodinane *(36)*.

Incorporation of the Dimers 1-10 into ODNs

The solid phase synthesis of the ODNs containing our modified dimers 1-10 and
their purification were performed using standard procedures *(11,37)*. The stability of
the dimers 1-10 under the conditions required for the solid phase synthesis was
verified by ^{1}H-NMR after treatment with aqueous NH$_3$ (25%) at 60° C for 12 hours.

**Thermal Denaturation of the Duplexes Formed Between Modified ODNs and
Complementary RNA Strands**

The influence of our new modifications 1-10 on the thermal stability of the duplexes
formed between the ODNs and their RNA complements was determined by
measuring the melting temperature (T$_m$) of the hybrids *(38)*. The ODNs shown in
Table I were used; the * represents a modified internucleosidic linkage (1-10). The
other linkages between the nucleosides are normal phosphodiester bonds.

Table I. Oligomer Sequence (5' → 3')

Entry	Sequences	T$_m$ (° C) Wild Type with RNA Complement
A	CTC GTA CCT *TTC CGG TCC	63.3
B	CTC GTA CT*T T*TC CGG TCC	61.8
C	GCG T*TT *TT*T T*TT *TGC G	50.2
D	TTT T*TC TCT CTC TCT	52.8

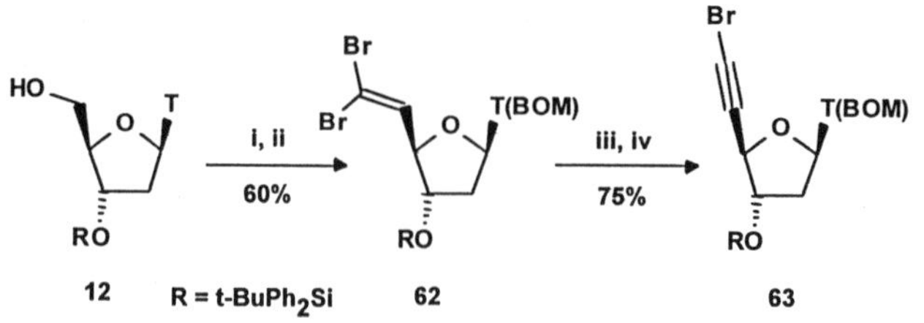

Scheme 12: i) BOMCl, DBU, CH$_3$CN. ii) oxalyl chloride, DMSO, CH$_2$Cl$_2$, -78° C, then Et$_3$N, Ph$_3$P=CBr$_2$. iii) TBAF, THF. iv) t-BuPh$_2$SiCl, imidazole, DMF.

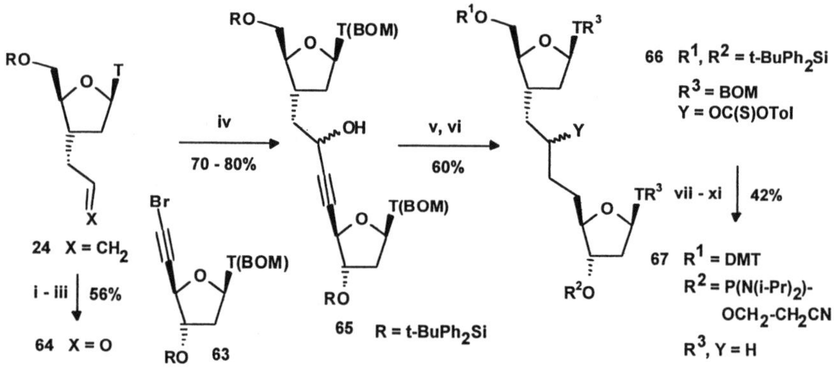

Scheme 13: i) see i) (Scheme 12). ii) NMMO, OsO$_4$, acetone/H$_2$O (4:1). iii) NaIO$_4$, dioxane/H$_2$O (3:1). iv) **63**, **64**, CrCl$_2$, NiCl$_2$, THF. v) H$_2$, Pd/C (5%), THF. vi) Tol-OCS-Cl, DMAP, Et$_3$N, CH$_2$Cl$_2$. vii) tributyltinhydride, AIBN, PhCH$_3$ (0.1 M), reflux. viii) H$_2$, Pd/C (10%), THF/MeOH (1:1). ix) TBAF, THF. x, xi) see vii, viii (Scheme 1).

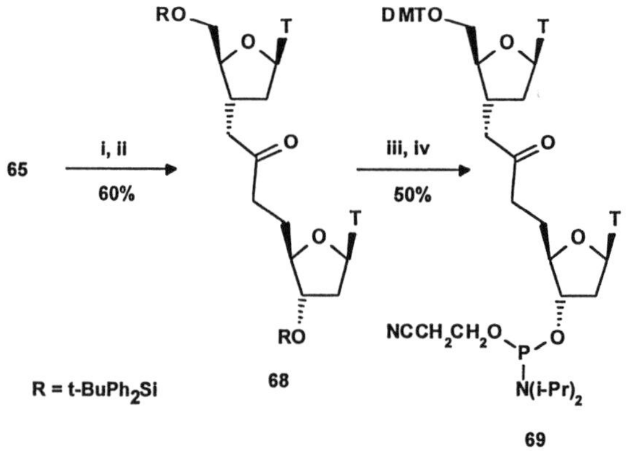

Scheme 14: i) H$_2$, Pd/C (10%), THF/MeOH (1:1). ii) *Dess-Martin* reagent, CH$_2$Cl$_2$. iii) TBAF, THF, AcOH. iv) see vii, viii) (Scheme 1).

We report here average values for the variation of the melting temperatures of the duplexes (T_m/modification) for the modifications **1-10** obtained from the duplexes A-D. These data are summarized in Table II.

The amide **1** modifications lead to a substantial destabilization (Table II: entries 1- 3). An increase of the size of the substituent on the nitrogen atom of the amide function further decreases the melting temperature of the duplex. Interestingly, amide **1** and amide **5** (entries 1, 9), which adopt similar geometries in the duplex with RNA, display also very similar T_ms. These results suggest that restricted rotation and steric hindrance near the upper sugar moiety destabilize the duplex. When the restricted rotation around the amide bond is located near the lower sugar unit as in **2** (entry 4), it also leads to destabilization. However, in this case the decrease of the T_m is only -1.6° C per modification. According to our molecular modeling study, amide **2** can adopt a geometry close to the one of the DNA strand in a duplex with RNA. In contrast to these destabilizing backbone modifications, amide **3** slightly increases the melting temperature (entry 5). The amides **3, 4** adopt similar geometries in the duplexes, and they display comparable T_m values. These results obtained for amides **1** and **5** and for amides **3** and **4** suggest that the geometrical factors are predominant in the stability of a duplex between a modified ODN and an RNA target.

Table II: Average ΔTm values per backbone modification

Entry	Modification		T_m/Mod.	Entry	Modification		T_m/Mod.
1	**1**	R = H	-2.8	11	**6**	R^1 = Me, R^2 = H	-3.6
2	**1**	R = Me	-3.2	12	**6**	R^1 = i-Pr, R^2 = H	-3.1
3	**1**	R = i-Pr	-3.6	13	**6**	R^1 = Et, R^2 = H	-3.8
4	**2**		-1.6	14	**6**	R^1 = H, R^2 = Me	-5.2
5	**3**	R = H	+0.4	15	**7**	R = H	-3.6
6	**3**	R = Me	-0.1	16	**7**	R = Me	-2.9
7	**3**	R = i-Pr	0.0	17	**7**	R = i-Pr	-2.0
8	**4**		-0.3	18	**8**	R = H	-4.4a)
9	**5**		-3.5	19	**8**	R = Me	-4.8a)
10	**6**	R^1, R^2 = H	-3.8a)	20	**9**		-2.5
				21	**10**		-4.2

a) No duplex formation >20° C when incorporated in the ODN C (Table I)

Moreover, we measured the T_m of the duplex formed between the ODN E (Table III) containing alternating phosphodiester-amide **3** linkages with its complementary RNA strand. A higher affinity for RNA was observed for the modified ODN than for the wild type. This result is consistent with the one when amide **3** is introduced as point modification in the backbone (Table II, entry 5).

The affinity of the corresponding ODN for an RNA target, as well as its resistance towards nucleases, is increased. The amide **1-5** modifications increase the resistance of the ODNs towards 3'-exonucleases (*39*) by a factor of 3-10. The resistance of the ODNs towards endo- and exonucleases under physiological conditions was highly increased in an alternating phosphodiester-amide **3** backbone (as in E).

Table III: **Melting temperature (Tm) with RNA complement**

Oligomer sequence (5' → 3')	Tm (^{o}C)	ΔTm (^{o}C)/modification
D TTT TTC TC TCT CTC TCT	52.8	
E T*T T*T T*C T*C T*C T*C T*CT	57.1	+0.6

* = 3'-CH$_2$-CO-NH-5'; For experimental details for Tm determination see (*38*)

When the amide bond is located in the middle of the backbone, an increase of the size of the substituent on the nitrogen atom does not destabilize substantially the duplex (entries 5-7). This is in agreement with our molecular modeling study. In the lowest energy conformation for amide **3**, there is space for a substituent as large as isopropyl without severe steric interactions.

All our amide modifications show at least the same base specificity (Watson-Crick base pairing rules) as the wild type. Mismatches opposite to an amide backbone in the RNA strand caused similar drops in T_m values as for the wild type.

The more rigid urea derivatives **6** (entries 10-14, Table II) displayed a severe destabilization of the duplexes. A substituent larger than hydrogen on the 5'-nitrogen (entries 10-13) has little effect on the T_m. In contrast, substituents on the 3'-nitrogen (entry 14) are not well tolerated.

A slight improvement of the thermal stability of duplexes formed with carbamate modified ODNs **7, 8** (entries 15-17, Table II) was observed compared to the urea analogs. Additional alkyl substituents on the 5'-nitrogen atom in **7** (entries 16, 17) had no effect on the T_ms. The thermal stability of ODNs with the isomeric carbamates **8** proved to be worse. No duplex formation above 20° C was observed with five modifications (entries 18, 19, Table II). The introduction of a rigid moiety and of steric hindrance in the vicinity of the 3'-carbon center of the upper sugar in the ureas **6** and in the carbamates **8** has a negative effect on the thermal stability of the duplexes.

These results make clear that restriction of the rotation has to occur preferentially right in the middle of the backbone. Adjusting the distance between the sugars and rigidity (preorganization) of the backbone is another important parameter to gain the desired thermal stability of the duplexes. The backbones **9** and **10,** which are much more flexible than the amides **3** and **4** displayed, decreased binding affinity (entries 20, 21, Table II).

Conclusions

We have identified two backbone replacements, namely the amides **3** and **4**, which show good binding affinity of the corresponding ODN for the RNA target. The amide **3** dimers are readily accessible and are introduced in biologically relevant sequences. An alternating phosphodiester-amide **3** (or amide **4**) backbone confers not only a good affinity for RNA but also a highly increased stability under physiological conditions. Further synthetic modifications of the amides **3** and **4**, which might be required to reach ultimately potent biological activity, are under current investigation.

Acknowledgments

We thank Drs. U. Pieles and D. Hüsken (Ciba-Geigy) for the synthesis and the purification of the ODNs as well as Dr. S. M. Freier (Isis Pharmaceuticals, Carlsbad, CA 92008, USA) for performing hybridization experiments. Fruitful discussions with Dr. H. Moser are gratefully acknowledged. We thank P. Hoffmann, A. Garnier, G. Rüegg, and T. Lochmann for technical assistance in the synthesis of the dimers and V. Drephal, M.-L. Piccolotto, and W. Zürcher for the ODNs syntheses, their purification, and the measurements of T_ms.

References and Notes

1. *Oligodeoxynucleotides. Antisense Inhibitors of Gene Expression*; Cohen, J. S., Ed.; CRC Press, Inc.: Boca Raton, FL, USA, 1989. (b) Uhlmann, E.; Peyman, A. *Chem. Rev.* **1990**, *90*, 543. (c) Crooke, S. T. *Annu. Rev. Pharmacol Toxicol.* **1992**, *32*, 329. (d) Cook, P. D. *Anti-Cancer Drug Design.* **1991**, *6*, 585.
2. Vasseur, J.-J.; Debart, F.; Sanghvi, Y. S.; Cook, P. D. *J. Am. Chem. Soc.* **1992**, *114*, 4006. (b) Reynolds, R. C.; Crooks, P. A.; Maddry, J. A; Akhtar, M. S.; Montgomery, J. A.; Secrist, J. A. *J. Org. Chem.* **1992**, *57*, 2983. (c) Jones, R. J.; Lin, K. Y.; Milligan, J. F.; Wadwani, S.; Matteucci, M. D. *J. Org. Chem.* **1993**, *58*, 2983. (d) Meng, B.; Kawai, S. H.; Wang, D.; Just, G.; Giannaris, P. A.; Damha, M. J. *Angew. Chem. Int. Ed. Engl.* **1993**, *32*, 729, and references cited therein. (e) Quaedflieg, P. J. L. M.; Van der Marel, G. A.; Kuyl-Yeheskiely, E.; Van Boom, J. H. *Recl. Trav. Chim. Pays-Bas.* **1991**, *110*, 435.
3. De Mesmaeker, A.; Lebreton, J.; Waldner, A.; Cook, P. D. *Backbone Modified Oligonucleotide Analogs*, International Patent WO 92/20823, 1992.
4. The names "amide **1**" to "amide **5**" refer to the order of synthesis of the modifications in our laboratories.
5. Jaroszewski, J. W.; Cohen, J. S. *Adv. Drug Delivery Rev.* **1991**, *6*, 235.
6. Lebreton, J.; De Mesmaeker, A.; Waldner, A.; Fritsch, V.; Wolf, R. M.; Freier, S. M. *Tetrahedron Lett.* **1993**, *34*, 6383.
7. Maillard, M.; Faraj, A.; Frappier, F.; Florent, J.-C.; Grierson, D.S.; Monneret, C. *Tetrahedron Lett.* **1989**, *30*, 1955.
8. Barton, D. H. R.; Géro, S. D.; Quiclet-Sire, B.; Samadi, M. *Tetrahedron Lett.* **1989**, *30*, 4969.

9. Harada, K.; Orgel, L. E. *Nucleosides Nucleotides* **1990**, *9*, 771.
10. Knorr, R.; Trzeciak, A.; Bannwarth, W.; Gillessen, D. *Tetrahedron Lett.* **1989**, *30*, 1927.
11. Sinha, N. D.; Biernat, J.; McManus, J.; Köster, H. *Nucl. Acids Res.* **1984**, *12*, 4539.
12. Krecmerova, M.; Hrebabecky, H.; Holy, A. *Collect. Czech. Chem. Commun.* **1990**, *55*, 2521.
13. De Mesmaeker, A.; Lebreton, J.; Waldner, A.; Fritsch, V.; Wolf, R. M.; Freier, S. M. *Synlett* **1993**, 733.
14. Fiandor, J.; Tam, S. Y. *Tetrahedron Lett.* **1990**, *31*, 597.
15. Samano, M. C.; Robins, M. J. *Tetrahedron Lett.* **1991**, *32*, 6293.
16. De Mesmaeker, A.; Waldner, A.; Lebreton, J.; Hoffmann, P.; Fritsch, V.; Wolf, R. M.; Freier, S. M. *Angew. Chem. Int. Ed. Engl.* **1994**, *33*, 226. (b) Lebreton, J.; Waldner, A.; Lesueur, C.; De Mesmaeker, A. *Synlett* **1994**, 137. (c) For an independent report, see: Idziak, I.; Just, G.; Damha, M. J.; Giannaris, P. A. *Tetrahedron Lett.* **1993**, *34*, 5417.
17. Dalcanale, E.; Montanari, F. *J. Org Chem.* **1986**, *51*, 567.
18. Lin, T.-S.; Prusoff, W. H. *J. Med. Chem.* **1978**, *21*, 109.
19. Maiti, S. N.; Singh, M. P.; Micetich, R. G. *Tetrahedron Lett.* **1986**, *27*, 1423.
20. Lebreton, J.; Waldner, A.; Fritsch, V.; Wolf, R. M.; De Mesmaeker, A. *Tetrahedron Lett.* in press.
21. Parkes, K. E. B.; Taylor, K. *Tetrahedron Lett.* **1988**, *29*, 2995.
22. Baldwin, J. E.; Kelly, D. R. *J. Chem. Soc., Chem. Commun.* **1985**, 682.
23. Jones, G. H.; Taniguchi, M.; Tegg, D.; Moffatt, J. G. *J. Org. Chem.* **1979**, *44*, 1309.
24. Hauser, F. M.; Hewawasam, P.; Rho, Y. S. *J. Org. Chem.* **1989**, *54*, 5110.
25. De Mesmaeker, A.; Lebreton, J.; Waldner, A.; Fritsch, V.; Wolf, R. M. *Bioorg. Med. Chem. Lett.* **1994**, *4*, 873.
26. Etzold, G.; Kowollik, G., Langen, P. *J. Chem. Soc., Chem. Commun.* **1968**, 422.
27. Devos, A.; Remion, J.; Frisque-Hesbain, A.; Colens, A.; Ghosez, L. *J. Chem. Soc., Chem. Commun.* **1979**, 1180. (b) De Mesmaeker, A.; Hoffmann, P.; Ernst, B. *Tetrahedron Lett.* **1989**, *30*, 3773.
28. Waldner, A.; De Mesmaeker, A.; Lebreton, J.; Fritsch, V.; Wolf, R. M. *Synlett* **1994**, 57.
29. Waldner, A.; De Mesmaeker, A.; Lebreton, J. *Bioorg. Med. Chem. Lett.* **1994**, *4*, 405.
30. Coull, J. M.; Carlson, D. V.; Weith, H. L. *Tetrahedron Lett.* **1987**, *28*, 745. (b) Stirchak, E. P.; Summerton, J. E.; Weller, D. D. *J. Org. Chem.* **1987**, *52*, 4202.
31. Lebreton, J.; De Mesmaeker, A.; Waldner, A. *Synlett* **1994**, 54. (b) De Mesmaeker, A.; Waldner, A.; Sanghvi, Y. S.; Lebreton, J. *Bioorg. Med. Chem. Lett.* **1994**, *4*, 395.
32. A synthesis of a base protected form of **9** was reported: Butterfield, K.; Thomas, E. J. *Synlett* **1993**, 411. (b) Compound **10** was reported in a patent application: Weis, A. L.; Hausheer, F. H.; Chaturvedula, P. V. C.; Delecki, D. J.; Cavanaugh, P. F.; Moskwa, P. S.; Oakes, F. T. *Compounds and Methods for Inhibiting Gene Expression*, International Patent WO 92/02534, 1992.
33. Ireland, R. E.; Norbeck, D. W. *J. Org. Chem.* **1985**, *50*, 2198. (b) Camps, F.; Coll, J.; Fabrias, G.; Guerrero, A.; Riba, M. *Tetrahedron Lett.* **1983**, *24*, 3387.
34. Takai, K.; Kuroda, T.; Nakatsukasa, S.; Oshima, K.; Nozaki, H. *Tetrahedron Lett.* **1985**, *26*, 5585.
35. Barton, D. H. R.; McCombie, S. W. *J. Chem. Soc. Perkin Trans I* **1975**, 1574.
36. Dess, D. B.; Martin, J. C. *J. Org. Chem.* **1983**, *48*, 4156.

37. Each oligonucleotide was prepared on an ABI 390 DNA synthesizer using standard phosphoramidite chemistry: Gait, M. J. *Oligonucleotide Synthesis: A Practical Approach*; IRL Press: Oxford, 1984. Each ODN was checked by mass spectroscopy (MALDI-TOF: Pieles, U.; Zürcher, W.; Schär, M.; Moser, H. E. *Nucl. Acids Res.* **1993**, *21*, 3191).

38. The thermal denaturation of DNA/RNA hybrides was performed at 260 nm. See ref. 16a) and Freier, S. M.; Albergo, D. D.; Turner, D. H. *Biopolymers* **1983**, *22*, 1107. All values are averages of at least three experiments. The absolute experimental error of the T_m values is ± 0.5 C.

39. The modifications were incorporated in the following sequence where * refers to the position of the modifications amide **1-5**: 5'-CGA CTA TGC AAT T*TC. The resistance was tested with 10% fetal calf serum at 37° C. See: Hoke, G. D.; Draper, K.; Freier, S. M.; Gonzales, C., Driver, V. B.; Zounes, M. C.; Ecker, D. J. *Nucl. Acids Res.* **1991**, *19*, 5743.

RECEIVED August 18, 1994

Chapter 3

Synthesis of Nonionic Oligonucleotide Analogues

Joseph A. Maddry, Robert C. Reynolds, John A. Montgomery, and John A. Secrist III

Organic Chemistry Department, Southern Research Institute, Birmingham, AL 35255

The design and synthesis of several classes of oligonucleotide analogs (OAs) incorporating improved, achiral, non-ionic phosphate linker surrogates are reported. Specifically, the synthesis and properties of monomers and oligomers based on five such linkers will be described, including lipophilic linkers based on various dialkylsiloxane moieties, especially diisopropylsiloxane; two moderately polar linkers (carboxylate ester and carboxamide) based on the appropriately 3'-substituted (hydroxy and amino), 5'-doubly-homologated nucleoside carboxylic acids; and polar surrogates (sulfonate esters and amides) prepared from the 3'-substituted, 5'-homologated nucleoside sulfonic acids. The potential advantages of oligonucleotides bearing non-ionic backbone linkages as clinical therapeutic agents are briefly discussed.

The enormous potential of antisense oligonucleotide analogs (OAs) for the treatment of disease has elicited great excitement among medicinal chemists, who are intrigued by their promise of high specificity and broad applicability. The major goal of this strategy is to inhibit gene expression by interference with the transcription, translation, or replication of a given gene, through complementary (normally Watson-Crick) hybridization between the target (sense) and OA (antisense) strands. This interest has prompted extensive research designed to improve the stability, bioavailability, and potency of OAs. Our interest in this area arose from the observation that the current generation of OAs suffered from several shortcomings that might ultimately preclude their introduction as clinically useful agents. Thus, antisense compounds composed of natural DNA or RNA sequences suffer from rapid enzymatic degradation and poor uptake; DNA analogs based on the phosphorothioate linker, while of improved stability, nonetheless are charged, resulting in inefficient cellular membrane permeability and rapid clearance from the plasma, and chiral, which can cause intractable purification problems; and methylphosphonates, though uncharged, still suffer from the chirality drawback and typically display low potency. It was our belief that useful drugs would emerge only through the use of achiral, non-ionic OAs (NOAs). Consequently, we embarked upon a program of design and synthesis to

0097–6156/94/0580–0040$08.00/0

investigate several classes of OAs incorporating improved phosphate linker surrogates. This chapter will describe the synthesis and properties of monomers and oligomers based on five such novel linkers (dialkylsiloxane, carboxylate ester, carboxamide, sulfonate ester, and sulfonamide) developed in our laboratories (*1-4*).

Silicon-Based NOAs

Initially we chose to pursue NOAs beaing a silicon atom in place of the phosphorus of the native linker. Silicon-derived linkers possess several potential advantages, besides their being non-ionic. The silicon atom is similar in size to phosphorus, and like its heavier counterpart in phosphate it bonds to four tetrahedrally arranged ligands; furthermore, these groups can be chosen to ensure an achiral environment. Considerations of synthetic feasability and hydrolytic stability led us away from moieties containing silicon directly bonded to four oxygen atoms, so we chose instead to pursue analogs containing dialkylsiloxane residues, where the alkyl groups could be selected to modulate linker stability and polarity, with the constraint that both alkyl groups must be identical to render the resulting phosphate mimic achiral. While our work was ongoing, similar efforts to prepare silicon-based OAs were reported by other laboratories (*5-6*).

Scheme 1 depicts our initial strategy for preparing siloxane-based NOAs. Suitably protected nucleosides bearing a free 3'-hydroxyl group were treated with a slight excess of the difunctional silylating agent, diphenyldichlorosilane. Subsequently, an equivalent portion of a 3'-blocked nucleoside was added, and the product then selectively deprotected in a subsequent step to permit chain elongation at the 5'-terminus. Oligomerization in the 3' → 5' direction was chosen to facilitate subsequent automation of the protocol. We employed fairly standard protecting groups for the nucleoside hydroxyls (dimethoxytrityl for the 5'-position, which was easily removed by mild acid treatment, and levulinyl at the 3'-position, cleaved by hydrazinolysis), and for simplicity began the development of the chemistry with thymidine, so no protection of the nucleoside base was necessary. Our choice of diphenylsiloxane as a phosphate replacement represented a compromise between the need for sufficient alkyl bulk at the silicon site to ensure hydrolytic stability, while retaining sufficient reactivity there to allow successive substitution by the two somewhat hindered nucleoside hydroxyl groups. In retrospect this choice was inappropriate, since nucleophilic reaction at phenyl-substituted silicon atoms was known to occur at a significantly faster rate than would be expected from steric considerations alone, approaching that of the congeneric methyl case; presumably, electronic effects are responsible for this acceleration.

From the outset, we realized the potential for symmetrically-substituted by-product formation (5'-5' and 3'-3' coupled products) resulting from the disubstitution of two equivalents of one of the nucleoside precursors, rather than stepwise addition of each of the two components. Thus the experimental design sought to minimize symmetrical coupling by conducting the reaction at reduced temperature under dilute conditions, and introducing the more hindered secondary alcohol as the initial reactant. Unfortunately, these techniques were only modestly successful, and no variant of the experimental conditions resulted in a high yield of the desired 5'-3' dimer with elimination of the disubstitution side reaction. These observations conform with the results subsequently described by other groups.

In another attempt to circumvent these problematic reactions, we examined

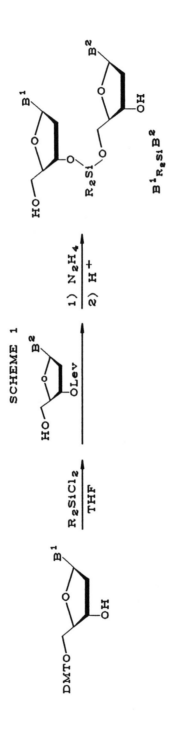

SCHEME 1

several dichlorosilanes whose reactivity was varied through modification of the alkyl substituents. Table I lists the various silanes that were examined. Though most of

Table I. Silanes $R_2SiX^1X^2$ examined for reactivity

R	X^1	X^2
C_6H_5	Cl	Cl
C_6H_5	$(CH_3)_2N$	$(CH_3)_2N$
C_6H_5	CH_3CO_2	CH_3CO_2
C_6H_5	Cl	$C_6H_{11}NH$
C_6H_5	Cl	$C_6H_5CH_2NH$
C_6H_5	Cl	$(CH_3)_2NH$
$(CH_3)_3C$	Cl	Cl
$CH_3(CH_2)_3$	Cl	Cl
CH_3CH_2	Cl	Cl
$c\text{-}C_6H_{11}$	Cl	Cl
$(CH_3)_2CH$	Cl	Cl
$(CH_3)_2CH$	CF_3SO_3	CF_3SO_3

these compounds were commercially available, others were prepared as needed by standard techniques, typically involving treatment of silicon tetrachloride with two equivalents of the appropriate Grignard reagent, followed by distillation. Also investigated were representative dialkyldiaminosilanes and one dialkyldiacetoxysilane; these reagents generally displayed selectivity no better than their chlorinated counterparts, and their reduced reactivity often mandated higher temperatures for complete conversion. Several differentially functionalized silanes were also prepared, in hopes that the two leaving groups would be sufficiently disparate in reactivity to facilitate the desired stepwise addition.

Though some of these coupling agents were examined only cursorily, we found that product distribution and yield were more successfully influenced through variation of the silicon alkyl groups then by replacement of chloride with alternative leaving groups. Thus, diisopropyldichlorosilane consistently gave the highest reproducible yield of the desired 3'-5'-linked, stepwise addition product, and so was employed for all subsequent studies. Though the ~40% coupling efficiency with this reagent is low, particularly if long oligomers were to be synthesized, we anticipated that coupling yields would markedly improve through automation of the sequence on a DNA synthesizer, where complications resulting from symmetrical disubstituion would be eliminated by virtue of the solid phase technique.

We next turned our attention to the corresponding chemistry of the remaining deoxynucleosides. Table II shows the oligomers in the dialkylsiloxane series that were synthesized. All four natural DNA bases were successfully incorporated into siloxane-linked dimers (though, for unrelated reasons, the 2'-benzoyl-3'-O-methylriboside was employed in the guanine case, rather than the deoxyriboside). A single example of a tetramer was also prepared (5'-A-C-C-C-3', with dibutylsiloxane linkages) but not fully

Table II. Dialkylsiloxane-linked deoxynucleoside dimers isolated

Sequence (5′→3′)	Silyl Linker
T-T	diphenyl, dibutyl, diisopropyl
A-C	dibutyl
C-C	dibutyl
A-G (3′-O-methyl)	diisopropyl

characterized. The resulting compounds, colorless, amorphous solids, had satisfactory stability, permitting facile deprotection and purification. One particular concern that emerged, however, was the extreme lipophilicity of these derivatives, even after deprotection, and especially the dramatic difference in apparent polarity (by tlc) between the dimeric and tetrameric species; extrapolating this trend to longer oligomers, it became obvious that this particular attribute was likely to interfere with hybridization to DNA or mRNA, where even if aqueous solubility of the silicon-based oligomer were not problematic, alternative conformational foldings were still likely that could effectively desolvate the hydrophobic backbone, or otherwise make duplex formation energetically unfavorable. Mediocre yields, polarity concerns, and the reports of parallel findings from other groups caused us to suspend work in the silicon series and focus on alternative phosphate surrogates (5-7).

Carboxyl-Based NOAs

In addition to the silicon-based compounds, we briefly pursued both carboxamide and carboxylate ester linkers. Oligonucleotide analogs containing these non-ionic linkers should have identical internucleoside spacing compared with the natural phosphate backbone, thus allowing base pairing and duplex formation with the complementary strand. Additionally, carboxylate ester and carboxamide linkers should not be nuclease substrates, although the possibilty that non-specific esterases or amidases might cleave such linkers was recognized.

Several permutations of the basic amide or ester link are possible and most have been reported since this work began (8-12). However, we intially synthesized linkers based on the known 6′-carboxyl homonucleosides shown in Scheme 2 (13). In the thymidine series, both benzyl and trimethylsilylethyl (TMSE) esters were pursued. The benzyl ester was useful because the chemistry was known and yields of pure 5′-deoxy-5′-thymidineacetic acid are relatively high (14). However, an advantage for TMSE esters was envisioned during a solid phase protocol, wherein ester coupling would be followed by removal of the TMSE blocking group with fluoride, allowing the next acyl activation and coupling to then proceed. Therefore, the TMSE esters were given a higher priority.

The 3′-acetyl protected nucleosides (1, B=A, N^2-dimethylaminomethyleneG, T) were oxidized via the Pfitzner-Moffat method and coupled directly with the TMSE-protected ylide (15) to give only the trans-oriented olefins in fair yields after workup. The olefins were reduced by standard catalytic hydrogenation to give good yields of 2 (B=A,G,T) (13-14). At this juncture, only the thymidine derivative was used to study dimerization conditions. Compound 2 was readily deprotected at the carboxyl group by treatment with tetra-N-butylammonium fluoride (1.0M in tetrahydrofuran) for between one and two hours to give the free 6′-carboxylate 3. Possibly, pulsing with higher concentrations in a solid phase, automated system could decrease the reaction time.

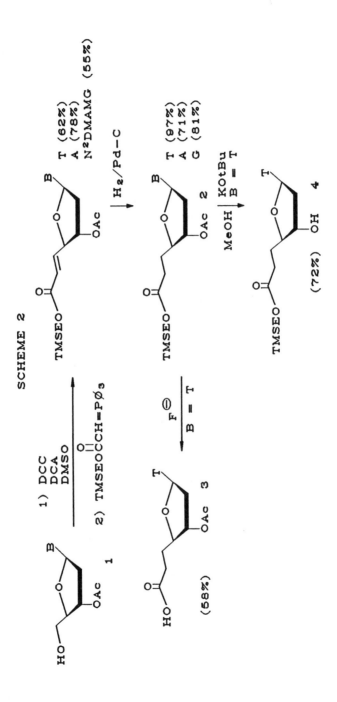

Additionally, the 3′-acetyl group in these nucleosides is much more labile than the other ester function and was easily removed by methoxide in methanol to give **4** in good yield. Finally, the ester dimer was prepared by coupling **3** and **4** using DCC as the coupling reagent (Scheme 3). Furthermore, an amide dimer has been prepared by coupling **3** with 3′-amino-5′-tritylthymidine using similar conditions (Scheme 3). Neither of these coupling reactions have been optimized. In addition to our concerns about the biological stability of these linkers, success with other linkers preempted this work.

Sulfonyl-Based NOAs

Because the sulfonyl moiety is isoelectronic to the natural phosphoryl internucleoside linkage of DNA and RNA, the development of oligonucleotide analogs based upon this unit is of great interest. Sulfonyl-based oligomers are achiral, and should be among the most polar of the non-ionic congeners. This hydrogen bonding potential should enhance solubility and binding to proteins such as RNase H (where even ionization of the acidic $N–H$ could occur, given sufficient basicity in the active site). Oligomers should also share the nuclease resistance of other analogs. Yet relative to charged species, superior lipid membrane permeability should promote cellular availability.

As in the case of carboxylic acid based linkages, sulfonate and sulfonamide linked NOAs require an analogous set of nucleoside monomers homologated at either the 3′ or 5′ position; synthetic expediency mandated elaboration of the primary carbon. Two strategies were considered for simultaneous homologation and introduction of the desired sulfonyl functionality: treatment of a 5′-nucleoside aldehyde with a sulfonyl-stabilized phosphorus ylide, and intramolecular displacement of a 5′-leaving group by the anion of the corresponding 3′-methylsulfonate ester. This latter route would lead to formation of a cyclic sulfonic acid ester (*i.e.*, a sultone), that could be hydrolyzed to the desired precursor. Although the latter method was unprecedented in nucleoside chemistry, an analogous reaction was known for simple sugars (*16*), and since this route avoided preparation of the 5′-formylnucleosides, it was selected for initial study. In the sulfonate series, an alternative strategy based on modification of carbohydrate sulfonates was reported while this work was in progress (*17*).

Scheme 4 shows the preparation of nucleoside sultones based on our extension of the Fraser-Reid methodology (*2*). 5′-*O*-tosyl-3′-*O*-mesylthymidine was treated with the ethylenediamine complex of lithium acetylide in DMSO, providing as the major product sultone **5** in 48% yield. Among the other materials isolated were the elimination products 5-methyl-1-(5-methyl-2-furanyl)-2,4(1*H*,3*H*)-pyrimidinedione and (*R*)-1-(2,5-dihydro-5-methylene-2-furanyl)-5-methyl-2,4(1*H*,3*H*)-pyrimidinedione. Sultone **5** could also be prepared from the 3′,5′-dimesylate in good but somewhat reduced yield (35%), though overall the simplified synthesis of the dimesylate compensated for the shortfall. None of the product resulting from nucleophilic attack upon the 5′ anion upon the 3′-mesylate was observed. Unfortunately, though **5** could be hydrolyzed by base to the desired deoxyribonucleoside sulfonic acid salt, this product was always accompanied by large amounts of the inseparable 3′ epimeric derivative, arising through the intermediacy of the 2,3′-anhydronucleoside; this latter species could be isolated under mild hydrolytic conditions.

We then turned our attention to sultone **6**, similarly prepared by two methods (Scheme 5) beginning with 1-(2-deoxy-β-D-*threo*-pentofuranosyl)thymine. In contrast to the previous case, the mixed tosylate/mesylate diester was clearly superior to the dimesylate in sultone formation, though again none of the product of inverse nucleophilic substitution was seen with the dimesylate. Treatment of **6** with a variety of nucleophiles generally gave good yields of the corresponding ring opened sulfonate salt; the 3′-azide,

SCHEME 3

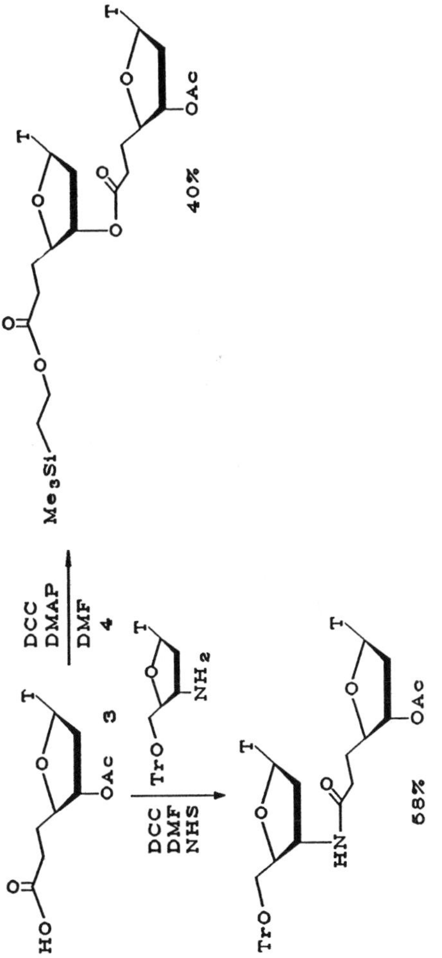

SCHEME 4

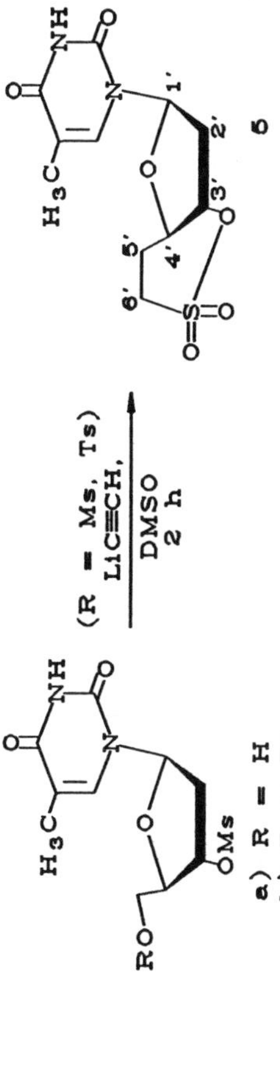

needed for preparation of sulfonamide-linked NOAs, was so obtained. Simple hydrolysis with aqueous ammonia, though slow, generated the requisite 3'-hydroxyl monomer for the ester series.   The corresponding deoxyuridine analog, 1-(2-deoxy-β-D-*threo*-pentofuranosyl)uracil, also underwent the reaction in good yield (53%, unpublished results), and from this sultone the corresponding 3'-hydroxyl and 3'-azide were prepared. The dimesylate of *N*-benzoyl-2'-deoxycytidine was also prepared, and produced the cyclized sultone in 33% yield (unpublished results); in addition, a second product was identified as the 3'-α-hydroxy-6'-sulfonic acid, produced by hydrolysis of the sultone. This reaction was not optimized, and the analogous reactions with the 3'-epimer or the mixed tosylate/mesylate were not investigated.

In the purine series, irrespective of the identity of the base or its state of protection, no conditions could be found that permitted sultone formation in either the dimesyl or tosyl/mesyl systems; upon treatment with lithium acetylide [or other] base, mixtures of materials were usually obtained, the preponderant product being the substituted furan that resulted from elimination of the two sulfonyl leaving groups, followed by aromatization (unpublished results).   Thus, we turned our attention to alternative, potentially superior methodology that would facilitate introduction of a variety of natural or modified purine or pyrimidine bases in either the ribo- or deoxyribo- series, using a common intermediate (Scheme 6).   This strategy employed sugar sultone 7, originally prepared by the Fraser-Reid group, which was then deprotected, functionalized as the 1,2-diacetate, and coupled to an appropriate base, which could then be further manipulated at the 3'-position to obtain the necessary precursors for sulfonate ester or sulfonamide NOAs.   Alternatively, 7 could be derivatized at position 3 prior to incorporation of the base.   In fact, both these sequences were successfully employed, and the resulting ribonucleoside analogs could then be deoxygenated at the 2'-position to yield the congeneric deoxynucleosides.   Both deoxyadenosine and deoxyguanosine precursors were prepared in this manner (unpublished results).

With the precursors in hand, coupling methodology became our next focus. Since coupled adducts would be readily accesible through the intermediacy of the nucleoside sulfonyl chlorides, our chief concern was development of sufficiently mild conditions for introduction of this functionality without concomitant elimination of the heterocyclic base. After ion exchange chromatography to convert the sulfonic acids to a form more readily soluble in organic media (typically *N,N*-diisopropylethylammonium salts), we found several satisfactory chlorinating agents, including dichlorotriphenylphosphorane, triphenylphosphine ditriflate, and triisopropylbenzenesulfonyl chloride, but found triphosgene to be superior (Scheme 7). Using this reagent, several examples of sulfonate ester and sulfonamide coupled dimers were prepared incorporating both purine and pyrimidine bases, including T—T (5'→3'; sulfonate and sulfonamide series) and T—A (sulfonamide). In the sulfonamide series, the 3'-terminal azido moiety served as an amino masking group, permitting extension of the oligomer after reduction. Reduction was accomplished with 10% palladium on carbon to provide the amine in good yield (3).

In summary, novel chemistry has been reported for the formation of antisense oligonucleotides bearing silicon-, carbon-, and sulfur-based phosphate surrogates that significantly extend the library of non-ionic linkers available for development of antisense OAs.

SCHEME 5

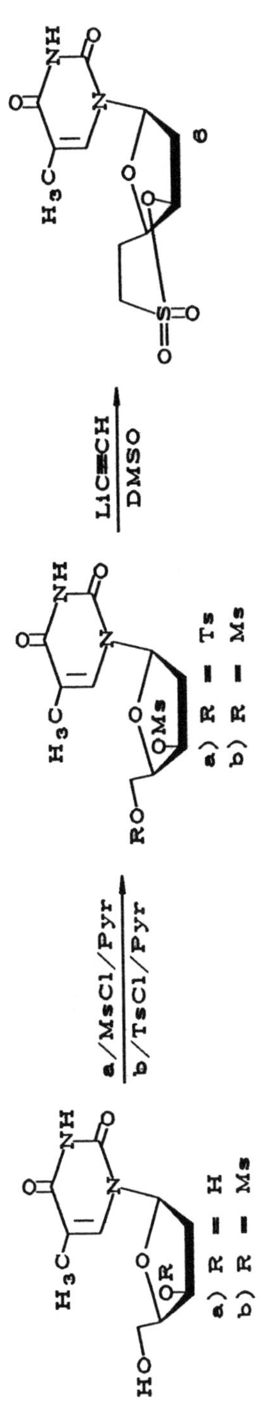

SCHEME 6

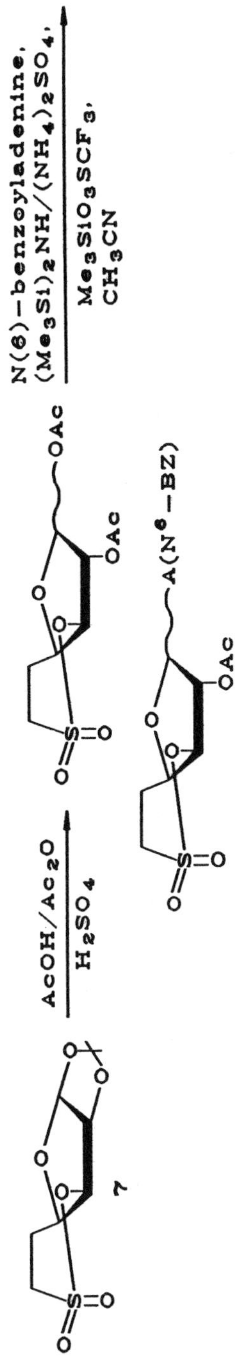

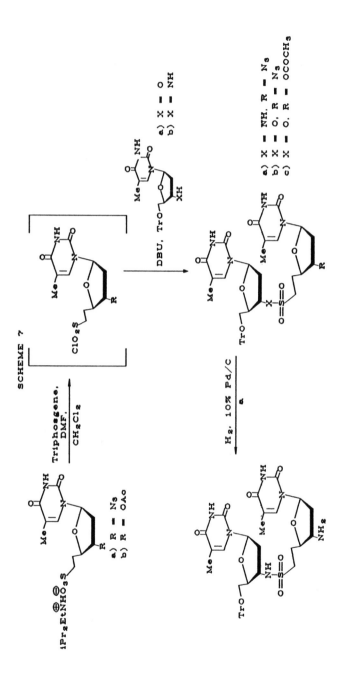

SCHEME 7

Acknowledgments

This research was supported by NIH grants U01-CA44082 and U01-AI26061-03. The authors also wish to acknowledge Drs. Anita Rathore, M. Shamim Akhtar, and, especially, Peter A. Crooks, for their vital contributions to this work.

Literature Cited

1. Secrist III, J.A.; Crooks, P.A.; Maddry, J.A.; Reynolds, R.C.; Rathore, A.S.; Akhtar, M.S.; Montgomery, J.A. *Nucleic Acids Res.* **1991**, Symp. Ser. No. 24, 5-8.
2. Crooks, P.A.; Reynolds, R.C.; Maddry, J.A.; Rathore, A.; Akhtar, M.S.; Montgomery, J.A.; Secrist III, J.A. *J. Org. Chem.***1992**, *57*, 2830-2835.
3. Reynolds, R.C.; Crooks, P.A.; Maddry, J.A.; Akhtar, M.S.; Montgomery, J.A.; Secrist III, J.A. *J. Org. Chem.***1992**, *57*, 2983-2985.
4. Glenmarec, C.; Reynolds, R.C.; Crooks, P.A.; Maddry, J.A.; Akhtar, M.S.; Montgomery, J.A.; Secrist III, J.A.; Chattopadhyaya, J. *Tetrahedron* **1993**, *49*, 2287-2298.
5. Cormier, J.F.; Ogilvie, K.K. *Nucleic Acids Res.* **1988**, *16*, 4583-4594.
6. Seliger, H.; Feger, G. *Nucleosides Nucleotides* **1987**, *6*, 483-484.
7. Sana, A.K.; Sardaro, M.; Waychunas, C.; Delecki, D.; Kutny, R.; Cavanaugh, P.; Yawman, A.; Upson, D.A.; Kruse, L.I. *J. Org. Chem.* **1993**, *58*, 7827-7831.
8. Idziak, I; Just, G; Damha, M.J.; Giannaris, P.A. *Tet. Lett.* **1993**, *34*, 5417-5420.
9. De Mesmaeker, A.; Lebreton, J.; Waldner, A.; Cook, P.D. *Backbone Modified Oligonucleotide Analogs.* International Patent WO 92/20, 823 (1992).
10. De Mesmaeker, A.; Lebreton, J.; Waldner, A.; Fritsch, V.; Wolf, R.M. *Amides as Backbone Replacements in Oligonucleotides.* 207th American Chemical Society National Meeting and Exposition Program (Abstract No. 38): San Diego, CA, March 13-17, 1994.
11. Lebreton, J.; DeMesmaeker, A.; Waldner, A.; Fritsch, V.; Wolf, R.M.; Freier, S.M. *Tet. Lett.* **1993**, *34*, 6383-6386.
12. DeMesmaeker, A.; Lerreton, J.; Waldner, A.; Fritsch, V.; Wolf, R.M. *Bioorg. Med. Chem. Lett.* **1994**, *4*, 873-878, and references therin.
13. Moffatt, J.G. In *Nucleoside Analogues. Chemistry, Biology, and Medical Applications*; Walker, R.T.; De Clercq, E.; Eckstein, F., Eds.; NATO Advanced Study Institute, Series A; Plenum Press: New York, 1979, Vol. 26; pp. 71-164.
14. Montgomery, J.A.; Thomas, H.J. *J. Org. Chem.* **1981**, *46*, 594-.
15. Oyama, K.; Nishimura, S.; Nonaka, Y.; Kihara, K.; Hashimoto, T. *J. Org. Chem.* **1981**, *46*, 5242-5244.
16. Fraser-Reid, B.; Sun, K.M.; Tsang, R.Y.-K.; Sinay, P.; Pietraszkiewicz, M. *Can. J. Chem.* **1981**, *59*, 260-263.
17. Musicki, B.; Widlanski, T.S. *J. Org. Chem.* **1990**, *55*, 4231-4233.

RECEIVED July 6, 1994

Chapter 4

Synthesis and Hybridization Properties of DNA Oligomers Containing Sulfide-Linked Dinucleosides

George Just and Stephen H. Kawai

Department of Chemistry, McGill University, Montréal,
Québec H3A 2K6, Canada

The replacement of the phosphodiester groups in nucleic acids by non-hydrolysable linkages, such as sulfide groups, represents an important strategy in the development of antisense inhibitors. Branched-chain nucleosides bearing mesylated β-hydroxyethyl groups at the α-3'-position were efficiently synthesized by three different routes. Subsequent coupling with 5'-deoxy-5'-thionucleosides afforded an array of sulfide-linked dinucleoside analogues whose furanose rings carry either a hydroxy, methoxy, or no substituent at the 2'-positions. After suitable modification, the dimers could be incorporated into DNA oligomers by standard automated methodology. The resulting sulfide-containing strands were found, in most cases, to bind complementary DNA and RNA and, in some cases, showed a surprising selectivity towards RNA over DNA. The hybridization properties of the equivalent sulfone systems were much poorer.

In response to the instability of natural DNA and RNA oligomers *in vivo*, as well as problems regarding cellular uptake, a great deal of recent work has been devoted to the development of nucleic acid analogues for use as specific inhibitors of gene expression -- the antisense strategy (*1-4*). In addition to oligonucleotides bearing modified phosphodiester linkages, a number of antisense systems in which the phosphate group has been replaced altogether have been described (*5-12*). Apart from evaluating the suitability of such modified oligomers as potential therapeutics, the study of their DNA and RNA hybridization properties addresses the fundamental question of the role played by the backbone (sugar-phosphate units in natural strands) in the formation and stabilization of nucleic acid helices.

One aspect of our research in the antisense field has focused on oligonucleotide analogues in which phosphodiester groups are replaced by dialkyl sulfides. Such linkages are susceptible neither to chemical nor enzymatic hydrolysis. The decision to

0097−6156/94/0580−0052$08.00/0

place the sulfur at the position shown was based on both synthetic and structural considerations. To familiarize ourselves with the latter, a scale model of a double helix was constructed. Careful study of the molecular models did not reveal any obvious unfavorable steric interactions upon replacement of the phosphate with the thioethylene group.

In this review, we describe three distinct approaches for the synthesis of 3'-deoxy-3'-*C*-hydroxyethyl nucleoside analogues carrying either a hydroxy, a methoxy, or no substituent at the 2'-position. These involved either modification of nucleosides or carbohydrates, or total synthesis. Coupling of the suitably activated "upper" monomer units **A** with 5'-deoxy-5'-thiolated "lower" monomer units **B** afforded dimers connected through a dialkyl sulfide linkage which, after modification and activation, could be incorporated into DNA by automated techniques. The hybridization properties of the resulting modified oligomers and their sulfone analogues to complementary DNA and RNA strands are described.

Synthesis of 3'-*C*-Substituted "Upper" Monomeric Units From Xylose.

Construction of Branched-Chain Sugars. Furanose derivatives bearing hydroxyethyl groups at the 3-position were readily prepared (*13*) from commercially available 1,2-*O*-isopropylidene-D-xylofuranose *via* tritylation and Moffatt oxidation of the sugar to the known ketone **1** (*14*). The two carbons of the branched-chain were introduced by way of a Horner-Wittig condensation of the anion of trimethyl phosphonoacetate with **1**, which afforded the two isomeric α,β-unsaturated esters **2** and **3** in a ratio of 3.8:1. Subsequent lithium aluminum hydride reduction resulted in completely stereoselective reduction of the α,β-unsaturated esters, affording the key 3-*C*-(2'-hydroxyethyl)furanose intermediate **4**. The Mitsunobu coupling of **4** with thiolacetic acid, followed by detritylation and reprotection with the more acid-stable *tert*-butyldiphenylsilyl group, provided the branched-chain thiosugar **5**.

The analogous system bearing the sulphur at the 5'-position was prepared by reversing the above sequence of steps. Alcohol **4** was first silylated to yield furanose **6**, followed by selective cleavage of the trityl group and Mitsunobu coupling with thiolacetate, which provided thiosugar **7**.

Acetolysis of Branched-Chain Sugars. A key step in our strategy towards the target nucleosides was the acetolytic conversion of the branched-chain acetonides to their corresponding 1,2-*O*-diacetate derivatives, required for the stereospecific introduction of the nitrogenous bases by Vorbrüggen's methodology. Although such acetolyses are generally quite straightforward in the case of simple sugars, the presence of the 3'-chain in the present systems gives rise to the possibility of competing side-reactions (*13*).

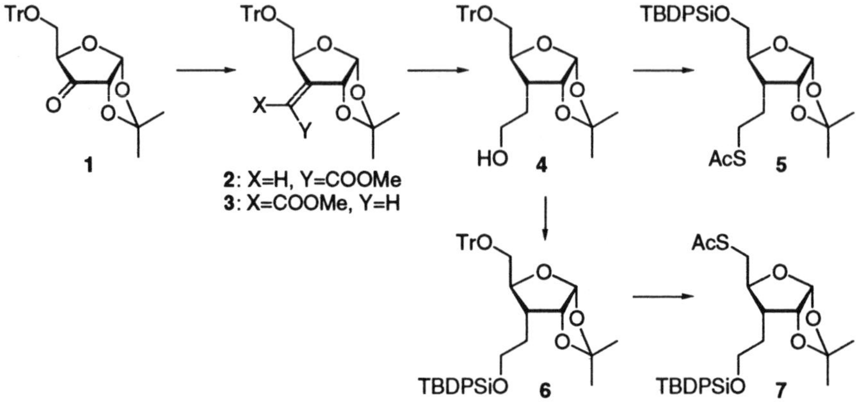

1

2: X=H, Y=COOMe
3: X=COOMe, Y=H

HO 4

AcS 5

TBDPSiO 6

AcS 7

The initial treatment of acetonide **5** with acetic acid containing acetic anhydride and *p*-toluenesulfonic acid at ambient temperature afforded two products with similar R$_f$. The desired 1,2-di-*O*-acetylfuranose **8** was thus obtained in 37% yield, primarily as the β-anomer. A slightly less polar compound, obtained in 42% yield, was found to be the *cis*-disubstituted thiolane structure **9**.

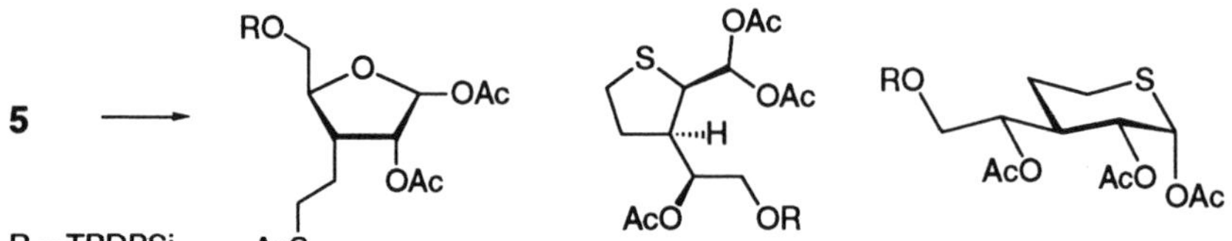

5 ⟶

R = TBDPSi   AcS   8

9

10

Whereas neither the nature of the acid nor the AcOH / Ac$_2$O ratio had much influence on the product ratio, the reaction temperature was found to have a profound effect on the course of the reaction. At lower temperatures (5-15°C), thiolane **9** was preferentially formed by a factor of 2:1. At 70°C, however, the desired β-furanose **8** was isolated in 70% yield, with only slight traces of **9** being formed. Small amounts of the α-anomer of **8** were also detected. Treatment of **5** with boron trifluoride etherate in acetic anhydride gave the acetylated thiopyranose **10** in a yield of 70%.

The acetolysis of isopropylidene sugar **7** also gave different products depending on the reaction temperature. When carried out at 75°C, the treatment afforded the desired 1,2-*O*-diacetylfuranose **11** as the sole product in 83% yield. As expected, the reaction performed at 15°C gave a much lower yield (8%) of furanose **11**, the major products being the open-chain aldehydrol derivatives **12** which were formed as a 5.7:1 mixture of separable 1(*S*)- and 1(*R*)-anomers in a combined yield of 80%. The reaction of acetonide **11** with boron trifluoride etherate in acetic anhydride afforded a complex mixture of products, from which thiolane **13** was isolated in 25% yield.

The formation of **12** from **7** provides a clue to the formation of the observed low-temperature acetolysis products. A cyclic oxonium ion formed by the scission of the exocyclic glycosyl bond is generally acknowledged to be the intermediate in the formation of 1,2-di-*O*-acetates, and this is no doubt the case for the formation of the desired furanose sugars **8** and **11**. At lower temperatures, however, endocyclic C-O bond cleavage is evidently favored, yielding an open-chain oxocarbonium intermediate **C**. The addition of an acetate would then provide **12** as a mixture of anomers at C-1. We believe that the C-1 acetoxy group in such a species may participate in the subsequent solvolysis of the isopropylidene moiety, yielding an acetoxonium ion **D** which bridges C-1 and C-2. Attack at C-2 by the sulfur of a thiolacetate group situated at either the 2'- (a) or 5-positions (b) would lead to the thiolanes **9** and **13**, respectively, and account for the double inversion at this center.

Acetolyses were eventually carried out on a number of branched-chain sugars lacking the thiolester group. In the case of 5-tritylated derivatives, the protecting group is cleaved under the reaction conditions and 1,2,5-triacetates are obtained. The reaction CSA / AcOH / Ac$_2$O carried out at 70°C for compounds **6** and **14** afforded the corresponding triacetates **16** and **17** in respective yields of 52 and 49%, in addition to 25-30% of the aldehydrol derivatives analogous to **12** (*15*). The key triacetate **18** was also prepared under these conditions from the corresponding acetonide **15** in 69% yield (*16*).

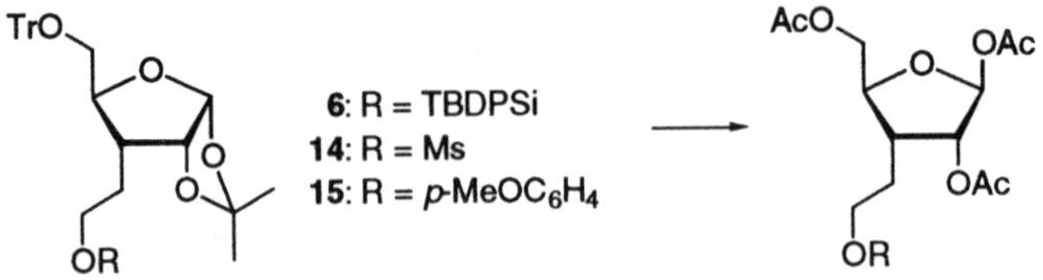

6: R = TBDPSi
14: R = Ms
15: R = *p*-MeOC$_6$H$_4$

16: R = TBDPSi
17: R = Ms
18: R = *p*-MeOC$_6$H$_4$

Nucleoside Formation and Activation. Vorbrüggen couplings of the acetylated furanoses and *per*-silylated bases were carried out in refluxing 1,2-dichloroethane employing trimethylsilyl triflate as the catalyst. The method was used to prepare a series of branched-chain nucleoside analogues **19** to **22** in yields, typically, of 85-90% and complete β-stereoselectivity (*15-17*). The overall yield of these nucleosides based on trityl ketone **1** is about 30-35%. This method, therefore, constitutes a versatile and high-yield route to 3'-deoxy-3'-*C*-(2"-hydroxyethyl) ribonucleoside analogues in which the base may be varied over a wide range.

Since our initial studies were to focus on 2'-deoxy-type systems, an activated nucleoside lacking a 2'-functionality was required. This was accomplished from compound **22**. After hydrolysis of the acetate groups, selective silylation and reaction with phenyl chlorothionoformate gave the nucleoside **24**. Radical reduction with tributyltin hydride provided the 2'-deoxy system, which was subjected to oxidative removal of the *p*-methoxyphenyl group employing ceric ammonium nitrate, followed by mesylation, which afforded the activated upper unit **25** (*16*).

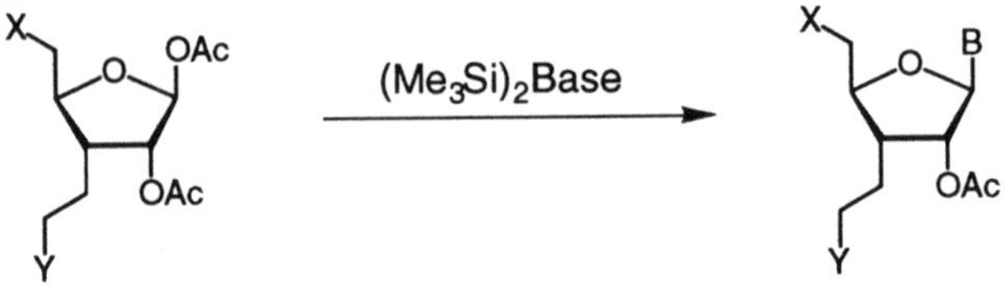

(Me$_3$Si)$_2$Base

8 ⟶ **19**: X = TBDPSiO, Y = SAc, B = Cyt, Ade[Bz]
11 ⟶ **20**: X = SAc, Y = TBDPSiO, B = Cyt, Ade[Bz]
16 ⟶ **21**: X = OAc, Y = TBDPSiO, B = Thy
18 ⟶ **22**: X = OAc, Y = *p*-MeOC$_6$H$_4$O, B = Thy

An appropriately protected upper unit suitable for the preparation of a *ribo*-system was obtained in a very straightforward manner by oxidative deprotection of **22** followed by mesylation to give monomer **23**.

The third type of upper unit to be prepared was a 2'-OMe derivative (Meng, B., Ph.D. Thesis, McGill University, 1994). This was again carried out from nucleoside **22** whose thymine ring was first protected against alkylation by treatment with mesitylenesulfonyl chloride and triethylamine, followed by 2,6-dichlorophenol and diazabicyclooctane to yield the *O*[4]-ether (*18*). The acetate groups were then cleaved, followed by selective silylation of the primary hydroxyl, which afforded **26**. The remaining 2'-hydroxyl group was then methylated using methyl iodide and sodium hydride in DMF. Removal of the protecting group from the pyrimidine was effected with *p*-nitrobenzaldoxime and tetramethylguanidine to give **27**. Oxidative deprotection followed by mesylation once more gave the target monomer **28**.

An alternative route to **28** involves elaboration of the sugar prior to nucleoside formation. Replacement of the trityl group in **15** by a benzyl ether, followed by acid-catalyzed methanolysis, afforded methyl glycoside **29**. Subsequent methylation employing methyl iodide and sodium hydride gave **30**, which was transformed to its anomeric acetate **31** by treatment with dimethylboron bromide followed by acetic acid-triethylamine. The tin tetrachloride-catalyzed Vorbrüggen coupling of **31** and *per*-silylated thymine proceeded to give nucleoside **32** with a β:α ratio of 1:1.5. When trimethylsilyl triflate was employed, the desired β-nucleoside was obtained stereoselectively in a yield of 48%.

From a Nucleoside. Since the procedures described above initially provide analogues of ribonucleosides which require further transformation to the corresponding 2'-deoxy derivatives, an alternate route which would provide directly the 2'-deoxynucleoside upper unit from thymidine was also devised (*16*).

The most expeditious way of preparing the required monomer **25** involves proceeding *via* the recently described (*19*) tributyltin-mediated radical allylation of the

3'-thionocarbonate derivative of thymidine **33**. In this manner, the 3'-branched chain may be introduced with high stereoselectivity. Oxidative scission of the olefinic bond of allylated nucleoside **34** with osmium tetroxide-sodium periodate yielded aldehyde **35**, which was reduced by sodium borohydride to give alcohol **36**. Mesylation provided the target system **25**.

By Total Synthesis. A third route to 2'-deoxy type upper units involves total assymmetric synthesis (*20*). 1-Benzyloxy-2(*S*)-hydroxybutyne **37** was obtained in four steps from *cis*-2-butene-1,4-diol in good yield and in >97% ee by known procedures (*21-23*). After acetalization with bromine and ethyl vinyl ether, the acetylenic bromoacetal was converted to the corresponding acetylenic ester, which, upon partial hydrogenation, afforded α,β-unsaturated ester **38** in 75-80% overall yield.

The subsequent tributyltin hydride-mediated radical cyclization proceeded with complete stereoselectivity and produced the dideoxyribofuranose **39** in high yield. The activation of acetal **39** could be achieved using dimethylboron bromide, through which the coupling with *bis*-(trimethylsilyl)thymine in dichloroethane could be carried out to give nucleoside **40**, for which the β-anomer predominated in a 3:1 ratio.

Synthesis of 5'-Deoxy-5'-Thionucleoside "Lower" Monomer Units

The second type of monomer unit required for formation of the sulfide-linked dimers is a suitably protected nucleoside bearing a 5'-thiol ("lower" units **B**). The 2'-deoxy system **43** was simply prepared in three steps from thymidine. The Mitsunobu coupling of thymidine and thiolacetic acid, carried out in THF/DMF mixture for solubility reasons, proceeded regioselectively to afford **41**. Subsequent silylation to **42** and deacetylation yielded the target thiol **43** in good yield (*16*).

The 2'-methoxy thionucleoside unit **46** was prepared from uridine (Wang, D., Ph.D. Thesis, McGill University, 1994). The 3' and 5'-positions of the free nucleoside were blocked by Markiewicz's tetra-*iso*-propyldisiloxane group and the pyrimidine then protected as its N^3-SEM derivative (SEM = -CH$_2$OCH$_2$CH$_2$SiMe$_3$) by treatment with 2-(trimethylsilyl)ethoxymethyl chloride and diisopropylethylamine to give **44**. The remaining hydroxyl was then quantitatively converted to its methyl ether using methyl iodide and silver oxide. Other conventional methylating conditions such as methyl iodide and sodium hydride led to decomposition products.

The disiloxane protecting group could then be selectively cleaved with three equivalents of tetra-*n*-butylammonium fluoride at ambient temperature without loss of the SEM-group, which was retained for solubility considerations, and whose removal requires heating to 55°C. Nucleoside **45** was then subjected to Mitsunobu coupling with thiolacetic acid, followed by treatment with methanolic ammonia, which yielded the free thiol **46**.

Synthesis and Hybridization Properties of DNA Containing Sulfide-Linked Dinucleosides

The first modified oligomer we studied was based on a fully 2'-deoxy-type dimer (*24*). The coupling of upper unit **25** with a 10% excess of the lower thiol unit **43** in DMF solution containing 1.5 equivalents of cesium carbonate afforded the sulfide-linked dimer **47** in 88% yield. Removal of the silyl groups with tetra-*n*-butylammonium fluoride gave, quantitatively, the diol **48**. Dimethoxytritylation of the primary hydroxyl to **49**, followed by phosporamidite formation by standard means, provided the activated species **50** in 75% yield. Prior to attempting incorporation into DNA, model studies were carried out employing dimer **47,** which indicated that the sulfide bond was stable to the oxidation conditions (I_2/pyridine/H_2O) used for the transformation of phosphite to phosphate triesters during the solid-phase synthetic cycle.

47: R = R' = TBDMSi
48: R = R' = H
49: R = DMTr, R' = H
50: R = DMTr,

The standard sequence used to study the binding properties of our sulfide-incorporating strands towards complementary DNA and RNA by thermal denaturation measurements was 5'-$G_pC_pG_pT^y_xT^y_pT^y_xT^y_pT^y_xT^y_pG_pC_pT$-3'; where p = phosphate; x = sulfide, sulfone, or phosphate linkage; and y = the 2'-substituents (other than H) of the nucleoside unit. Phosphate (p) links will henceforth not be indicated. A control DNA strand **DNA-I**, 5'-$GCGT_6GCT$-3' as well as its complementary DNA and RNA strands **DNA-II**, 5'-$AGCA_6CGC$-3' and **RNA-II**, 5'-r($AGCA_6CGC$)-3' were prepared. Control melts of the fully natural duplexes **DNA-I:DNA-II** and **DNA-I:RNA-II** gave T_m values of 59.5° and 52.5°C, respectively. (All T_m values reported herein were obtained for 1M NaCl / pH 7 phosphate buffer.)

The modified dodecamer **S-I**, having the sequence $GCG(T_sT)_3GCT$, was prepared by standard automated solid-phase methods. A fifth bottle containing a solution of activated dimer **50** in acetonitrile was attached to the synthesizer and the coupling time for this unit increased to 10 minutes. DNA strands which incorporate the T_sT dimer could thus be prepared with coupling yields of >98%. T_m values of 37.2° and 36.0°C were measured for duplexes **S-I:DNA-II** and **S-I:RNA-II**, respectively. Oligomers analogous to **S-I,** which incorporate only one or two sulfide-linked units (sequences of type $GCGTT(T^y_sN^y)TTGCT$ and $GCG(T^y_sN^y)TT(T^y_sN^y)GCT$), were also prepared and their hybridization properties evaluated. It was found that the decrease in T_m for these systems varied linearly with the number of sulfides present in the strand, the average values for ΔT_m/sulfide being 7.4° and 5.5°C for binding to **DNA-II** and **RNA-II**, respectively.

We next proceeded to prepare the mixed ribo-deoxyribo dimer **54** and incorporate it into our DNA sequence (*25*). The coupling of the mesylated upper unit **23** with thionucleoside **43** gave the sulphur-linked dimer **51** in very good yield. Deacetylation in methanolic ammonia to yield **52**, followed by dimethoxytritylation and immediate reacetylation of the 2'-position afforded **53**. Desilylation and phosphoramidite formation provided **54**.

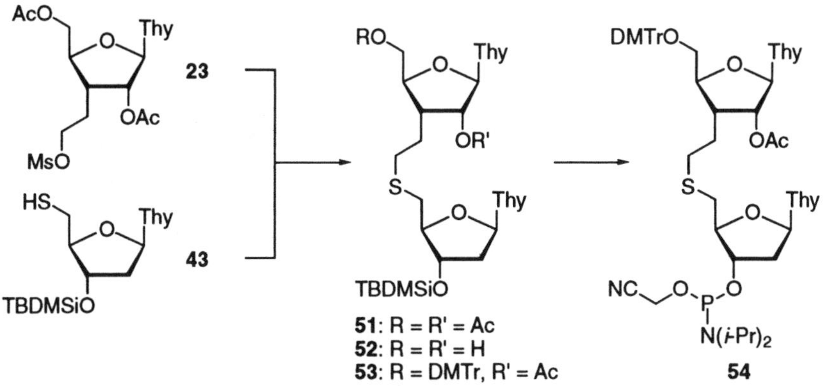

51: R = R' = Ac
52: R = R' = H
53: R = DMTr, R' = Ac

The modified dodecamer **S-II**: GCG(T^{OH}_sT)$_3$GCT, which contains three sulfide-linked ribofuranosylthymine-thymidine dimers, was synthesized using phosphoramidite **54** by the same automated methodology. The results of the thermal denaturation studies clearly indicated that strands containing the (T^{OH}_sT) dimer could discriminate between complementary RNA and DNA over a wide range of salt

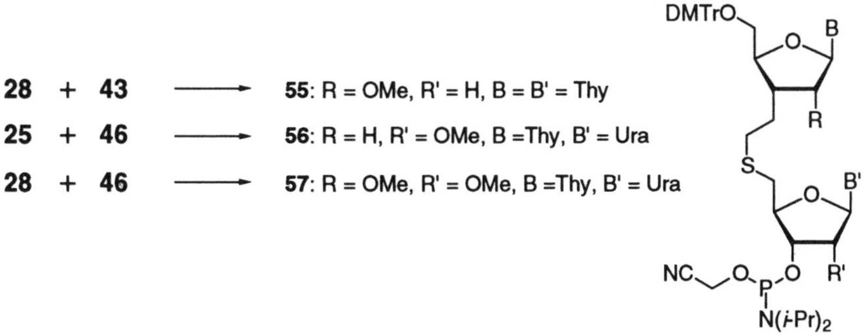

28 + 43 ⟶ 55: R = OMe, R' = H, B = B' = Thy

25 + 46 ⟶ 56: R = H, R' = OMe, B = Thy, B' = Ura

28 + 46 ⟶ 57: R = OMe, R' = OMe, B = Thy, B' = Ura

concentrations. While hybridization with a T_m of 31.5°C occurred for the **S-II:RNA-II** pair, no cooperative binding was observed between **S-II** and **DNA-II**. Study of the oligomers bearing one and two T^{OH}_sT units also showed a linear dependence of the melting temperature on the number of sulfides, with average ΔT_m/sulfide values for the system of 14.2° and 7.3°C for DNA and RNA, respectively. The binding selectivity for RNA exhibited by **S-II** is quite surprising; only a few oligonucleotide analogues exist that can discriminate between DNA and RNA, only complexing with the latter, and these structures are very different from the natural structures   (26-28).

These results prompted us to prepare oligomers incorporating sulfide dimers bearing methoxy groups at the 2'-positions of either or both of the sugar rings (Wang, D., Ph.D. Thesis, McGill University, 1994). The coupling of the upper monomeric units **28** and **25** with the lower thiol units **43** and **46** by the usual method, gave three sulfide dimers of type T^{OMe}_sT, T_sU^{OMe} and $T^{OMe}_sU^{OMe}$. Deprotection, followed by dimethoxytritylation and phosphatidylation, gave the activated dimers **55, 56,** and **57** ready for DNA incorporation.

	S-I	S-II	S-III	S-IV	S-V
S + DNA-II:	37.2° (8.8°)	NB (14.2°)	NB (15.1°)	26.2° (11.2°)	24.8° (13.1°)
S + RNA-II:	36.0° (7.0°)	31.5° (7.3°)	30.5° (8.0°)	35.5° (6.1°)	43.8° (3.8°)

The phosphoramidites were then introduced into the standard sequence, giving dodecamers **S-III, S-IV,** and **S-V** (of type GCG(T^y_sN^y)$_3$GCT). As can be seen from the T_m data tabulated above (the values in parentheses indicate the average ΔT_m/sulfide based on studies of strands containing one to three T^y_sN^y units; NB = no observed binding), dodecamer **S-III** exhibited binding properties very similar to those observed for **S-II**. Placing the methoxy group in the other ribose ring (**S-IV**) had a similar effect on the T_m values. However, when both sugar rings of the dimers carried a methoxy group at the 2'-position, the binding to RNA improved with an average drop in T_m per sulfide of only 3.8°C.

Sulfones

We next proceeded to investigate the effect of transforming the sulfide-linkages in our oligomers to sulfone groups (Meng, B., Ph.D. Thesis, McGill University, 1994). The diol sulfide **48** could be quantitatively oxidized using buffered Oxone reagent to the corresponding sulfone, which, after dimethoxytritylation and phosphitylation, afforded the activated dimer **58**. This unit could be incorporated into the standard dodecamer sequence; however, treatment of the synthesized strand with ammonia during the final deprotection step failed to provide the desired oligomer **S-VI**. Instead, a series of shorter fragments were obtained. Model studies carried out on sulfone **59** indicated that, even upon brief exposure to ammonia at room temperature, degradation occurred with thymine being isolated from the reaction mixture. It is most likely that ammonia causes a β-elimination reaction.

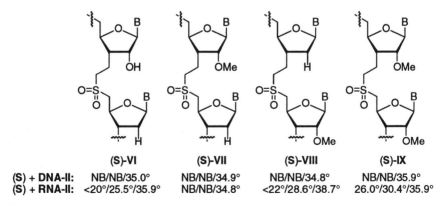

We, therefore, oxidized purified strands **S-II, S-III, S-IV,** and **S-V** with Oxone reagent to yield the corresponding trisulfones **S-VI, S-VII, S-VIII,** and **S-IX,** respectively. Although the mobilities of these sulfone-incorporating oligomers were very similar to those of the starting trisulfide strands, treatment with ammonia was found to degrade them to the smaller fragments, indicating that the desired oxidation had taken place. All four sulfone strands, **S-VI** to **S-IX,** showed no binding to complementary DNA and only very poor binding to complementary RNA. Strands containing one or two sulfone-bridges were also prepared and their hydridization properties evaluated. As can be seen from the T_m data shown below (melting temperatures given for oligomers incorporating three, two, and one sulfone linkage, respectively), the binding of these types of systems to complementary DNA and RNA is very weak, a single insertion resulting in drops in T_m of about 25°C for binding to **DNA-II** and about 15°C for **RNA-II**. Apparently, whatever steric interference is caused by the sulfide linkage is accentuated when replaced by a sulfone group.

	(S)-VI	(S)-VII	(S)-VIII	(S)-IX
(S) + DNA-II:	NB/NB/35.0°	NB/NB/34.9°	NB/NB/34.8°	NB/NB/35.9°
(S) + RNA-II:	<20°/25.5°/35.9°	NB/NB/34.8°	<22°/28.6°/38.7°	26.0°/30.4°/35.9°

Conclusion

We have shown that sulfide-linked dinucleoside analogues may be readily prepared from monomeric units, which are in turn accessible by a number of very efficient routes. These thioether dimers may then be incorporated into natural DNA by standard automated techniques. The modified oligomers thus obtained retain good binding to complementary RNA and (in most cases) DNA. In view of the efficiency of the sulfide forming step, an obvious extension to this work is the preparation of oligonucleotide analogues for which all of the internucleoside linkages consist of thioethylene groups. The preparation of the monomeric intermediates for one such system has already been reported by our group (*16*).

It should also be mentioned that a modified oligomer, which only bears internucleoside sulfides, would constitute an ideal binding moiety for an artificial restriction enzyme. Since thioethers should not be cleaved by agents that catalyze phosphodiester hydrolysis, the attachment of such a catalytic group to a polysulfide oligomer would yield a system well-designed for the site-specific hydrolysis of RNA or DNA without the risk of self-cleavage.

Acknowledgments

This work was supported by the Natural Sciences and Engineering Research Council of Canada, the Ministère de l'éducation du Québec (FCAR), and BioChem Pharma Inc. Oligomer syntheses and thermal denaturation studies were carried out in collaboration with Prof. M. J. Damha (Department of Chemistry, McGill University). We are grateful to Prof. O. A. Mamer (McGill University Biomedical Mass Spectrometry Unit) for the measurement of mass spectra.

Literature Cited

1. Englisch, U.; Gauss, D. H. *Angew. Chem., Int. Ed. Engl.* **1991**, *30*, 613.
2. Uhlmann, E.; Peyman, A. *Chem. Rev.* **1990**, *90*, 544.
3. Dolnick, B. J. *Biochem. Pharm.* **1990**, *40*, 671.
4. *Oligonucleotides: Antisense Inhibitors of Gene Expression;* Cohen, J. S., Ed.; Top. Mol. Struct. Biol. 12; MacMillan Press: London, 1989.
5. Engholm, M.; Buchardt, O.; Nielson, P. E.; Berg, R. H. *J. Amer. Chem. Soc.* **1992**, *114*, 1895.
6. Meier, C.; Engels, J. W. *Angew. Chem., Int. Ed. Engl.* **1992**, *31*, 1008.
7. Huie, E. M.; Krishenbaum, M. R.; Trainor, G. L. *J. Org. Chem.* **1992**, *57*, 4569.
8. Huang, Z.; Schneider, K. C.; Benner, S. A. *J. Org. Chem.* **1991**, *56*, 3869.
9. Musicki, B.; Widlanski, T. S. *Tetrahedr. Lett.* **1991**, *32*, 1267.
10. Matteucci, M. *Nucleosides Nucleotides . ***1991**, *10*, 231.
11. Cormier, J. F.; Ogilvie, K. K. *Nucleic Acids Res.* **1988**, *16*, 4583.
12. Stirchak, E. P.; Summerton, J. E.; Weller, D. D. *J. Org. Chem.* **1987**, *52*, 4202.
13. Kawai, S. H.; Chin, J.; Just, G. *Carbohydr. Res.* **1991**, *211*, 245.
14. Sowa, W. *Can. J. Chem.* **1968**, *46*, 1586.
15. Kawai, S. H.; Just, G. *Nucleosides Nucleotides . ***1991**, *10*, 1485.
16. Kawai, S. H.; Wang, D.; Just, G. *Can. J. Chem.* **1992**, *70*, 1573.
17. Kawai, S. H.; Chin, J.; Just, G. *Nucleosides & Nucleotides . ***1990**, *9*, 1045.
18. Nyilas, A.; Chattopadhyaya, J. *Acta Chem. Scand.* **1986**, *B40*, 826.
19. Chu, C. K.; Doboszewski, B.; Schmidt, W.; Ullas, G. V. *J. Org. Chem.* **1989**, *54*, 2767.

20. Just, G.; Lavallée, J.-F. *Tetrahedr. Lett.* **1991**, *32*, 3469.
21. Takano, S.; Samizu, K.; Sugihara, T.; Ogasawara, K. *J. Chem. Soc., Chem. Commun.* **1989**, 1344.
22. Yadav, J. S.; Deshpande, P. K.; Sharma, G. V. M. *Pure Appl. Chem.* **1990**, *62*, 1333.
23. McCombie, S. W.; Shankar, B. B.; Ganguly, A. K. *Tetrahedr. Lett.* **1989**, *30*, 7029.
24. Kawai, S. H.; Wang, D.; Giannaris, P. A.; Dahma, M. J.; Just, G. *Nucl. Acids Res.* **1993**, *21*, 1473.
25. Meng, B.; Kawai, S. H.; Wang, D.; Just, G., Giannaris, P. A.; Damha, M. J. *Angew. Chem., Int. Ed. Engl.* **1993**, *32*, 729.
26. Fujimori, S.; Shudo, K.; Hashimoto, Y. *J. Amer. Chem. Soc.* **1990**, *112*, 7436.
27. Giannaris, P. A.; Dahma, M. J. *Nucl. Acids Res.* **1993**, *21*, 4742.
28. Adams, A. D.; Petrie, C. R.; Meyer, R. B., Jr. *Nucl. Acids Res.* **1991**, *19*, 3647.

RECEIVED July 25, 1994

MODIFICATIONS OF SUGAR MOIETIES

Chapter 5

4'-Thio-RNA: A Novel Class of Sugar-Modified β-RNA

Laurent Bellon[1], Claudine Leydier, Jean-Louis Barascut, Georges Maury, and Jean-Louis Imbach

Laboratoire de Chimie Bio-Organique, Unité de Recherche Associée au Centre National de la Recherche Scientifique 488, Universite Sciences et Techniques du Languedoc, Place E. Bataillon, 34095 Montpellier Cedex 05, France

A new series of sugar-modified oligoribonucleotides, namely 4'-thio-β-D-oligoribonucleotide, is described herein. The synthesis of the 4'-thio-D-ribonucleotide derivatives and their assembling on solid support using phosphoroamidite chemistry are reported together with nuclease resistance and binding affinities studies. 4'-thio-RNA exhibit high resistance towards enzymatic degradation and stable hybridization with complementary RNA strands. Finally, interaction of 4'-thio-oligouridylates with HIV-1 Reverse Transcriptase are presented and discussed.

The existence of naturally occurring antisense RNA transcripts (1-3) in conjunction with the recent discovery of the catalytic roles of RNA, creating (4) or breaking (5) phosphodiester linkages, has led to a growing interest in RNA molecules. Therefore, synthetic RNA oligonucleotides were prepared, but their antisense potency in biological medium have been hampered by their poor nuclease stability. Only a few structural modifications of oligoribonucleotides, i.e., 2'-O-Me-RNA (6-7), phosphorothioate-RNA (8), or α-RNA (9), have been proposed to improve resistance towards enzymatic degradation. This can be related to the more difficult synthetic approach imposed by the 2'-functionality together with the necessity to synthesize the modified constitutive synthons.

Looking at ribozyme molecules, we have been interested in designing a new oligoribonucleotidic series with potential catalytic activity. Considering the wild-type RNA structure, we decided to keep the 2'-hydroxyl group available for catalysis and to retain β-configuration and unmodified phosphate backbone for mimicry of folded structures and solubility reasons.

[1]Current address: Isis Pharmaceuticals, 2292 Faraday Avenue, Carlsbad, CA 92008

This led us to consider the isosteric replacement of the annular oxygen atom of the ribofuranose ring by a sulfur atom in the hope of improving nuclease resistance and binding affinities of the corresponding 4'-thio-oligonucleotides.

Such 4'-thio-β-D-oligoribonucleotides (4'-S-RNA) can be obtained by solid support assembling of the 4'-thio-β-D-ribonucleotides building blocks using phosphoroamidite methodology (*10*). This approach required the previous synthesis of the 4'-thio-β-D-ribonucleoside precursors from the corresponding nucleobases and the suitable 4-thio-D-ribofuranose.

Synthesis Chemistry

A number of 4'-thionucleosides have been reported as early as three decades ago (*11-21*), and more recently, a new interest in the 4'-thionucleosidic field came out with the work of several teams (*22-36*) describing various syntheses of 2'-deoxy-4'-thio-D-ribonucleosides derivatives.

Although numerous syntheses of 4-thiosugars including 4-thio-D-ribofuranose have been described (*13-20*), the multi-step synthetic pathways used to obtain 4-thio-D-*erythro*-pentofuranose have led to poor overall yields. Consequently, our initial effort was focused on the design of a strategy that could give rise to a better access to the desired 4-thio-D-ribofuranose. This synthetic route (Figure 1) is based on the introduction of a sulfur atom at the C-1 position of a L-Lyxose derivative intermediate, **6**, followed by a nucleophilic displacement of the previously activated 4-hydroxyl group with inversion of configuration (*37, 38*).

As shown in Figure 1, the expected acetyl-2,3,5-tri-O-benzyl-4-thio-D-ribofuranose, **9**, can be obtained either from D-ribose, **1**, or L-lyxose, **4**, with 21% or 35% overall yields, respectively. The L-Lyxose route is more straightforward and less time-consuming.

The synthesis of the necessary 4'-thio-β-D-ribonucleosides were performed using appropriate glycosylation reactions (*39, 40*) (Figure 2) with the exception of N9-(4-thio--D-ribofuranosyl)-guanine which could not be obtained in satisfactory yields.

As an anomeric mixture was obtained in all cases, an anomeric configuration of each nucleoside was carefully determined using NOE and NOESY 2-dimensional proton N.M.R. experiments. However, pure N$_4$-benzoyl-4'-thio-β-D-cytidine, **13**, could be reached within the same overall yield either by the direct condensation route or from protected 4'-thio-β-uridine, **10**, *via* the thio-amide analogue, **11** (Leydier, C.; Bellon, L.; Barascut, J.-L.; Deydier, J.; Maury, G.; Pelicano, H.; Elalaoui, M. A.; Imbach, J.-L. *Nucleosides Nucleotides*, **1994**, 000) (Figure 3).

The 5'- and 3'-hydroxyl functions of the different 4'-thio-ribonucleosides were respectively protected with DmTr and TBDMS groups as previously described (*41, 42*). The 3'-methoxy-N,N-diisopropyl-phosphoroamidite-4'-thio-D-ribonucleoside derivatives were obtained according to similar procedure used in the oxygenated series.

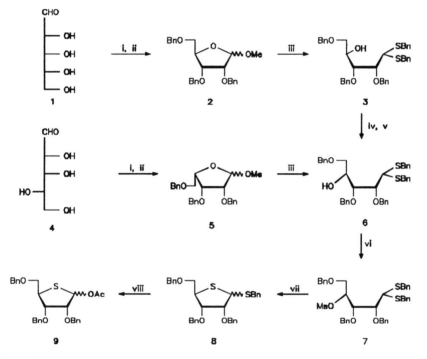

Figure 1. i = MeOH, HCl ; ii = BnBr, KOH, THF ; iii = BnSH, BF$_3$, Et$_2$O ; iv = Ph$_3$P, DEAD, p-NO$_2$PhCO$_2$H, THF ; v = K$_2$CO$_3$, MeOH ; vi = MsCl, Pyr.; vii = NBu$_4$I, BaCO$_3$, Pyr.; viii = Hg(OAc)$_2$, AcOH.

B = Th, Ur, Cy^{Bz}, Ad.

Figure 2. i = BSA, TMSTf, CH$_3$CN (for Ad : i = TMSTf, CH$_3$CN, MS 4A) ;
ii = BBr$_3$, CH$_2$Cl$_2$ (for Ad, ii = BzCl, Pyr. then BBr$_3$, CH$_2$Cl$_2$).

Figure 3. i = Lawesson's Reagent ; ii = NH$_3$, MeOH ; iii = BzCl, Pyr. ;
iv = BBr$_3$, CH$_2$Cl$_2$.

Various 4'-thio-ß-<u>D</u>-oligoribonucleotides (42); 4'-S(U-U-U-U-U-U), 4'-SU_6, 14, 4'-S(U-U-U-U-U-U-U-U-U-U-U-U), 4'-SU_{12}, 15, and a 4'-thiododecamer complementary to the splicing acceptor site of the *tat* HIV gene, 4'-S(ACACCCAAUUCU), 16; were designed in order to evaluate their nuclease resistance and hydrogen bonding properties. All these syntheses required the suitable 5'-DmTr-4'-thio-uridine-solid support (42) [Long Chain Alkylamines on Controlled Pore Glass beads (LCA-CPG)] prepared according to the procedure of Damha et al. (43) with a loaded yield of 21 µmol.g^{-1}. Standard RNA elongation cycle allowed high coupling yields (> 98% as monitored from the dimethoxytrityl cation release). The 4'-thiooligomers were deprotected according to a standard protocol including a thiophenol and a tetrabutyl-ammonium fluoride treatment and finally purified by reverse phase HPLC (42).

Preliminary Evaluations as Potential Antisense Agents

Nuclease Resistance. Using some of these previous 4'-thiooligoribonucleotides, we studied their resistance towards enzymatic degradation comparatively with the corresponding β-RNA sequences. As a first example, the substrate activities of 4'-SU_6, 14, and its oxygenated parent U_6 were studied comparatively with respect to various purified nucleases as well as in culture medium (Table I). It appears from Table I that 4'-SU_6 is much more resistant to the four considered degradating enzymes compared to the wild type. Our results show that the most active nuclease is the snake venom phosphodiesterase in accordance with previous data (44). As it is well known that 3'-end protection of an oligomer confers an enhancement of the enzymatic stability towards 3'-exonucleases (45), we synthesized 4'-SU_6-nprOH, in which a 3'-phosphodiester propanol termini is introduced by means of the "Universalis" solid support (46). This modified hexamer exhibited an improved half-life time of 250 minutes in the presence of snake venom phosphodiesterase.

Table I. Half-Life Time (Min) of Oligoribonucleotides U_6 and 4'-SU_6 in the Presence of Various Nucleases and in Culture Medium

Enzyme or Culture Medium	Half-Life Time (min)	
	U_6	4'-SU_6
5'-Exonuclease : Calf spleen phosphodiesterase	17	3900
3'-Exonuclease : Snake venom phosphodiesterase	1	76
Endonuclease S_1	120	930
Ribonuclease A	<1	670
RPMI 1640 + 10% heat inactivated fetal calf serum	<1	182

We then evaluated the enzymatic stability of 4'-S(ACACCCAAUUCU), **16**, in cell culture medium at 37°C (Figure 4) using on-line ISRP cleaning HPLC technique (*45*). RPMI 1640 complemented with 10% heat inactivated fetal calf serum was chosen because of its frequent use in hybrid translation arrest experiments *in vitro*.

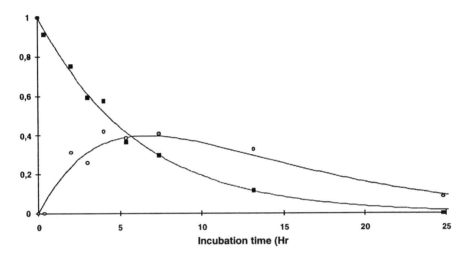

Figure 4. Enzymatic Degradation Kinetics of 4'-S(ACACCCAAUUCU) (■) and Its First Metabolite (○)in RPMI 1640 Culture Medium Complemented with 10% Heat Inactivated Fetal Calf Serum (RPMI was purchased from GIBCO BRL UK)

The different hydrolysis rate constants k_1 and k_2 were determined according to the simple models 4'-S(ACACCCAAUUCU) $\xrightarrow{k_1}$ 11-mer $\xrightarrow{k_2}$ 10-mer, (k_1 = 2.72.10^{-3}, k_2 = 2.29.10^{-3}). The corresponding half-life times are summarized in Table II and compared to the results obtained with other modified dodeca-oligonucleotides of the same sequence (*47*).

It is apparent from Table II that, in conditions where all the oligonucleotide analogs previously analyzed (*47*) were half degraded within 1 hour (entry 1-3), the 4'-thio-RNA derivative, **16**, was found to be resistant to enzymatic cleavage (entry 4). Moreover, due to their high nuclease resistance, it is noteworthy that all the sterile conditions required during the handling and the storage of natural RNA are not necessary for 4'-S-RNA. 4'-Thio-oligomers were recovered intact after being stored frozen for several months at - 20 °C.

Table II. **Half-Life Time of Oligonucleotide Analogues and Their First Degradation Product in RPMI 1640 Culture Medium supplemented with 10% Heat Inactivated Fetal Calf Serum**

Oligonucleotide / Series	$T_{1/2}$ 12-Mer (Min)	$T_{1/2}$ 11-Mer (Min)
5'-ACACCCAATTCT-3' (β-DNA.)	31	13
5'-ACACCCAAUUCU-3' (2'-OMe-β-RNA).	68	58
5'-ACACCCAAUUCU-3' (2'-O-allyl-β-RNA).	60	ND
5'-ACACCCAAUUCU-3' (4'-S-β-RNA) **16**.	255	300

Hydrogen-Bonding Properties. In many cases, the hybridization abilities of antisense oligonucleotides targeted against messenger RNA have been assessed on the basis of Tm measurements with modified complementary oligodeoxynucleotides (DNA) rather than the less accessible oligoribonucleotides (RNA). Thus we decided to evaluate the hybridization properties of the previously synthesized 4'-thio-dodecamer 4'-SU$_{12}$ versus the complementary DNA and RNA sequences d(C$_2$A$_{12}$C$_2$) and poly(A), respectively (42).

Upon increasing the temperature of pre-cooled equimolar 4'-SU$_{12}$ / Poly(A) or 4'-SU$_{12}$ / d(C$_2$A$_{12}$C$_2$) mixture, hyperchromicity was observed from which melting temperature was determined (Table III). A comparison of the Tm data obtained from melting curves at high (1 M NaCl) and low (0.1 M NaCl) salt concentrations (Table III) indicates in both cases that the homoduplex 4'-SU$_{12}$ / Poly(A) is more stable than the heteroduplex 4'-SU$_{12}$ / d(C$_2$A$_{12}$C$_2$). This confirms that RNA/RNA duplexes are more stable than RNA/DNA duplexes.

Table III. **Tm Values (Tm ±0.5°C) of 4'-SU$_{12}$ Hybrids with Complementary RNA or DNA Strands**

Duplex	[NaCl]	Tm °C
4'-SU$_{12}$ / Poly(A)	1 M	46
4'-SU$_{12}$ / d(C$_2$A$_{12}$C$_2$)	1 M	27
4'-SU$_{12}$ / Poly(A)	0.1M	33
4'-SU$_{12}$ / d(C$_2$A$_{12}$C$_2$)	0.1M	5

Moreover, it appears that the Tm value (46°C) determined for the duplex 4'-SU_{12} / Poly(A) is higher than the calculated melting temperature of the duplex U_{12} / A_{12} (Tm = 32°C, 1M NaCl, concentration of oligonucleotide 3μM) (*48*), suggesting that the introduction of a sulfur atom in the 4'-thiooligoribonucleotide improved its thermal stability in duplexes compared to the natural product.

Stoichiometric experiments performed with 4'-SU_{12} and its RNA complementary sequence Poly(A) using a molar NaCl buffer allowed to ascertain a one to one 4'-thio-RNA / RNA duplex formation (Figure 5) (Bellon, L.; Leydier, C.; Barascut, J.-L.; Imbach, J.-L.; unpublished results).

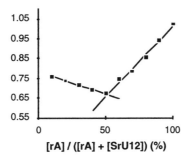

Figure 5. Stoichiometric Measurement of 4'-SU_{12} / Poly(A).
Conditions: 1M NaCl, 10mM Sodium Cacodylate

We then determined the annealing properties of 4'-S(ACACCCAAUUCU), **16**, with synthetic DNA and RNA complementary strands and compared these results with literature data (Table IV) (*47*).

Table IV. Thermal Stability (Tm ±0.5° C) of Hybrids Formed Between
Antisense Oligomers ACACCCAAT(U)T(U)CT(U) and DNA or RNA Targets
AGAAT(U)T(U)GGGT(U)GT(U)

Entry	Antisense Oligomers	Vs 12-Mer DNA Sense	Vs 12-Mer RNA Sense
1	ß-DNA	47.6	46.1
2	ß-S-DNA	37.9	36.1
3	ß-2'OMe-RNA	39.0	58.4
4	ß-2'Oallyl-RNA	35.7	54.0
5	ß-Me-P-DNA	40.3	27.4
6	4'-S-ß-RNA **16**	29.6	56.9

(Condition: 3 μM for each synthetic strand in 0.1 M NaCl, 10 mM Na cacodylate pH 7.0.)

Tm data clearly indicate that the duplex formed between 4'-S-RNA and RNA is much more stable than with DNA by about 27° C (entry 6). If we compare duplexes formed with several antisense oligomers and sense RNA, the 4'-S-RNA exhibits high Tm values comparable to the two 2'-$\underline{O}$-substituted-RNA (entry 3 and 4).

These annealing properties observed under nearly physiological conditions for the duplex 4'-S-RNA / RNA in conjunction with the high nuclease resistance of these 4'-thio-analogs suggest that this new 4'-S-RNA series might be used as efficient antisense molecules.

Interaction with HIV-1 Reverse Transcriptase

Homopolymeric oligonucleotide analogs have been successfully used as non-sequence-specific inhibitors of HIV-1 Reverse Transcriptase. (RT). As an example, phosphorothioate oligodeoxynucleotides (49) or α-oligonucleotides (50), either annealed or non-annealed to a template, competitively inhibit the incorporation of dTMP at the 3'-end of dT_{14} bound to Poly(A) (51, 52). To date, very few RNA homopolymeric oligonucleotide primer analogs have been studied for their possible negative effects on RT activity, even though the natural primer is t-RNA Lys 3 (53). Accordingly, we have studied the binding affinities of the 4'-thio-uridylates 4'-SU_6 and 4'-SU_{12}, hybridized or not with poly (A), versus RT using fluorescence spectrophotometry experiments (Table V). The Kd values extracted from intrinsic fluorescence variation of RT (54) in the presence of the different species are reported in Table V (Column 2). Both oligomers SrU_6 and SrU_{12} showed a marked affinity for the enzyme, whereas their corresponding hybrids with Poly(A) exhibited slightly lower affinities.

Table V. Interaction of 4'-SU_6, and 4'-SU_{12}, with HIV-1 RT

Ligand	Intrinsic Fluorescence Kd (nM)^a	Displacement of fFuorescent 36.F19 Kd (nM)
4'-SU_6	4 ± 0.5	b
Poly(A).4'-SU_6	14.5 ± 0.5	260
4'-SU_{12}	11 ± 1.2	b
Poly(A).4'-SU_{12}	19 ± 4	310

Dissociation constants of complexes as determined by fluorescence change measurements at 25°C. a) Average value from 4 experiments. Relative experimental errors were estimated to be less than 10 %. b) No change in fluorescence intensity.

We then evaluated the capacity of the 4'-thio-homopolymers, annealed or not, to displace a fluorescent hybrid (36F.19) bound to the polymerase active site of the

RT (*55*) (Table V, Column 3). None of the 4'-thiooligomers studied competed with the fluorescent hybrid for binding to the polymerase site. This strongly suggests that these oligomers bind the enzyme in a distinct secondary site. As exemplified by the Kd values, the corresponding duplexes formed with poly (A) displaced 36F.19 from its complex only at high concentrations, which is consistent with very low binding affinities of the 4'-thio-RNA hybrids for the polymerase active site.

Steady-state kinetics (*42*) of the interaction of 4'-SU_{12} with poly (A) revealed that no priming of the polymerization by this 4'-thio-hexamer occurred. Neither 4'-SU_{12} nor its duplex inhibited polymerase activity of the enzyme (data not shown).

Conclusion

We have synthesized for the first time three 4'-thiooligoribonucleotides, namely 4'-SU_6, 4'-SU_{12}, and 4'-S(ACACCCAAUUCU). We have shown that such oligomers possess a high nuclease resistance as compared to the wild type RNA against various nucleases and in cell culture medium. In addition, we studied their capacity to bind to complementary nucleic acid sequences, and we found that they can form stable duplexes with Poly (A) or with synthetic RNA complementary strands under nearly physiological conditions. In contrast, a strong destabilization effect was noticed when these oligomers were associated with the corresponding DNA. With respect to the interaction of these oligomers with reverse transcriptase, all available evidence is consistent with the existence of binding to RT of 4'-SU_6 and 4'-SU_{12}, annealed or not with Poly(A), at a site that is different from the polymerase site of the enzyme. The binding of the oligomers to this secondary site does not inhibit DNA polymerization catalyzed by the enzyme.

All these data firmly indicate that this new stable chimeric ß-RNA series merits further study to evaluate its potency to interfere with the nucleic acid cellular machinery.

Literature Cited

1. Green, P. J.; Pines, O.; Inouye, M. *Annu. Rev. Biochem.*. **1986**, *55*, 569-597.
2. Khochbin, S.; Lawrence, S. J. *EMBO J.* **1989**, *8*, 4107-4114.
3. Krystal, G. W.; Armstrong, B. C.; Battey, J. F. *Mol. Cell Biol.* **1990**, *10*, 4180-4191.
4. Cech, T. R. *Annu. Rev. Biochem.* **1990**, *59*, 543-568.
5. Altman, S. *J. Biol. Chem.* **1990**, *265*, 20053-20056.
6. Inoue, H.; Hayase, Y.; Imura, A.; Iwai, S.; Miura, K.; Ohtsuka, E. *Nucl. Acids Res.* **1987**, *15*, 6131-6148.
7. Sproat, B. S.; Lamond, A. I.; Beijer, B.; Neuner, P.; Ryder, V. *Nucl. Acids Res.* **1989**, *17*, 3373-3386.
8. Morvan, F.; Rayner, B.; Imbach, J.-L. *Tetrahedron Lett.* **1990**, *31*, 7149-7152.
9. Debart, F.; Rayner, B.; Degols, G.; Imbach, J.-L. *Nucl. Acids Res.* **1992**, *20*, 1193-1200.
10. Beaucage, S. L.; Caruthers, M. H. *Tetrahedron Lett.* **1981**, *22*, 1859-1862.

11. Whistler, R. L.; Dick, W. E.; Ingle, T. R.; Rowell, R. M.; Urbas, B. *J. Org. Chem.* **1964**, *29*, 3723-3725.
12. Urbas, B.; Whistler, R. L. *J. Org. Chem.* **1966**, *31*, 813-816.
13. Reist, E. J.; Gueffroy, D. E.; Goodman, L. *J. Am. Chem. Soc.* **1964**, *86*, 5658-5663.
14. Nayak, U. G.; Whistler, R. L. *Liebigs Ann. Chem.* **1970**, *741*, 131-138.
15. Fu, Y. L.; Bobek, M. *J. Org. Chem.* **1976**, *41*, 3831-3834.
16. Hoffman, D. J.; Whistler, R. L. *Biochemistry.* **1970**, *9*, 2367-2372.
17. Reist, E. J.; Gueffroy, D. E.; Goodman, L. *Chem. & Ind.* **1964**, 1364-1365.
18. Bobek, M.; Whistler, R. L.; Bloch, A. *J. Med. Chem.* **1970**, *13*, 411-413.
19. Reist, E. J.; Fisher, L. V.; Goodman, L. *J. Org. Chem.* **1968**, *33*, 189-192.
20. Whistler, R. L.; Doner, L. W.; Nayak, U.G. *J. Org. Chem.* **1971**, *36*, 108-110.
21. Ototani, N.; Whistler, R. L. *J. Med. Chem.* **1974**, *17*, 535-537.
22. Secrist III, J. A.; Riggs, R. M.; Tiwari, K. N.; Montgomery; J.A. *J. Med. Chem.* **1992**, *35*, 533-538.
23. Secrist III, J. A.; Tiwari, K. N.; Riordan, J. M.; Montgomery, J. A. *J. Med. Chem.* **1991**, *34*, 2361-2366.
24. Tiwari, K. N.; Montgomery, J. A.; Secrist III, J. A. *Nucleosides Nucleotides* . **1993**, *12*, 841-846.
25. Dyson, M. R.; Coe, P. L.; Walker, R. T. *Carbohydr. Res.* **1991**, *216*, 237-248.
26. Dyson, M. R.; Coe, P. L.; Walker, R. T. *J. Chem. Soc.* **1991**, 741-742.
27. Dyson, M. R.; Coe, P. L.; Walker, R. T. *Nucleic Acids Res. Symposium Series.* **1991**, *24*, 1-4.
28. Dyson, M. R.; Coe, P. L.; Walker, R. T. *J. Med. Chem.* **1991**, *34*, 2782-2786.
29. Basnak, I.; Hancox, E. L.; Connolly, B. A.; Walker, R. T. *Nucleic Acids Res. Symposium Series.* **1993**, *29*, 101-102.
30. Uenishi, J.; Motoyama, M.; Nishiyama, Y.; Wakabayaschi, S. *J. Chem. Soc. Chem. Commun.* **1991**, 1421-1422.
31. Uenishi, J.; Kawanami, H.; Kubo, Y. *Nucleic Acids Res. Symposium Series.* **1993**, *29*, 37-38.
32. Uenishi, J.; Motoyama, M.; Takahashi, K. *Nucleic Acids Res. Symposium Series.* **1992**, 77-78.
33. Tber, B.; Fahmi, N.; Ronco, G.; MacKenzie, G.; Villa, P.; Ville, G. *Collect. Czech. Chem. Commun.* **1993**, *58*, 18-21.
34. Bredenkamp, M. W.; Holzapfel, C. W.; Swanepoel, A. D. *Tetrahedron Lett..* **1990**, *31*, 2759-2762.
35. Bredenkamp, M. W.; Holzapfel, C. W.; Swanepoel, A. D. *S. Afr. J. Chem.* **1991**, *44*, 31-33.
36. Clement, M. A.; Berger, S. H. *Med. Chem. Res.* **1992**, *2*, 154-164.
37. Bellon, L.; Barascut, J.-L.; Imbach, J.-L. *Nucleosides Nucleotides.* **1992**, *11*, 1467-1479.
38. Bellon, L.; Leydier, C.; Barascut, J.-L.; Imbach, J.-L. *Nucleosides Nucleotides.* **1993**, *12*, 847-852.
39. Vorbruggen, H.; Krolikiewicz, K.; Bennua, B. *Chem. Ber.* **1981**, *114*, 1234-1255.

40. Genu, C.; Gosselin, G.; Puech, F.; Henry, J.-C.; Aubertin, A. M.; Obert, G.; Kirn, A.; Imbach, J.-L. *Nucleosides Nucleotides.* **1991**, *10*, 1345-1376.
41. Bellon, L.; Morvan, F.; Barascut, J.-L.; Imbach, J.-L. *Biochem. Biophys. Res. Commun.* **1992**, *184*, 797-803.
42. Bellon, L.; Barascut, J.-L.; Maury, G.; Divita, G.; Goody, R.; Imbach, J.-L. *Nucl. Acids Res.* **1993**, *21*, 1587-1593.
43. Damha, M. J.; Giannaris, P. A.; Zabarylo, S. V. *Nucl. Acids Res.* **1990**, *18*, 3813-3821.
44. Debart, F.; Rayner, B.; Imbach, J.-L.; *Tetrahedron Lett.* **1990**, *31*, 3537-3540.
45. Pompon, A.; Lefebvre, I.; Imbach, J.-L. *Biochem. Pharm..* **1992**, *43*, 1769-1775.
46. Morvan, F.; Genu, C.; Rayner, B.; Gosselin, G.; Imbach, J.-L. *Biochem. Biophys. Res. Commun.* **1990**, *172*, 537-543.
47. Morvan, F.; Porumb, H.; Degols, G.; Lefebvre, I.; Pompon, A.; Sproat, B. S.; Rayner, B.; Malvy, C.; Lebleu, B.; Imbach, J.-L. *J. Med. Chem.* **1993**, *36*, 280-287.
48. Borer, P. N.; Dengler, B.; Tinoco, I.; Uhlenbeck, O. C. *J. Mol. Biol.* **1974**, *86*, 843-853.
49. Stein, C. A.; Cohen, J. S. In *Oligonucleotides, Topics in Molecular and Structural Biology*; Cohen, J. S., Ed.; Mac Millan Press, **1989**, *Vol. 42*; pp 97-117.
50. Rayner, B.; Malvy, C.; Paoletti, J.; Lebleu, B.; Paoletti, C.; Imbach, J.-L. In *Oligonucleotides, Topic in Molecular and Structural Biology*; Cohen J. S., Ed.; Mac Millan Press, **1989**, *Vol 42*; pp 119-136.
51. Majumdar, C.; Stein, C. A.; Cohen, J. S.; Broder, S.; Wilson, S. *Biochemistry.* **1989**, *28*, 1340-1346.
52. Maury, G.; Rayner, B.; Imbach, J.-L.; Muleer, B.; Restle, T.; Goody, R. S. *Nucleosides Nucleotides.* **1991**, *10*, 325-327.
53. De Vico, A. L.; Sarngaharan, M. G. *J. Enz. Inhibition.* **1992**, *6*, 9-34.
54. Maury, G.; Elaloui, A.; Morvan, F.; Muller, B.; Imbach, J.-L.; Goody, R. S. *Biochem. Biophys. Res. Comm.* **1992**, *186*, 1249-1256.
55. Muller, B.; Restle, T.; Reinstein, J.; Goody, R. S. *Biochemistry.* **1991** *30*, 3709-3715.

RECEIVED June 16, 1994

Chapter 6

Hexopyranosyl-Like Oligonucleotides

Piet Herdewijn[1], Hans De Winter[2], Bogdan Doboszewski[1],
Ilse Verheggen[1], Koen Augustyns[1], Chris Hendrix[1],
Tula Saison-Behmoaras[3], Camiel De Ranter[2], and Arthur Van Aerschot[1]

[1]Laboratory of Medicinal Chemistry, Rega Institute,
Minderbroedersstraat 10, B-3000 Leuven, Belgium
[2]Laboratory of Analytical Chemistry and Medicinal Fysicochemistry,
Katholieke Universiteit Leuven, Van Evenstraat 4,
B-3000 Leuven, Belgium
[3]Laboratoire de Biophysique, Muséum National D'Histoire Naturelle,
rue Cuvier 43, F-75231 Paris, France

Oligonucleotides containing monomers 1-(2,3-dideoxy-β-D-erythro-hexopyranosyl)thymine; 1-(2,4-dideoxy-β-D-erythro-hexopyranosyl)-thymine; 1-(3,4-dideoxy-β-D-erythro-hexo-pyranosyl)-thymine; 1-(2,3-dideoxy-α-D-erythro-hexopyranosyl)thymine; 1-[2,3-dideoxy-3-C-(hydroxymethyl)-α-L-threo-pentopyranosyl]-thymine; and 1,5-an-hydro-2,3-dideoxy-2-(thymin-1-yl)-D-arabinohexitol were modeled and synthesized. Only oligonucleotides constructed of 1,5-anhydro-hexitol nucleosides are able to hybridize with natural DNA.

The synthesis of oligonucleotides with hexopyranose nucleoside building blocks, started during the second half of the eighties, was based on well known physicochemical considerations. Regardless of solvent effects, when two single-stranded oligonucleotides join each other to form a duplex, the formation of the duplex is driven by enthalpy and entropy factors. Stacking interactions and hydrogen bonding (enthalpy changes) stabilize helix formation; loss in conformational freedom (entropy changes) disfavors duplex formation. The loss in entropy, however, is small when starting from conformational rigid structures (i.e., when the single-stranded oligonucleotide is built with conformationally rigid building blocks and when this single-stranded oligonucleotide itself displays a structure that is complementary to its target). A pyranose oligonucleotide has, theoretically, a free-energy advantage over a furanose oligomer, assuming that the same intermolecular forces (enthalpy changes) determine duplex formation. These oligonucleotides should be able to physically block the transcription process.

The big challenge was to find a pyranose-oligomer that was able to hybridize with natural furanose-DNA under the normal Watson-Crick pairing selectivity. Or, better said, the pairing priorities do not have to be exactly the same, but the

0097−6156/94/0580−0080$08.00/0
© 1994 American Chemical Society

selectivity of pairing with dA, dG, dC, and T is important, either in the parallel or antiparallel direction.

In an effort to find out why nature selected pentoses and not hexoses as sugar building blocks for nucleic acids, Dr. Eschenmoser and his collaborators investigated the synthesis of β-D-2,3-dideoxyglucopyranosyl (6'→4') oligonucleotides (homo DNA) (*1*) and of β-D-ribopyranosyl (4'→2') oligonucleotides (pRNA) (*2*). The selection of these building blocks is based on retrosynthetic considerations of structures, which may have occurred naturally. These oligonucleotides, however, do not show base pairing with natural DNA backbones so that their usefulness as potential therapeutic agents is highly limited. However, we learned from this work that "the Watson-Crick rules are not only a consequence of the chemical properties of the four bases but also of the specific furanose structure of the natural DNA backbone". This information further stimulated us to investigate several different pyranose-like structures.

Synthesis and Properties of Oligonucleotides Containing 2,3-Dideoxy-β-D-Erythro-Hexopyranosyl, 3,4-Dideoxy-β-D-Erythro-Hexopyranosyl, and 2,4-Dideoxy-β-D-Erythro-Hexopyranosyl Nucleoside Building Blocks

Our research started with the synthesis of oligonucleotides having normal pyranose nucleoside building blocks incorporated and with a 4'-(T^1), 2'-(T^2) or 3'-(T^3) positioned secondary hydroxylgroup.

1-(2,3-Dideoxy-β-D-erythro-hexopyranosyl)thymine (T^1) was synthesized by reaction of bis(trimethylsilyl)thymine and 3,4,6-tri-O-acetyl-D-glucal (*3, 4*) in the presence of trimethylsilyl triflate followed by catalytic hydrogenation and deacetylation (Scheme 1) (*5*). The same molecule was also synthesized by condensation of methyl 2,3-dideoxy-α-D-erythro-hexopyranose with thymine in the presence of HMDS, TMSCl, and $SnCl_4$ in 65% yield (*1*).

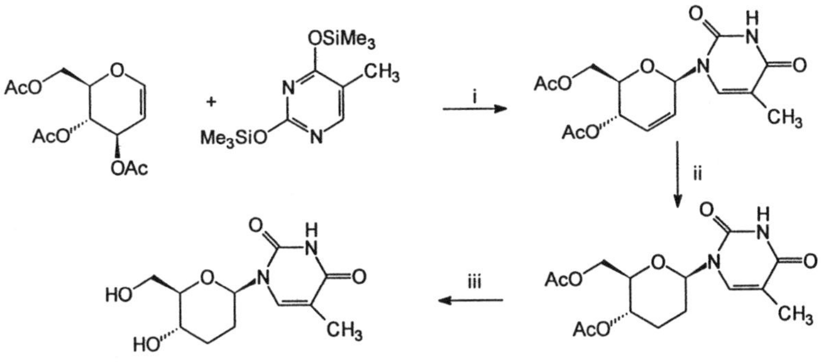

Scheme 1: Synthesis of 1-(2,3-dideoxy-β-D-erythro-hexopyrano-syl)thymine. (T^1) i: TMSOTfl., CH_3CN, 1 h, RT, 45%; ii: H_2, 10% Pd/C, MeOH, 2 h, RT, 85%; iii: NH_3, MeOH, 16 h, RT, 80%.

1-(3,4-Dideoxy-β-D-erythro-hexopyranosyl)thymine (T^2) could be obtained as depicted in Scheme 2. Methyl α-D-glucopyranoside was selectively benzoylated at position 2 and 6 (*6*) and the trans vicinal diol functionality was converted to a double bound (*7*), which was hydrogenated (*8*). The methyl glycoside was converted to the 1-acetoxy sugar and condensed with silylated thymine. The benzoyl protecting groups were removed in the usual way (*9*).

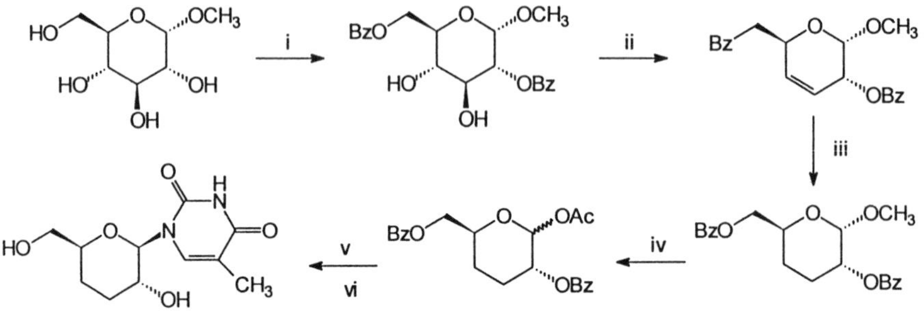

Scheme 2: Synthesis of 1-(3,4-dideoxy-β-D-erythro-hexopyrano-syl)thymine (T^2). i: (Bu$_3$Sn)$_2$O, PhCH$_3$, BzCl, 8 h, RT, 80%; ii: Ph$_3$P, imidazole, 2,4,5-triiodoimidazole, PhCH$_3$, 3 h, Δ, 85%; iii: H$_2$, Pd/C, Et$_3$N, EtOH, 95%; iv: AcOH, Ac$_2$O, H$_2$SO$_4$, 24 h, RT, NaOAc, 65%; v: SnCl$_4$, CH$_3$CN, 1.5 h, 40° C, 88%; vi: NH$_3$, MeOH, RT, 3 days, 100%.

The first part of the reaction scheme (Scheme 3) leading to 1-(2,4-dideoxy-β-D-erythro-hexopyranosyl)thymine (T^3) was originally described by Corey (*10*) and Falck (*11*). Tri-O-acetyl-D-glycal is converted to methyl 2-deoxy-α-D-glu-copyranoside by methoxymercuration and reduction of the chloromercury derivative. Tritylation of the primary hydroxyl group was followed by a stereoselective epoxi-dation and regioselective opening of the epoxide with lithium aluminium hydride. Detritylation of methyl 2,4-dideoxy-α-D-erythro-hexopyranoside with a catalytic amount of p-toluenesulphonic acid in methanol followed by benzoylation gave methyl 3,6-di-O-benzoyl-2,4-dideoxy-D-erythro-hexopyranoside as a mixture of the α- and β-anomer. When the detritylation was performed with formic acid or with 80% HOAc, the 1,6-anhydro compound was obtained. Treatment of the anomeric mixture with 80% HOAc followed by acetylation afforded 62% of the 1-acetoxy sugar, which was used in the sugar-base condensation reaction. The outcome of the sugar-base condensation reaction leading to the nucleoside analogue is very suscep-tible to the reaction conditions used. Sugar-base condensation reaction with thymine or N^4-benzoylcytosine at room temperature overnight in the presence of BSA, TMSOTfl gives preponderantly the thermodynamically   more stable β-anomer. Shorter reaction times give more of the α-anomer. Condensation with N^2-acetyl-O^6-diphenylcarbamoyl guanine also afforded predominantly the β-anomer. Very long reaction times are necessary to synthesize the adenine analogue. Here no α-anomer was isolated, and the yield was quite low (30%) (*12, 13*).

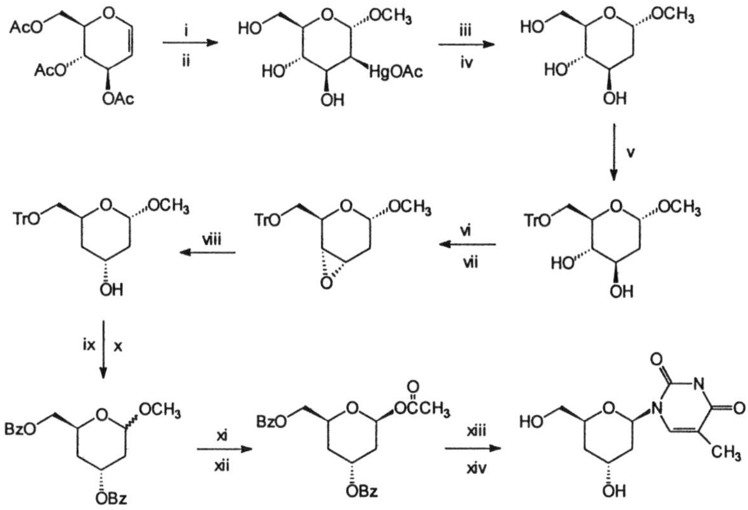

Scheme 3: Synthesis of 1-(2,4-dideoxy-β-D-erythro-hexopyrano-syl)thymine (T³). i: NaOMe, MeOH, 1.5 h, RT; ii: Hg(OAc)₂, 2.5 h; RT, 72%; iii: NaCl, MeOH, 20 min, RT; iv: NaBH₄, ¹PrOH, 30 min, 0° C, HCl, 91%; v: TrCl, pyr, 16 h, RT, 85%; vi: NaH, HMPT, 30 min, RT; vii: (¹Pr)₃C₆H₂SO₂ imidazole, THF, -10° C, 4 h, 80%; viii: LiAlH₄, Et₂O, 1 h, RT, 95%; ix: p-toluensulphonic acid 0,05%, MeOH, 24 h, RT; x: BzCl, pyr, 4 h, RT, 90%; xi: 80% HOAc, 5 h, 80° C; xii: Ac₂0, pyr, 72 h, RT, 62% together with 7% of the α-ano-mer; xiii: thymine, BSA, TMSOTfl, C₂H₄Cl₂, 16 h, RT, 71% together with 9% of the α-anomer; xiv: NH₃, MeOH, 4 days, RT, 88%. Nu-cleosides with an adenine base moiety are designated as **A³**, with cy-tosine as **C³** and with guanine as **G³**.

Oligonucleotides containing **T¹**, **T²**, and **T³**; together with oligonucleotides containing 2,4-dideoxy-β-D-erythro-hexopyranose nucleosides with adenine, cytosine, and guanine moieties; were synthesized on solid support using the phosphoramidite approach. Monomers used to synthesize the oligonucleotides are depicted in Scheme 4a.

Loading of the long chain alkylamino-controlled pore glass was carried out using reagents shown in Scheme 4b. The loading capacity varies between 25 and 30 µmol/g. Enzymatic degradation of the synthesized oligonucleotides using snake venom phosphodiesterase and alkaline phosphatase demonstrates the correct ratio of unmodified over modified nucleosides.

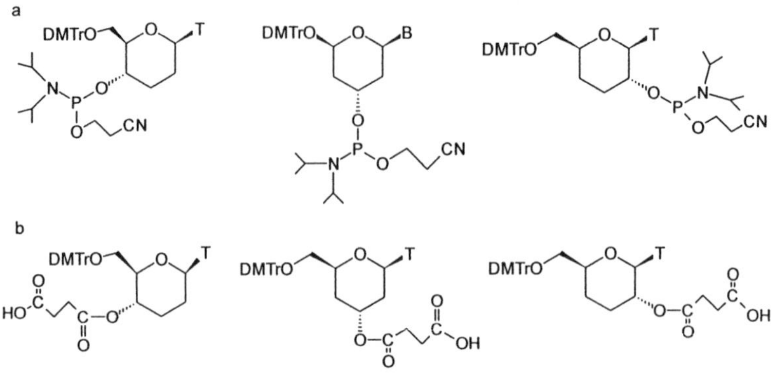

a

b

Scheme 4: B: thymin-1-yl; N^4-benzoylcytosin-1-yl; N^6-benzoy-ladenin-9-yl; N^2-isobutyryl-guanin-9-yl.

Melting curves of the synthesized oligonucleotides annealed to unmodified $(dA)_{13}$ were recorded in 0.1 M NaCl, 0.02 M potassium phosphate pH 7.5, 0.1 mM EDTA at a concentration of 4 µM for each oligonucleotide (Table I). Substitution with one or two 2,4-dideoxyhexose nucleosides (T^3) at either end resulted in a melting temperature that was approximately the same as for the unsubstituted oligonucleotide. These terminal hexoses are still base paired and take part in the cooperative melting of the duplex. Substitution in the middle, however, has a more pronounced effect, resulting in a destabilization corresponding to 1.5 times the nearest neighbor interaction. This means that the destabilization is of the same magnitude as the least stable base-base mismatch and slightly lower than what would be expected for an internal loop containing two unpaired bases.

Table I: Enzymatic Stability (14) and T_m Values (9) of T^1, T^2, and T^3 Containing Oligonucleotides Hybridized with Natural $(dA)_{13}$

	t 1/2 (snake venom PD)	T_m (° C)	$\Delta G^{25° C}$ (kJ/mol)
T_{13}	1 (2 min)	32.3	41.4
$T^3T_{11}T^3$	2.1	33.9	42.9
$T^3T^3T_9T^3T^3$	4.5	29.9	38.2
$T_6T^3T_6$	2.6	25.6	33.2
$T^2T_{11}T^2$	2.9	31.4	39.6
$T^2T^2T_9T^2T^2$	stable (1 h)	22.5	30.2
$T_6T^2T_6$	3.5	17.3	23.2
$T^1T_{11}T^1$	2.4	32.2	40.8
$T^1T^1T_9T^1T^1$	17	26.7	34.4
$T_6T^1T_6$	2.4	17.5	24.2

The effect of one substitution in the middle is, therefore, limited to one base pair. This destabilization is only considered here with respect to normal Watson-Crick A-T base pairing. As we will learn later, alternative base pairing could give more stable duplexes with 2,4-dideoxyhexopyranose nucleotides ($\mathbf{T}^3$) (*14*).

The effect on duplex stability of a substitution with $\mathbf{T}^1$ or $\mathbf{T}^2$ is more dramatic. The ΔG values for oligonucleotides having two 3,4-dideoxyhexose nucleotides at either end correspond to a calculated value for an inner 11-mer with two dangling ends, indicating that one modified nucleoside is base paired and the other not (*9*). The destabilization caused by a substitution with $\mathbf{T}^2$ in the middle of a sequence indicates the formation of an internal loop of at least four unpaired bases. The results obtained with $\mathbf{T}^2$ are very much like the results obtained with $\mathbf{T}^1$, so that one can conclude that the magnitude of destabilization of the duplex increases in the order $\mathbf{T}^3 \!\ll\! \mathbf{T}^1 \!<\! \mathbf{T}^2$ (Figure 1). Completely modified 13-mers of $\mathbf{T}^1$, $\mathbf{T}^2$, or $\mathbf{T}^3$ are not capable of base pairing with an unmodified $(dA)_{13}$.

Substitution with one hexose at the end gives only a slight increase in stability against 3′-exonuclease, but substitution with two hexoses at either end resulted in a larger protection against enzymatic degradation (*14*). However, the stability against 3′-exonuclease decreases in the opposite order ($\mathbf{T}^2 \!>\! \mathbf{T}^1 \!\gg\! \mathbf{T}^3$) as the hybridization capacity.

The incorporation of a 2,4-dideoxyhexose nucleoside into a natural oligonucleotide, however, alters base-pairing specificity. This can be deduced from the melting points of the oligonucleotides depicted in Table II, which were synthesized in a program to select oligonucleotides active against Ha-ras oncogene expression (*14*). These oligonucleotides are complementary to the mRNA sequence of the activated Ha-ras oncogen of the human T24 bladder carcinoma. The mutation site (G-T transversion in the 12th codon) is located in the middle of the sequence. The oligonucleotides are targeted partly to the loop and partly to the RNA-RNA duplex of the hairpin formed by the Ha-ras mRNA (*15*).

Table II: Physical Properties of Oligonucleotides

(3′ → 5′)	5′d(GGCGCCGTC GGTG)3′		5′d(GGCGCCGGC GGTG)3′		5′r(GGCGCCGUC GGUG)3′ ≠	
	T_m (° C)	ΔG^{25} (kJ/mol)	T_m (° C)	ΔG^{25} (kJ/mol)	T_m (° C)	ΔG^{25} (kJ/mol)
d(CCGCGGCAGCCAC)	69.4	84.4	66.4	78.9	67.1	73.9
d(CC^{3}G^{3}CGGCAGCCAC)	64.2	76.2	61.1	69.1	63.8	71.5
d(CCGCGGCA^{3}GCCAC)	62.6	77.7	65.4	80.1	62.6	71.0

≠: flanked by (dT)$_3$ at both sides.

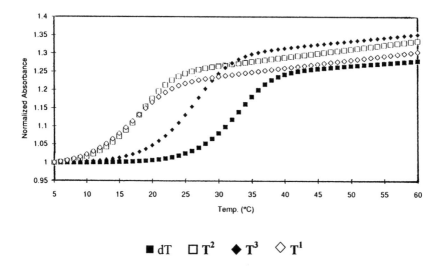

Figure 1: Comparative Melting Curves of $(dT)_{13}.(dA)_{13}$ with T^1, T^2, or T^3 Incorporated in the Middle of the Oligomer

It is clear from this table that an A^3-G mismatch is more stable than an A^3-T Watson-Crick base pair and that base pairing specificity alters from A-T>A-G>A-A>A-C to A^3-G>A^3-T>A^3-A>A^3-C when the natural furanose was substituted for a 2,4-dideoxyhexose. It seems that the A^3-G mismatch can adapt better to the slightly altered backbone structure caused by the incorporation of the hexopyranosyl nucleoside. Moreover, the 2,4-dideoxyhexose nucleoside is slightly better accepted in an RNA-DNA duplex (less pronounced decrease in melting point by introduction of the pyranose nucleoside) than in a DNA-DNA duplex, which further encouraged us to evaluate the constructs in a biological model (*14*).

The oligonucleotide protected at the 3′-end with two hexose nucleosides was only 2% degraded after 19 hour incubation at 37° in cell culture medium supplemented with 7% heat inactivated fetal calf serum.

When we investigated the RNase H (Rabbit Reticulocyte) mediated cleavage of Ha-ras mRNA (*16*), it was clear that the enzyme respects the normal Watson-Crick base pairing in the neighborhood of the cleavage site as no cleavage of the normal Ha-ras mRNA (A-G mismatch) was observed. The introduction of two pyranose nucleosides positioned eight bases upstream of the cleavage site, however, induced stronger cleavage of the mutated Ha-ras mRNA than the natural oligonucleotide. This double modified oligonucleotide was also able to inhibit the growth of T24 cells, while the others were not (*16*). The observed cleavage and inhibitory effects with modified oligonucleotides may be due to the formation of a more stable duplex with the target and increased nuclease resistance. A control oligonucleotide containing an acyclic nucleoside at the 3′-end was found to hybridize well with the target, however, failed to demonstrate the activity. The improved activity of the oligonucleotide, having two pyranose nucleosides incorporated five bases upstream of the mutation site (in comparison with the activity of the non-modified oligonucleotide), should be explained by the combined effect of an increased nuclease resistance and an increased RNase H activation. Its 100% selectivity could be due to the fact that the oligonucleotide is targeted to the mutation site, which is located in the middle of the sequence so that the stem-loop structure is efficiently disrupted (*16*), and due to the fact that an A^3-G mismatch is not recognized by RNase H although it is more stable than the A^3-T base pairing.

Modeling of Pyranose-Like Oligonucleotides

Because the hydroxyl group and the hydroxymethyl group may be attached onto the tetrahydropyran ring in different positions, and because it is very difficult to predict which combination will result in strong hybridizing properties, molecular modeling studies were initiated. The study had the purpose to investigate which pyranose oligonucleotide is able to form stable duplexes with natural single-stranded DNA. In order to keep the number of structures being modeled to a tractable quantity, we have restricted ourselves to the modeling of only the pyranose nucleosides with one hydroxymethyl- and one hydroxyl group distributed over the positions 2′, 3′, 4′ and 5′ and with a thymine base moiety at the 1′-position in a β orientation. These restrictions already gave 56 possible combinations. The work itself started with an investigation of the conformational preference of pyranose nucleosides using X-ray crystallography and molecular mechanical force field calculations (AMBER) (*17*). The X-ray studies give a "standard" pyranosyl nucleoside conformation, and this was used as a starting conformation in our selection strategy based on a combination of energy minimization and high temperature dynamics. The stability of dsDNA duplexes consisting of a strand of pyranosyl nucleosides with promising hybridization capabilities and a strand of natural DNA was investigated during 50 and 200 ps molecular dynamics simulations in water.

      Conformational analysis of 13 different pyranose nucleosides with a pyrimidine base moiety by X-ray diffraction studies (Figure 2) revealed that the base moiety is always oriented equatorially with respect to the sugar part ($^4C_1$ puckering for the β-nucleoside) and that the pyranosyl sugar ring adopts a slightly flattened chair conformation. The C_1'-O_5' bond length of the sugar is slightly shorter than the C_5'-O_5' distance, and the χ torsion angle is locked in the -anticlinal region ($\chi = 180°$ to $270°$). Molecular mechanic calculations and molecular dynamics simulation on pyranosyl nucleosides using the AMBER force field (*17*) confirm the X-ray observations.

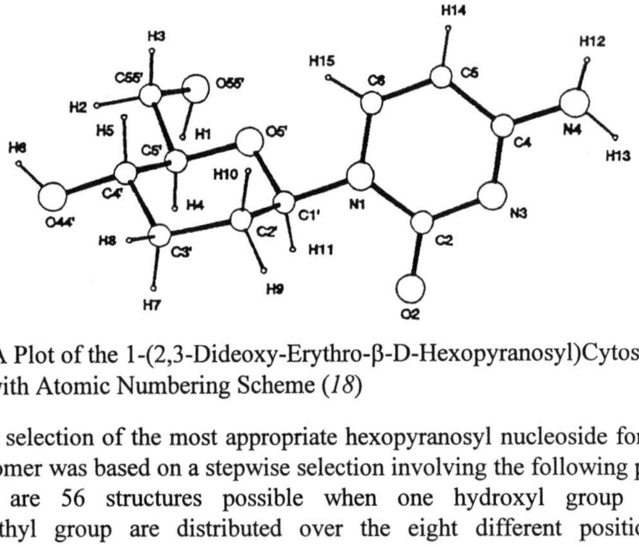

Figure 2: A Plot of the 1-(2,3-Dideoxy-Erythro-β-D-Hexopyranosyl)Cytosine X-ray Structure with Atomic Numbering Scheme (*18*)

      The selection of the most appropriate hexopyranosyl nucleoside for building up the oligomer was based on a stepwise selection involving the following processes. (a) There are 56 structures possible when one hydroxyl group and one hydroxymethyl group are distributed over the eight different positions of a pyranosyl-like sugar ring with a base substituent in the C-1' β-configuration. (b) A manual selection procedure using Dreiding models retrieved 15 pyranosyl structures with promising hybridizing capabilities (Table III). (c) The ability of tetramers of the 15 pyranose nucleosides to form stable double helical complexes with unmodified $(dA)_4$ strands was investigated by exploration of the conformational space of the 15 modified pyranosyl thymidine tetramers using high temperature molecular dynamics and energy minimization techniques. The energy difference between the helix conformation and the lowest energy conformation should at least be similar or less than the comparable energy difference for the unmodified $(dT)_4$ tetramer. As the chain direction goes from the free hydroxymethyl (5') to the free hydroxyl end (3'), T^1 to T^{10} were modeled as antiparallel oriented helices, T^{11} to T^{14} as parallel chains.

Table III: The 15 β-Pyranosyl Nucleosides as Subunits for the Construction of Tetramers

Code	Position							
	2′		3′		4′		5′	
	A	B	C	D	E	F	G	H
T^1	-	-	-	-	-	OH	CH$_2$OH	-
T^2	-	OH	-	-	-	-	CH$_2$OH	-
T^3	-	-	-	OH	-	-	CH$_2$OH	-
T^4	-	-	-	-	-	-	CH$_2$OH	OH
T^5	-	-	-	OH	CH$_2$OH	-	-	-
T^6	-	-	-	-	CH$_2$OH	-	-	OH
T^7	-	OH	-	-	CH$_2$OH	-	-	-
T^8	CH$_2$OH	-	-	OH	-	-	-	-
T^9	CH$_2$OH	OH	-	-	-	-	-	-
T^{10}	-	-	OH	-	-	CH$_2$OH	-	-
T^{11}	-	-	-	-	-	-	OH	CH$_2$OH
T^{12}	-	-	-	-	-	CH$_2$OH	OH	-
T^{13}	-	-	-	-	OH	CH$_2$OH	-	-
T^{14}	-	-	-	CH$_2$OH	OH	-	-	-
T^{15}	-	-	-	CH$_2$OH	-	-	OH	-

Models of the tetramer double helices were constructed by template, forcing the base atoms of each pyranosyl strand onto the corresponding d(T)$_4$ base positions in the Arnott geometry (*19*) of d(A)$_4$.d(T)$_4$, while the sugar and phosphate backbone atoms were allowed to relax. The energy minimization of the tetramers leads in some cases (T^3, T^9, and T^{12}) to totally distorted duplexes, characterized by the absence of intrachain base stacking, missing interchain hydrogen bonding, and almost perpendicular propeller twisting within the base pairs.   In most cases, however, the double helices are more or less normal shaped, except that often only one interchain hydrogen bond instead of the expected two are observed between the terminal base pairs. The high temperature molecular dynamics conformational search (MD calculations for 200 ps at 1000 K and sampling frequency of one conformation every 10 ps followed by MD at 300 K for 10 ps and minimization until the rms derivative of the function gradient was less than 0.1 kcal/mol Å) is summarized in Table IV. Although it has been proven that PNA forms triple helices with DNA, the polyamide nucleic acid (PNA-T_4) (20) built as a heteroduplex with $(dA)_4$ was incorporated in this study. In addition, the three already synthesized oligonucleotides (T^1, T^2, and T^3) can be considered as negative controls, as it was known in advance that incorporation of those hexose nucleosides gives less stable duplexes. As can be seen from the table, the energy needed for forcing the modified tetramer chains into a helix conformation is only for the T^5 and T^{14} (Scheme 5) tetramers, comparable to the 27.4 kcal/mol needed for the formation of the unmodified $(dT)_4$ chain.

Table IV: Conformational Search at 1000 K on the 15 Modified Pyranosyl Nucleoside Tetramers. Energies are reported in kcal/mol.

	E_{low}	E_{helix}	ΔE
T^1	-141.6	-96.6	45.0
T^2	-162.2	-112.8	49.4
T^3	-174.8	-140.6	34.2
T^4	-158.9	-95.5	63.4
T^5	-144.4	-122.2	22.2
T^6	-140.5	-96.9	43.6
T^7	-158.3	-102.9	55.4
T^8	-139.0	-68.7	70.3
T^9	-155.4	-93.8	61.6
T^{10}	-127.6	-90.1	37.5
T^{11}	-156.8	-99.5	57.3
T^{12}	-147.3	-98.8	48.5
T^{13}	-149.1	-116.2	32.9
T^{14}	-134.6	-117.3	17.3
T^{15}	-179.3	-140.2	39.1
$(dT)_4$	-111.0	-83.6	27.4
$(PNA)_4$	-191.4	-168.2	23.2

E_{low}: energy of the conformation with the lowest energy minimum
E_{helix}: helix energy
ΔE: destabilization energy

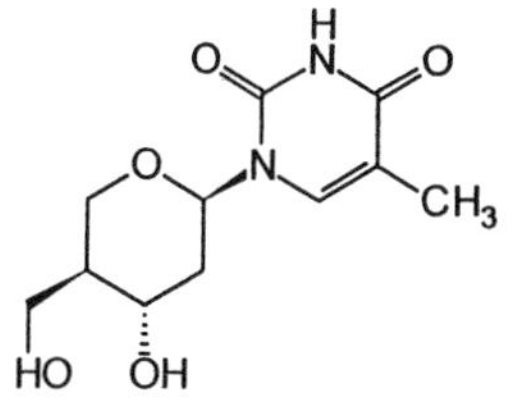

antiparallel
T^5

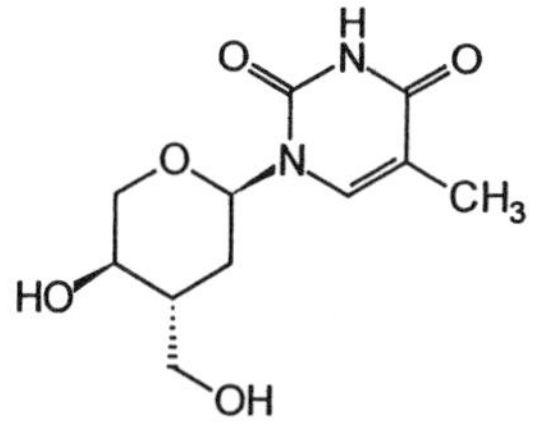

parallel
T^{14}

Scheme 5: Structures of T^5 and T^{14} predicted as candidates for oligo-nucleotide synthesis.

The helical conformation of T^1, T^2, and T^3 is less stable and requires a substantial energy sacrifice, which is in agreement with the observed melting points. The destabilization energy of the PNA is less than the ΔE of the natural dsDNA tetramer. These calculations, however, should be interpreted carefully since they are performed on the single T4 and A4 strands, and interchain interactions such as base pair formation and electrostatic interactions are thus neglected. Therefore, the most promising duplexes were more extensively examined using molecular dynamics in an aqueous environment to account for solvent and entropy effects.

(d) Aqueous molecular dynamics simulation on six of the DNA complexes as hexamers was performed: $(dA)_6$ with $(T^5)_6$, $(T^{14})_6$, $(T^1)_6$, $(T^3)_6$, and $(PNA)_6$ and $(dT)_6$. All systems were supplied with 10 Na^+ counterions initially positioned on the phosphate bisectors 3 Å from the P atom (except for the polyamide system, which was completed with only five counterions), and the double helices were subsequently soaked into a large water bath constructed of TIP3P water molecules, which were a snapshot from a Monte Carlo simulation of water (*21*). Finally, removal of sterically disallowed waters, or waters more than 9 Å distant from any solute atom, led to double helices immersed in a water droplet of about 800 water molecules, enough to provide in excess of two to three complete hydration shells for the DNAs. Periodic boundary conditions were not applied. The SHAKE algorithm (*22*) was used to constrain all bond lengths to their equilibrium value and to allow a time step of 2 fs. A non-bonded cut-off distance of 14 Å was used to compromise between reliability of the model and computational cost. Polyamide and pyranosyl sugar charges were derived through fitting the STO-3G electrostatic potentials (*23*) calculated by GAUSSIAN 80-UCSF (*24*); the phosphate groups, thymine bases, and oligo(dA) strand charges were taken from Weiner et al. (*25*). The terminal amino and carboxyl group of PNA were ionized.

Molecular dynamics in water for 50 ps reveals that higher functions of the sugar moiety of the deoxyadenosines are in the high-energy C-4′ endo region with hexamers containing T^1, T^3, and T^5, rather than with hexamers constructed of T^{14} and PNA. Moreover, the pyranosyl sugars of T^1 and T^3 spent almost 35% of their total time in boat and twist boat conformations. The α, β, γ, and δ backbone torsion angles of the six duplexes are comparable. However, the ε, ζ, and χ angles differ considerably. The ε and ζ angles of T^{14} and PNA correspond quite well with each other and differ somewhat from those of the natural duplex. On the other hand, there is a better fit between χ values for PNA and the natural duplex than with T^{14}. The average propeller twist for the base pairs of $(dA)_6.(dT)_6$ is 45°, while it is only 23° for $(dA)_6.(T^{14})_6$ with the other four helices distributed between them. The values for the helix repeat angles overlap each other within standard deviations and are in agreement with experimental data (9.5 to 10.5 residues for one turn). Only the unmodified oligomer, PNA, and T^{14} are able to maintain their hydrogen bond motif during the whole simulation period while remaining in a helix-like conformation.

In the $(dA)_6.(T^5)_6$ heteroduplex, a gap of three base pairs deep at the 3'-end of the adenine strand is formed after 35 ps simulation. The $(dA)_6.(T^{14})_6$ double helix remains intact over the course of the 50 ps simulation, and this was confirmed after an extended simulation period of 200 ps (Figure 3).

Although we are aware of the limitations of the present molecular dynamics simulations in predicting the stability of nucleic acids double helices (the stability of the $(dA)_6.(T^{14})_6$ duplex during the total 200 ps of MD simulation with very strong hydrogen bonds), no unusual ring puckering and backbone torsion angles in good agreement with the corresponding angles of a DNA helix encourages us to start the synthesis of T^{14} oligonucleotides.

Synthesis and Properties of 1,3,4-Substituted Pyranosyl-Like Oligonucleotides

The modeling studies, as well as our previous work on 1,5-anhydrohexitol nucleosides, stimulated us to start the synthesis of oligonucleotides with a 1,3,4-arrangement of the base moiety, the hydroxymethyl group, and the secondary hydroxyl group. The structures of these molecules are depicted in Scheme 6. In these molecules, the oxygen function of the tetrahydropyran ring is situated on the three possible positions left.

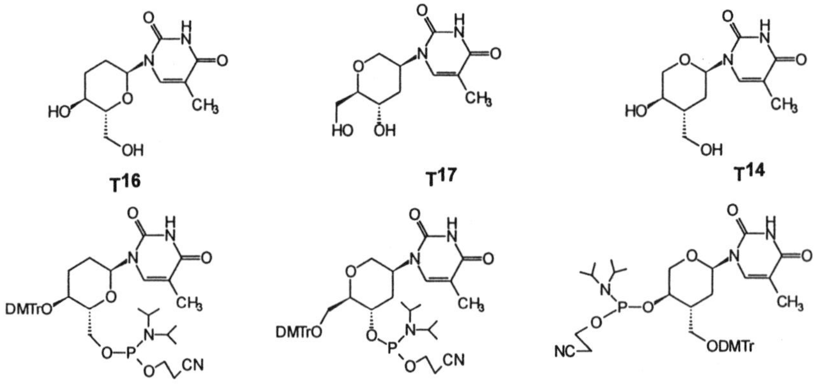

Scheme 6: 1,4-Substituted tetrahydropyran nucleosides and their phosphoramidite derivatives used as building blocks for oligonucleotide synthesis.

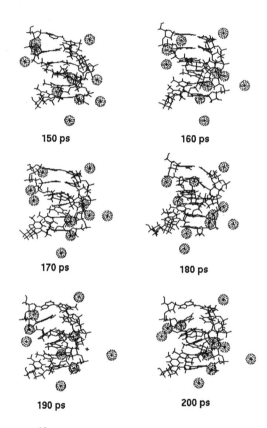

150 ps

160 ps

170 ps

180 ps

190 ps

200 ps

Figure 3: The $(dA)_6 \cdot (T^{14})_6$ Simulations Shown at 10 ps Intervals Along the 150-200 ps Trajectory

Compound T^{16} was selected because of its structural resemblance to T^{14}. When one takes into account that an anomeric-positioned base moiety of pyranosyl nucleosides has an equatorial orientation, then the structure of T_{14} and T_{16} should be similar (Scheme 7).

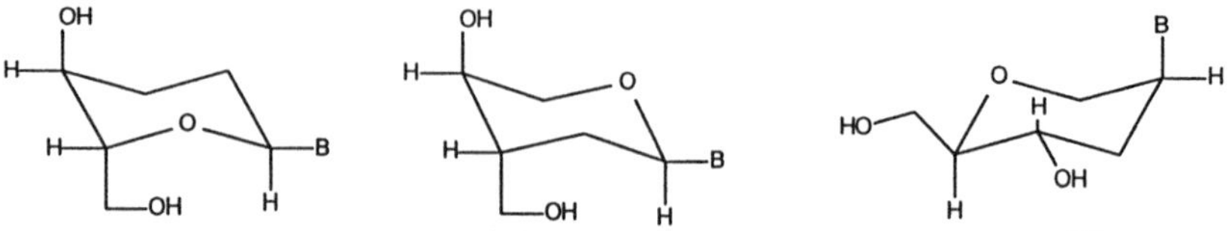

T^{16} T^{14} T^{17}

Scheme 7: Conformation of the 1,4-pyranosyl-like nucleosides.

The phosphoramidites, which were used to synthesize the oligonucleotides, are also depicted in Scheme 6. From the structures of the phosphoramidites, it can be deduced that T^{17} should be incorporated in the antiparallel way, while T^{14} has to be incorporated in a parallel manner. T^{16} was also incorporated in natural DNA, which justified the synthesis of its reversed phosphoramidite building block.

The phosphoramidite 1-(2,3-dideoxy-α-D-erythro-hexopyranosyl) thymine (T^{16}) was synthesized according to Scheme 8. The starting material was obtained as depicted in Scheme 1. The condensation of D-glucal with silylated thymine gives an equimolecular mixture of the α and β isomer, which was separated before reduction of the double bond was carried out.

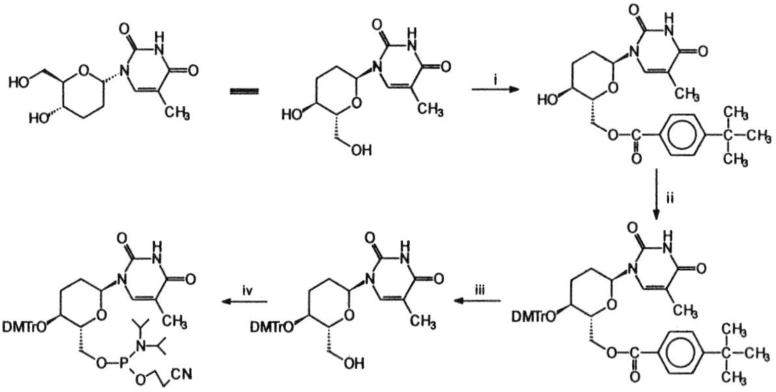

Scheme 8: Synthesis of the phosphoramidite building blocks of T^{16}. i: ^{t}Butylbenzoyl chloride (1.2 eq), pyr, 4° C, 2 h, 85%; ii: DMTrCl (3 eq), pyr, 24 h; iii: NH$_3$, MeOH, 48 h, 65%; iv: (iPr)$_2$NPClOCH$_2$CH$_2$CN (1.5 eq), (iPr)$_2$NEt (3eq), CH$_2$Cl$_2$, 3 h, RT, 55%.

Scheme 9 shows the synthesis of the anhydrohexitol nucleoside T^{17}. Acetobromo-α-D-glucopyranose, obtained from anhydrous D-glucose (*26*), was reductively dehalogenated with Bu_3SnH to 2,3,4,6-tetra-O-acetyl-1,5-anhydro-D-glucitol (*27*). After removal of the protecting groups, the 4- and 6-hydroxyl groups were protected with a benzylidene function. Selective toluoylation at position 2 was carried out with dibutyltin oxide and toluoyl chloride. The hydroxyl group at position 3 was removed by conversion to the 2,4-dichloro phenylthiocarbonate derivative followed by reduction with Bu_3SnH, AIBN. Removal of the toluoyl group in position 3 was followed by a Mitsunobu reaction with N^3-benzoylthymine in the presence of Ph_3P and diethyl azodicaboxylate. Finally the protecting groups were removed to yield 1,5-anhydro-2,3-dideoxy-2-(thymin-1-yl)-D-arabinohexitol (*28, 29*). ^1H NMR and X-ray analysis proved the base moiety of the nucleoside is situated in an axial orientation in the anti conformation. This confirms the conformation depicted in Scheme 7.

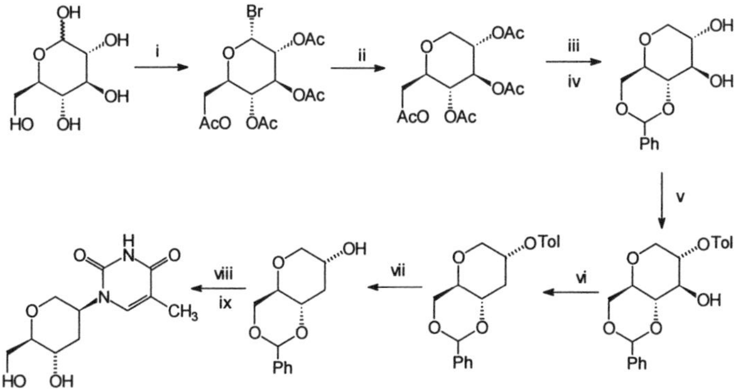

Scheme 9: Synthesis of 1,5-anhydro-2,3-dideoxy-2-(thymin-1-yl)-D-arabinohexitol (T^{17}). i: Ac_2O, HBr, HOAc, 10 h, RT, 60%; ii: Bu_3SnH, Et_2O, RT; KF, H_2O, 86%; iii: NaOMe, MeOH, 2 h, RT; iv: PhCH=O, $ZnCl_2$, $PhCH_3$, 2 days, RT, 75%; v: Bu_2SnO, C_6H_6, dioxane, p-toluoyl chloride, 5 h, RT, 78%; vi: $SCCl_2$, DMAP, CH_2Cl_2, 1.5 h, RT, 2,4-dichlorophenol, 2 h, Bu_3SnH, AIBN, $PhCH_3$, 16 h, 80° C, 75%; vii: NaOMe, MeOH, 4 h, RT, 80%; viii: N^3-benzoylthymine, $(Ph)_3P$, DEAD, dioxane, 16 h, RT; ix: NH_3, MeOH, 16 h, RT; 80% HOAc, 5 h, 80° C, 50%.

T^{14} could be obtained as shown in Scheme 10 (*30*). Condensation of protected D-xylopyranose with silylated thymine and deprotection afforded mainly 1-(β-D-xylopyranosyl)thymine. Selective protection of the 4'-hydroxyl group was followed by a reductive removal of the 2'- and 3'-hydroxyl group. The hydroxymethyl group was introduced by reaction with (bromomethyl)dimethylsilyl chloride, radicalar cyclization, and a Tamao oxidation. Protection of this hydroxyl group followed by epimerization at 4′-carbon atom using Mitsunobu conditions and full deprotection yielded the desired 1-[2′,3′-dideoxy-3′-C-(hydroxymethyl)-α-L-threo-pentopyranosyl] thymine. When the hydroxymethyl group is introduced first on a 2,3-dideoxy-2,3-unsaturated methylpyranoside followed by a sugar-base condensation reaction, only the α-anomer was formed (31). The conformation of T^{14} as depicted in Scheme 7 was confirmed by X-ray analysis. The physicochemical characteristics of the oligonucleotides containing T^{14}, T^{16}, and T^{17} are shown in Table V.

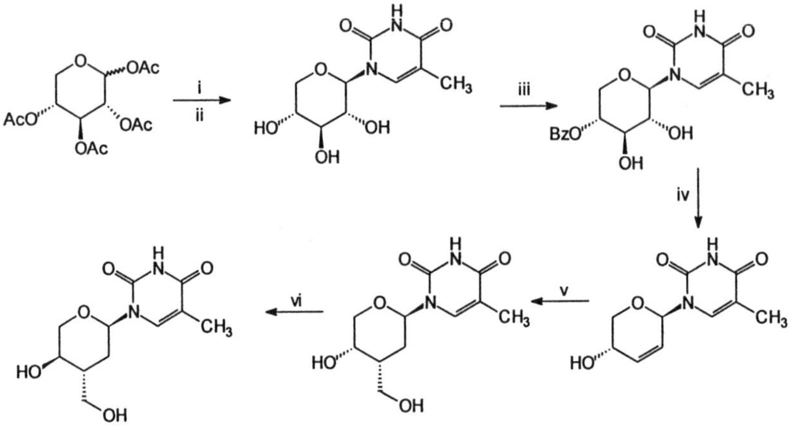

Scheme 10: Synthesis of 1-[2′,3′-dideoxy-3′-C-(hydroxymethyl)-α-L-threo-pentopyranosyl] thymine (T^{14}). i: silylated thymine, TMSOTfl, $C_2H_4Cl_2$, 50°, 20 h, 62%; ii: NaOMe, MeOH, 16 h, RT, 75%; iii: Bu_2SnO, MeOH, 2 h, Δ; benzoyl chloride, DMF-dioxane, 16 h, RT, 71%; iv: $(Ph)_3P,I_2$, imidazole, CH_3CN-$PhCH_3$, 80°, 2 h; Zn, 80°, 2 h and NH_3, MeOH, 16 h, RT, 70%; v: $(BrCH_2)(CH_3)_2SiCl$, imidazole, DMF, 0° to RT, 2 h; Bu_3SnH, AIBN, $PhCH_3$, 2 h, Δ; KF, $KHCO_3$, H_2O_2, DMF, RT, 16 h, 31%; vi: $(Ph)_3P$, DEAD, PhCOOH; dioxane, m-xylene, 16 h, RT and NH_3, MeOH, 16 h, RT, 23%.

Table V: Melting Temperatures and Free Energies for
Oligonucleotides Containing T^{14}, T^{16}, and T^{17}

	T_m (° C)	ΔH (kJ/mol)	$\Delta G^{25°}$ (kJ/mol)
T_{13}	32.3	367	41.4
$T^{16}.T_{10}.T^{16}.T$	27.3	318	35.1
$T^{16}.T^{16}.T_8.T^{16}.T^{16}.T$	20.8	319	28.3
$T^{16}.T_5.T^{16}.T_4.T^{16}.T$	21.2	299	28.9
$(T^{16})_{12}.T$	-	-	-
$(T^{14})_{12}.T$	-	-	-
$T_6.T^{17}.T_6$	26.9	349	35.0
$T^{17}.T^{17}.T_9.T^{17}.T^{17}$	26.5	324	34.3
$(T^{17})_{13}$	45.1 (1)	ND	ND
$A_6.A^{17}.A_6$	30.6	361	39.6
$A^{17}.A^{17}.A_9.A^{17}.A^{17}$	29.1	342	37.3
$(A^{17})_{13}$	20.5	ND	ND
$(A^{17})_{13}.(T^{17})_{13}$	76.3	ND	ND

(1) The melting point is measured at 284 nm. When measured at 260 nm the
melting curves show a minimum at 47° C.

Although incorporation of T^{16} in the 6′ → 3′ direction in a parallel
oligonucleotide is not identical with incorporation of T^{16} in the 3′ → 6′ direction in
an antiparallel oligonucleotide, the latter strategy was followed with incorporation of
T^{16} in a natural oligonucleotide. This could give a first idea about the destabilization
of 1,4-substituted pyranose-like nucleosides. As can be seen from the table,
incorporation of T^{16} gives a reduction in duplex stability, which is slightly less than
the incorporation of T^3. Nevertheless, the completely modified oligothymidylate did
not hybridize with a natural poly A. Unfortunately the oligonucleotide consisting of
T^{14} building blocks, which was predicted by the modeling studies as being the prime
candidate for strong hybridization, likewise, did not form duplexes with poly A. The
other possibility that the duplex is extremely stable cannot be excluded by now.

The results with T^{17} are more promising. The destabilization effect by
incorporating several 1,5-anhydroxyhexitol nucleosides in an $A_{13}.T_{13}$ antiparallel
duplex is slightly better than the results obtained with T^3. An important difference is
that with T^3 no hypochromicity was detected when evaluating the melting behavior
of a fully modified $(T^3)_{13}$ mer, and with T^{17} there is a clear transition. The
interactions at 0.1 M NaCl, however, are more complex than within a natural dsDNA
duplex and differ between a $(T^{17})_{13}.A_{13}$ duplex and a $(A^{17})_{13}.T_{13}$ combination. The
duplex formed between polyA^{17} and polyT^{17} is very stable. A striking difference

between the nucleoside T^{17} and all the other pyranose-like nucleosides with an anomeric base moiety is the axial orientation of the base moiety. The reason why the modeling process failed and the properties of the associations formed between an oligonucleotide constructed of the 1,5-anhydrohexitol nucleoside and a natural oligonucleotide are currently under investigation. Preliminary results show that one of the reasons for the failure of the modeling calculations might be that a conformational search of 200 ps using high temperature MD is most probably inadequate. While the 200 ps calculations give a destabilization energy that increases along T^{14}<PNA<dT (Table IV), this order is completely reversed after a 1000 ps run: dT<PNA<T^{14} (data not shown). In addition, MD simulations in vacuum boundary conditions are most probably inaccurate for highly charged systems such as oligonucleotides. It is evident from these modeling experiments that a lot of research still needs to be done before ending up with an accurate model to predict the hybridizing capabilities of oligonucleotides.

Literature Cited

1. Böhringer, M.; Roth, H.-J.; Hunziker, J.; Göbel, M.; Krishnan, R.; Giger, A.; Schweizer, B.; Schreiber, J.; Leumann, C.; Eschenmoser, A. *Helvetica Chimica Acta* **1992**, *75*, 1416-1477.
2. Pitsch, S.; Wendeborn, S.; Jaun, B.; Eschenmoser, A. *Helvetica Chimica Acta* **1993**, *76*, 2161-2183.
3. Ueda, T.; Watanabe, S.-I. *Chem. Pharm. Bull.* **1985**, *33*, 3689-3695.
4. Herscovici, J.; Montserret, R.; Antonakis, K. *Carbohydrate Research* **1988**, *176*, 219-229.
5. Augustyns, K.; Van Aerschot, A.; Urbanke, C.; Herdewijn, P. *Bull. Soc. Chim. Belg.* **1992**, *101*, 119-130.
6. Ogawa, T.; Matsui, M. *Tetrahedron* **1981**, *37*, 2363-2369.
7. Garegg, P. J.; Samuelsson, B. *Synthesis* **1979**, 813-814.
8. Garegg, P. J.; Johansson, R.; Samuelsson, B. *J. Carbohydrate Chemistry* **1984**, *3*, 189-195.
9. Augustyns, K.; Vandendriessche, F.; Van Aerschot, A.; Busson, R.; Urbanke, C.; Herdewijn, P. *Nucleic Acids Research* **1992**, *20*, 4711-4716.
10. Corey, E. J.; Weigel, L. O.; Chamberlin, A. R.; Lipshutz, B. *J. Am. Chem. Soc.* **1980**, *102*, 1439-1441.
11. Yang, Y.-L.; Falck, J. R. *Tetrahedron Letters* **1982**, *23*, 4305-4308.
12. Augustyns, K.; Van Aerschot, A.; Herdewijn, P. *Bioorganic & Medicinal Chemistry Letters* **1992**, *2*, 945-948.
13. Augustyns, K.; Rozenski, J.; Van Aerschot, A.; Janssen, G.; Herdewijn, P. *J. Org. Chem.* **1993**, *58*, 2977-2982.
14. Augustyns, K.; Godard, G.; Hendrix, C.; Van Aerschot, A.; Rozenski, J.; Saison-Behmoaras, T.; Herdewijn, P. *Nucleic Acids Research* **1993**, *21*, 4670-4676.
15. Lima, W. F.; Monia, B. P.; Ecker, D. J.; Freier, S. M. *Biochemistry* **1992**, *31*, 12055-12061.

16. Duroux, I.; Godard, G.; Boidot-Forget, M.; Schwab, G.; Hélène, C.; Siason-Behmoaras, T. *EMBO J.* (in press).
17. Pearlman, D. A.; Case, D. A.; Caldwell, J. C.; Seibel, G. L.; Singh, U. C.; Weiner, P.; Kollman, P. A. Amber 4.0, University of California, San Francisco **1991**.
18. De Winter, H.; Blaton, N.; Peeters, O.; De Ranter, C.; Van Aerschot, A.; Herdewijn, P. *Acta Crystallogr. Sect. B48* **1992**, 95-103.
19. Arnott, S.; Campbell-Smith, P. J.; Chandresekharan, R. *CRC Handbook of Biochemistry*; CRC: Boca Raton, FL, 1984, *Vol. 2*, pp. 411-422.
20. Nielsen, P. E.; Egholm, M.; Berg, R. H.; Buchardt, O. *Science* **1991**, *254*, 1497-1500.
21. Jorgensen, W.; Chandrasekhar, J.; Madura, J.; Impey, M.; Klein, R. *J. Chem. Phys.* **1983**, *79*, 926-935.
22. van Gunsteren, W. F.; Berendsen, H. J. C. *Mol. Phys.* **1977**, *34*, 1311-1327.
23. Singh, U. C.; Kollman, P. A. *J. Comp. Chem.* **1984**, *5*, 129-145.
24. Singh, U. C.; Kollman, P. A. *GAUSSIAN 80-UCFS*, QCPE 446.
25. Weiner, S. J.; Kollman, P. A.; Nguyen, D. T.; Case, D. A. *J. Comp. Chem.* **1986**, *7*, 230-252.
26. Ravindranathan Kartha, K. P.; Jennings, H. J. *J. Carbohydrate Chemistry* **1990**, *9*, 777-781.
27. Kocienski, P.; Pant, C. *Carbohydrate Research* **1982**, *110*, 330-332.
28. Van Aerschot, A.; Verheggen, I.; Herdewijn, P. *Bioorganic & Medicinal Chemistry Letters* **1993**, *3*, 1013-1018.
29. Verheggen, I.; Van Aerschot, A.; Toppet, S.; Snoeck, R.; Janssen, G.; Balzarini, J.; De Clercq, E.; Herdewijn, P. *J. Med. Chem.* **1993**, *36*, 2033-2040.
30. Doboszewski, B.; Herdewijn, P. *Tetrahedron Lett.* **1994** (submitted).
31. Augustyns, K.; Rozenski, J.; Van Aerschot, A.; Busson, R.; Claes, P.; Herdewijn, P. *Tetrahedron* **1994**, *50*, 1189-1198.

RECEIVED August 8, 1994

Chapter 7

α-Bicyclo-DNA: Synthesis, Characterization, and Pairing Properties of α-DNA-Analogues with Restricted Conformational Flexibility in the Sugar–Phosphate Backbone

M. Bolli[1], P. Lubini[2], M. Tarköy[1], and C. Leumann[1]

[1]Institute of Organic Chemistry, University of Bern,
CH-3012 Bern, Switzerland
[2]Laboratory of Organic Chemistry, Federal Institute of Technology,
CH-8092 Zurich, Switzerland

The synthesis of a rigid nucleoside analogue, where the centers C(3')
and C(5') of the natural deoxyribonucleosides are connected by an
ethylene bridge, is reported. The design of these nucleosides is based
on the hypothesis, that oligonucleotides derived therefrom, have single
strand structures that are more preorganized for duplex formation and,
thus, have the potential of forming more stable duplexes with natural
RNA and DNA complements. In both, the α– and β–anomeric forms of
these bicyclonucleosides, the presence of the ethylene bridge forces
their furanose unit into the 2'-endo/1'-exo conformation, thus
resembling in this part of the molecule a nucleotide unit in B-type
DNA. UV-melting curves of oligo-α-bicyclonucleotides of the bases
adenine and thymine indicate effective base-pairing to complementary
β-DNA and RNA sequences. CD-spectra indicate structures for the
hybrid duplexes that are comparable to that of natural DNA, but
significantly different from that of the hybrid β/α-duplexes in oligo-
deoxyribooligo-nucleotides. α-Bicyclo-DNA sequences are more stable
to hydrolysis catalyzed by snake venom phosphodiesterase compared to
natural DNA.

Due to the success of using natural oligonucleotides as selective inhibitors of protein
biosynthesis (*1-2*), and due to the advent of modern synthetic methods for
oligonucleotide synthesis (*3*), a large number of DNA-analogues have been
synthesized within the last two decades. These analogues were essentially designed to
display the following properties; (i): formation of thermodynamically more stable
duplexes or triplexes with single stranded RNA or double stranded DNA under
conservation of the base pairing selectivity; (ii): higher stability against cellular
nucleases; and (iii): more efficient penetration of cell membranes in order to reach
their targets in the cytoplasm or the nucleus, respectively. So far, most of the

0097–6156/94/0580–0100$08.00/0

approaches focussed on the modification of the phosphodiester groups (*4-6*). Recently, however, the number of reported nucleoside and oligonucleotide analogues with sugar components different from those of natural DNA and RNA is rapidly growing. Research in this direction started with the study of α-DNA, a DNA analogue that is built from α-deoxyribonucleosides (*7-17*). In a theoretical article, *Séquin* proposed in 1972 that α-DNA can base-pair to a complementary β-DNA strand in a parallel orientation exhibiting Watson-Crick base-pairing (*7*). The study was based on model building using *Dreiding* stereomodels. More than ten years later this prognosis was verified by NMR-analysis of such heteroanomeric duplexes (*11, 16*). Although a desirable x-ray structural analysis is still missing, the NMR-analyses revealed that the overall geometry of an α,β-duplex is (with the exception of strand orientation) similar to that of a DNA duplex of the B type. Meanwhile, oligonucleotides from α-ribonucleosides (*18-19*), from nucleosides containing other 2'-deoxy-pentofuranosyl sugar components (e.g. xylose (*20-21*)), or from natural sugar-modified nucleosides (e.g. 2'-O-alkylribonucleosides (*22*)) have been reported too.

While a growing body of oligonucleotide analogues has become available, the specific structural and energetic features governing the affinity and base-pairing selectivity in duplex and triplex formation are still not very well understood. In an effort to elucidate the effect of reducing torsional flexibility in the backbone of an oligonucleotide chain upon its base-pairing selectivity and -energy, we designed and synthesized a new type of nucleosides that we call bicyclonucleosides (Figure 1). These nucleosides differ from the natural nucleosides by an ethylene bridge introduced between the centers C(3') and C(5'). The configurations at these centers were chosen as to match the conformation of a natural nucleotide unit in duplex-DNA as closely as possible. We have shown previously, that the structural consequence of the additional ethylene bridge upon the conformation of the furanose part in the β-bicyclonucleosides with respect to the natural deoxynucleosides consists in the restriction of the torsional angles δ and γ to the anticlinal range (*23-24*). Bicyclonucleosides, therefore, show exclusively S-type conformation in their furanose units (C1'-exo, C2'-endo) and thus are preorganized for the formation of duplexes resembling that of B-DNA. While we have already reported on the synthesis, structure and pairing properties of β-bicyclo-(oligo)nucleotides (*25-26*) we will focus in the following on the synthesis, structural characterization and pairing properties of sequences of α-bicyclo-DNA containing the nucleobases adenine and thymine.

Synthesis of the Bicyclo-Sugar Unit

In our synthetic plan, we decided to construct the nucleosides **1-8**, starting from a common bicyclic carbohydrate precursor, rather than via modification of the parent natural 2'-deoxyribonucleosides. We envisaged building up the carbon framework of the bicyclic sugar **14** (Scheme 1) by a *Horner-Wittig* reaction of a suitably protected cis-2,3-dihydroxy-cyclopentanone with phosphonoacetate as the C-2 unit. While the resulting double bond was thought to act as an anchor for the stereoselective introduction of the tertiary hydroxyl group at the bridgehead, the oxygen atoms of the secondary hydroxyl group and of the furanose ring in **14** were already incorporated in the chiral ketone (±)**9**, the synthesis of which has been reported earlier (*27*). Since this material was used as the racemate, separation of enantiomers had to be achieved by resolution of an intermediate.

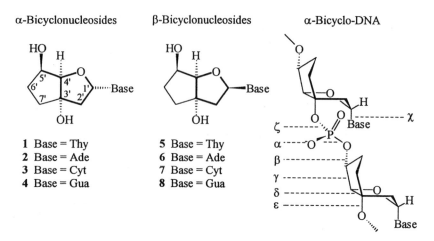

Figure 1. The bicyclonucleosides containing the 4 natural DNA-bases.

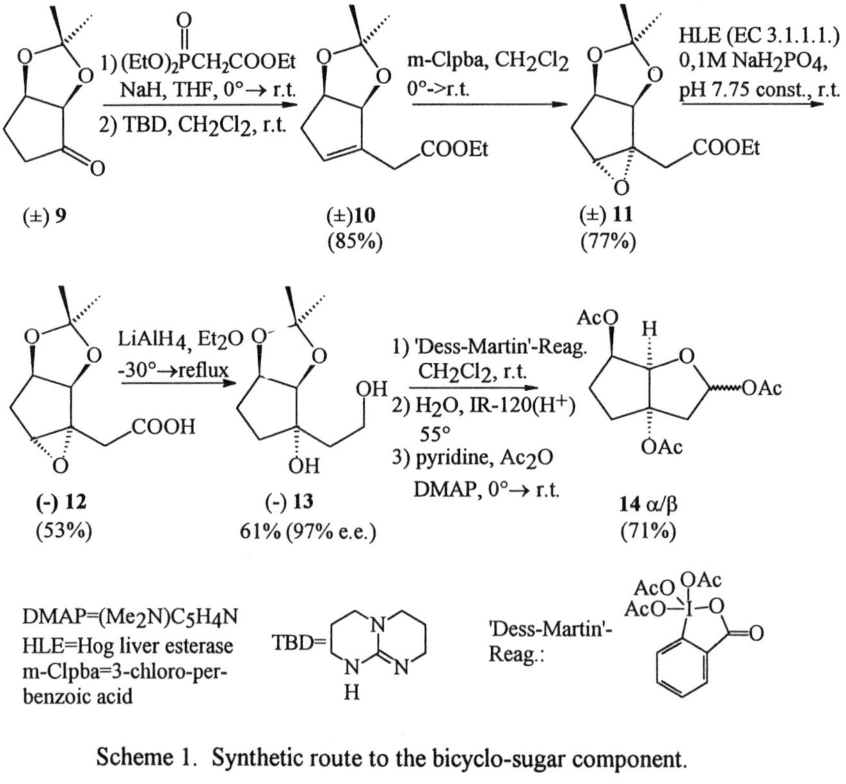

Scheme 1. Synthetic route to the bicyclo-sugar component.

Reaction of (±)**9**, (Scheme 1) with the anion of triethyl phosphonoacetate in THF resulted in the almost quantitative formation of the corresponding E/Z-mixture of α,β-unsaturated esters that were converted directly in a base catalyzed tautomerization reaction to the thermodynamically more stable, deconjugated β,γ-unsaturated ester (±)**10**. Introduction of the tertiary hydroxyl function was realized by an epoxidation/reduction pathway. Epoxidation of (±)**10** with a peracid proceeded, as expected, stereoselectively from the convex side of the bicyclic ring system and afforded the desired exo-epoxyester (±)**11**. Only minor amounts (5-10%) of the corresponding endo diastereoisomer were observed. The enantiomeric epoxide (±)**11** could readily and efficiently be resolved by hog liver esterase (E.C.3.1.1.1.) catalyzed hydrolysis. The acid (-)**12** (optical rotation sense assigned tentatively), isolated after ca. 50 % conversion of (±)**11** was directly reduced to the diol (-)**13** that could be obtained almost enantiopure (97% e.e.), and the absolute configuration of which has been established by x-ray analysis of a camphanic acid derivative thereof. The synthesis of the bicyclic sugar **14** α/β was finished in three steps, involving selective oxidation of the primary alcohol function in (-)**13** to the aldehyde with the "Dess-Martin" periodinane (*28*), followed by acid catalyzed cleavage of the ketal function and peracetylation of the spontaneously cyclized sugar derivative.

Synthesis of the Bicyclonucleosides

The remaining task in the synthesis of the bicyclonucleosides **1-8** was the formation of the nucleosidic bond between the bicyclo sugar derivatives **14** α/β and the four nucleobases adenine, guanine, cytosine and thymine. We chose the method of *Vorbrüggen* and coworkers (*29-31*), which is based on a Lewis acid induced formation of the nucleosidic bond between a sugar component and a persilylated (and where necessary acyl-protected) nucleobase. Using this method, the corresponding bicyclonucleoside derivatives **15-18** were readily obtained as anomeric mixtures in α,β-ratios ranging roughly from 1:2 - 2:1 in yields of 73-82% (Scheme 2).The point of attachment of the bases (N[1] for pyrimidine- and N[9] for purine bases) in the so formed nucleosides were unambiguously assigned by NMR-spectroscopic methods. While only N[1]-connection of the base in the pyrimidine nucleosides and N[9]-connection in the adenine containing nucleoside were observed, N[7] and N[9] mixtures were encountered in the nucleosidation reaction with the acyl protected guanine base. The separation of the anomeric mixtures, however, turned out to be difficult, and was even after saponification of the ester groups only successful in the case of the guanine nucleosides **21**. The discovery, that introduction of a bulky silyl protecting group into the secondary hydroxyl function of the α/β-mixtures **15, 16, 17** lead to enhanced chromatographic separability of the anomeric species, came to our rescue at this point. The reaction sequence: i):introduction of the tert.-butyldimethylsilyl group in the secondary hydroxyl function of the α/β-mixtures **15, 16, 17**; ii): separation of anomers by flash chromatography, and; iii): desilylation to the (base protected) nucleosides **1, 19, 20** (and the corresponding β-anomers) finally gave us the desired anomerically pure material. The relative configurations at the anomeric centers in the α-nucleosides **1, 19, 20**, as well as in the corresponding β-nuclesides were unambiguously assigned by ^{1}H-NMR NOE spectroscopy.

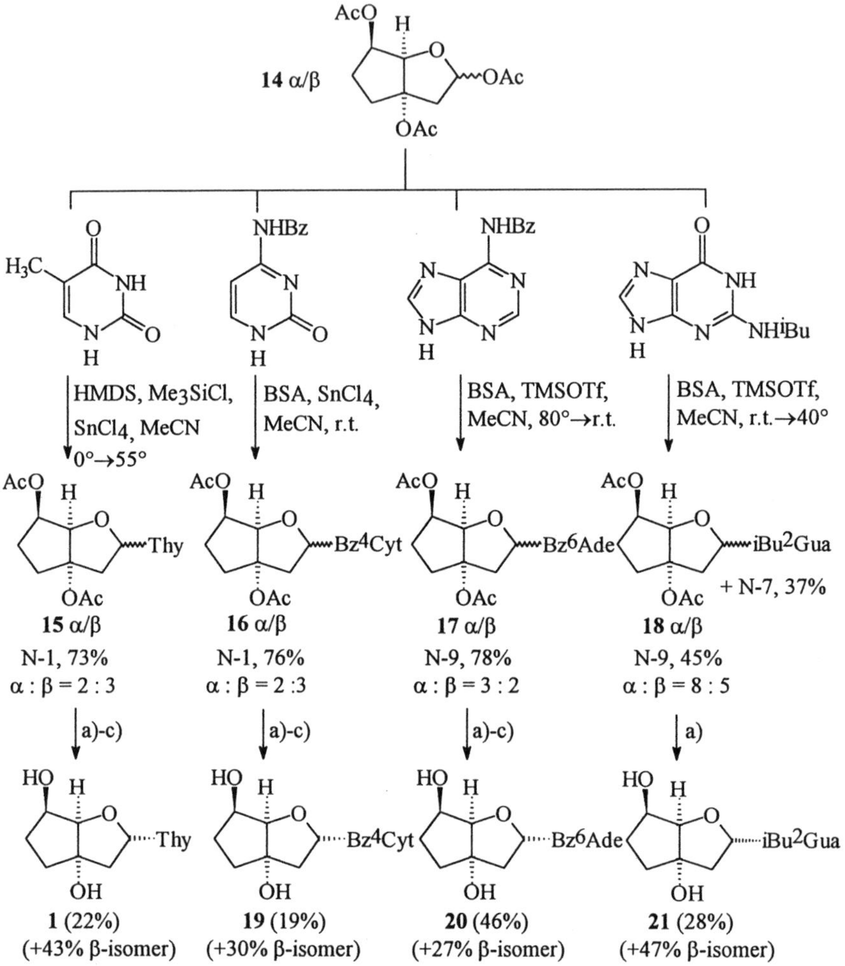

Scheme 2. Synthesis of the α- and β-bicyclonucleosides.

a) 0.2M NaOH in THF/MeOH/H$_2$O 5:4:1, 0-2°. b) (t Bu)Me$_2$Si(CF$_3$SO$_3$), pyridine, 0°, chromatographic separation of β-anomers. c) Bu$_4$NF·3H$_2$O, 55°: HMDS=(Me$_3$Si)$_2$NH; BSA=MeC(OSiMe$_3$)=NSiMe$_3$; TMSOTf=CF$_3$SO$_3$SiMe$_3$.

X-Ray Analysis of 19. Structural and conformational insight into the series of the α-anomeric bicyclonucleosides was obtained in the case of the α-bicyclocytidine derivative **19** by x-ray analysis. The crystallographic data of **19**, indicate two molecules A and B within the asymmetric unit of the crystal. In their furanose unit, molecule A shows a $C_{1'}$-exo conformation whereas molecule B adopts a $C_{2'}$-endo conformation (Figure 2, Table I).

Table I. Torsion angles of 19 and 5 (*23*) from x-ray analysis

Torsion angles	19 (molecule A)	19 (molecule B)	5
γ (O5'-C5'-C4'-C3')	146.7	133.9	149.3
δ (C5'-C4'-C3'-O3')	124.5	136.2	126.5
χ (O4'-C1'-N1-C2)	172.5	169.1	-112.7
v_0 (C4'-O4'-C1'-C2')	-32.0	-18.7	-42.4
v_1 (O4'-C1'-C2'-C3')	32.7	28.8	43.1
v_2 (C1'-C2'-C3'-C4')	-21.4	-26.9	-27.3
v_3 (C2'-C3'-C4'-C4')	3.3	16.9	3.2
v_4 (C3'-C4'-O4'-C1')	18.0	1.0	23.9

With respect to the furanose unit, both molecules are within the S-conformational range, which in the natural nucleoside series is responsible for DNA duplexes of the B-type. The secondary hydroxyl substituent, however, occurs in a pseudoequatorial position in both conformers with the effect of torsion angle γ being in the antiperiplanar (ap) range (Table I). Such an arrangement for torsion angle γ is observed in Z-DNA duplexes (G-nucleosides) but does not occur in duplexes of the A- and B-type (*32*). The torsion angle χ, describing the conformation around the nucleosidic bond, adopts in both molecules values near 170°. Although the base is still in the anti range as in DNA duplexes of the A and B type, the torsion angle χ differs by ca. 70° from that observed in the nucleoside residues in duplex-DNA of the B-type and by ca. 20° from those in duplexes of the A-type. The fact that the cytosine base plane in both molecules is almost parallel to the $C_{4'}$-$O_{4'}$-bond of the sugar can be interpreted in terms of avoiding unfavorable interactions with the the C3'-hydroxyl substituent.

From the comparison of the structures of the α-bicyclonucleoside **19** with that of the β-bicyclonucleoside **5** one can see that the change of configuration at $C_{1'}$ does not influence the conformation of the bicyclic core system. The furanose ring pucker as well as the conformation of the carbocyclic ring are of the same type in both, α- and β-nucleosides. This is in contrast to what is observed in the series of the natural (deoxy)ribonucleosides. The preferred conformations of the furanose unit there are $C_{2'}$-exo, $C_{3'}$-exo and $C_{4'}$-endo in the α-anomers and $C_{3'}$-endo and $C_{2'}$-endo in the β-anomers.(*33*).

The preferred conformation of the furanose unit of nucleosides in solution can be determined from ^{1}H-NMR coupling constant analysis. In the case of the bicyclonucleosides this analysis is rendered more difficult because of the missing hydrogen atom at C3'. From the data given in Table II it becomes evident that all α-bicyclonucleosides investigated show almost equal coupling constants between the vicinal protons H1'-H2'α, H1'-H2'β and between H4'-H5', indicating that the influence of the nature of the base on the conformation of the sugar is neglectable. Values for the torsion angles γ (130°-150°) are readily obtained by applying the Karplus relation (*34*), and are in agreement with those obtained in the x-ray analysis of **19**.

Table II. Selected chemical shifts (δ) and coupling constants (*J*) for α-bicyclo-nucleoside-derivatives (recorded in D_2O)

entry	δ [ppm]				*J* [Hz]		
	H-C(1')	H_α-C(2')	H_β-C(2')	H-C(4')	$^3J_{(1',2'\beta)}$	$^3J_{(1',2'\alpha)}$	$^3J_{(4',5')}$
1	6.15	2.38	2.58	4.37	4.3	7.0	5.3
2	6.32		2.71-2.80	4.32	3.9	6.1	5.2
3	6.20	2.29	2.57	4.36	3.6	7.1	5.2
4	6.17	2.48	2.52	4.09	4.3	6.4	4.9

Synthesis of Building Blocks for Oligonucleotide Synthesis

The α-bicyclonucleoside derivatives **1** and **20** (Scheme 3) served as starting materials for the preparation of the building blocks for solid phase oligonucleotide synthesis according to the phosphoramidite method of *Caruthers* (*35*). Tritylation of **1** and **20** with 4,4'-(dimethoxy)triphenylmethyl trifluoromethanesulfonate, which is easily prepared and obtained in crystalline form by reaction of dimethoxytrityl chloride with silver triflate in THF, under standard conditions yielded the O5'-protected derivatives **22** and **23** in yields of 79-91%. As expected, no tritylation of the tertiary hydroxyl group was observed. Subsequent elaboration of the phosphoramidite building blocks **12-15** was effected by reaction of **22** and **23** with chloro-allyloxy-diisopropylamino-phosphine in THF and proceeded in the same manner as in the case of the natural deoxynucleosides, irrespective of the different steric environment of the reacting center (secondary vs. tertiary OH-group). The phosphoramidites **24** and **25**, thus obtained as mixtures of diastereoisomers (ca. 1:1 ratio as determined by ^{1}H-, ^{31}P-NMR) could easily be purified by conventional column chromatography and stored at -20° for months without observable decomposition.

Nucleoside-derivatized solid support **28** and **29** was prepared following standard protocols in nucleic acid chemistry (*36*) by reaction of the nitrophenol esters **26** and **27** with long chain alkylamine CPG. Loading capacities of 32±2 μmol nucleoside/g CPG were obtained (trityl assay). Esters **26** and **27** were synthesized in yields of 84-85% by reaction of the tritylated precursors **22** and **23** with succinic anhydride and esterification of the corresponding succinates with p-nitrophenol using DCC as the coupling reagent.

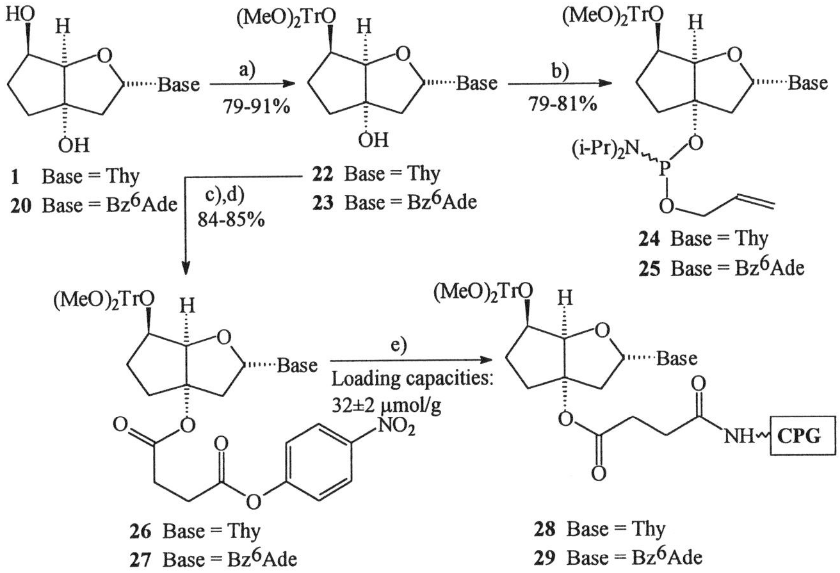

a) (MeO)$_2$Tr$^+$CF$_3$SO$_3^-$, (2.8-3.0 eq.), pyridine, r.t.; b) {(H$_5$C$_3$O)[(i-Pr)$_2$N]}PCl, (2.0-4.8 eq.),
(i-Pr)$_2$EtN, (4.0-9.6 eq.), THF, r.t.; c) succinic anhydride (10 eq.), pyridine, DMAP (5 eq.),
r.t.; d) p-nitrophenol (1.4 eq.), dioxane, DCC (3 eq.), r.t.; e) Long chain alkylamine-CPG,
DMF, dioxane, Et$_3$N, r.t.

Scheme 3.   Preparation of the building blocks for automated oligonucleotide
synthesis.

Synthesis of Oligo-α-Bicyclodeoxynucleotides

The synthesis of the α-bicyclo-oligodeoxynucleotides α-bcd(A$_{10}$), α-bcd(T$_{10}$) and α-bcd(T$_{20}$) was performed on a DNA-synthesizer (*Pharmacia Gene Assembler Special*) on a 1.3 μmol scale. The synthesis cycle is completely compatible with that of the natural DNA-synthesis. The only parameters changed were the detritylation and coupling time which were slightly extended relative to that for the synthesis of natural DNA-oligomers. Reagent and phosphoramidite concentrations as well as the phosphoramidite/tetrazol ratio remained unchanged. Coupling efficiencies of >98% per step (trityl assay) were obtained. After the chain assembly was complete (trityl off mode), removal of the phosphate protection according to known procedures (*37*) followed by application of standard methods for the detachment from the solid support and cleavage of the base-protecting groups yielded the crude α-bicyclo-DNA sequences that were further purified by ion-exchange hplc, followed by reversed phase-hplc. Quality control and structural analysis of homogeneous oligomer fractions were performed by capillary gel electrophoresis and matrix-assisted laser-desorption ionization time of flight mass spectroscopy (MALDI-ToF-MS) (*38*).

Although having phosphodiester linkages to tertiary alkoxy groups, the oligo-bicyclodeoxynucleotides are chemically stable and no signs of degradation (hplc control) under the conditions used for the investigation of their pairing properties (e.g. melting curves, mixing curves) were observed.

Complexes of α-bcd(T$_{10}$), and α-bcd(A$_{10}$) with Complementary RNA

We have shown earlier, that an oligomer of α-thymidine, 20 nucleotides in length, in contrast to its natural β-isomer, forms an antiparallel oriented monomolecular hairpin duplex with T-T-base pairs in the stem (*39*). This duplex melting was easily monitored by the corresponding UV-melting curve at 260 nm, and a T$_m$-value of 52° (concentration invariant, 10 mM MgCl$_2$) was found. We therefore investigated, whether duplex formation does also occur in the case of its bicyclic analogue α-bcd(T$_{20}$). UV-melting curves of α-bcd(T$_{20}$) at different wavelengths, however, did not show any signs of a cooperative structural transition (data not shown) indicating no T-T-base pairing in the case of α-bcd(T$_{20}$). With complementary natural DNA and RNA, however, α-bcd(T)- and α-bcd(A)-sequences readily form complexes. UV-melting curves of stoichiometric (1:1) mixtures of α-bcd(T$_{10}$):poly(A) and α-bcd(A$_{10}$):poly(U) are of sigmoidal shape indicating a cooperative melting process (Figure 3). The corresponding T$_m$-values (1M NaCl) of 36° in the former and 56.5° in the latter case indicate considerably stronger pairing of α-bcd(A$_{10}$) with its RNA-complement than α-bcd(T$_{10}$), as was already observed in the case of the β-series. α-bcd(T$_{10}$), β-bcd(T$_{10}$) and natural d(T$_{10}$) at comparable oligonucleotide and salt concentration (1M NaCl) have almost the same affinity to poly(A) which is reflected by the corresponding T$_m$-values that differ only by 2° for the three different duplexes. The pairing of α-bcd(A$_{10}$) to poly(U) (T$_m$=56°, 1M NaCl) is comparable to that of β-d(A$_{10}$) (T$_m$=56°, 1M NaCl) to poly(U). The strongest pairing with poly(U) was observed in the case of β-bcd(A$_{10}$) (T$_m$=70°, 1M NaCl) as the complementary strand.

The stoichiometry of complex formation in the α-bicyclo-DNA/RNA-hybrids

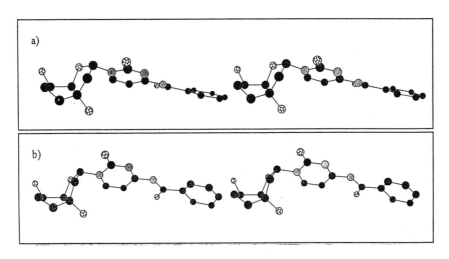

Figure 2.   X-ray structure (stereoscopic views) of the bicyclocytidine derivative
19: a) molecule A; b) molecule B.

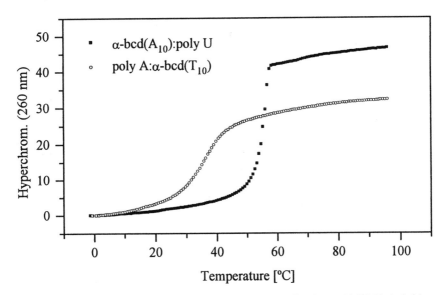

Figure 3.   UV-melting curves (260 nm) of the two bicyclo-DNA/RNA hybrids.
c(base-pair) = 55µM in 10 mM NaH_2PO_4, 1M NaCl, pH 7.0.

under discussion was investigated by the method of Job (40). Under the salt conditions used (1M NaCl) the UV-mixing curve of α-bcd(T_{10})/poly(A) shows two straight lines with an intersection point at 54 mol-% U indicating only duplex formation (Figure 4a). The deviation from the ideal value (50%) is most probably due to the uncertainty in the exact determination of the concentration of poly(A). With respect to the preferred pairing stoichiometry, this pairing system therefore behaves as β-bcd(T_{10})/poly(A) and α-d(T_8)/poly(A) (13). Entropic considerations (molecularity) may contribute to the fact that no triplex formation under high salt conditions is observed. The mixing curve of the alternative hybrid α-bcd(A_{10})/poly(U) also shows two straight lines, the intersection point of them, however, being at 67 mol-% U, indicating triplex formation of the U-A-U-type even at 1:1 stoichiometry. The fact, that at equimolar base-to-base ratio, triplex and unpaired purine single strand coexist rather than purely duplex, is also observed in the system β-d(A_{10})/poly(U). The situation is different in the case of β-bcd(A_{10})/poly(U), where duplex and triplex both exist at the corresponding molar base-ratios.

CD-spectra of both hybrid-RNA complexes (Figure 5a) are seemingly different from each other, the triplex α-bcd(A_{10}):2poly(U) showing the most intense ellipticity. The spectrum of the duplex α-bcd(T_{10}):poly(A) exhibits appreciable similarities to the spectra of β-d(T_{10}):poly(A) but is quite different from that of α-d(T_8):poly(A) (13). The differences are mainly located at wavelength near 280 nm, where the latter system shows negative ellipticities and the former positive ellipticities. Different duplex structures with respect to strand polarity and/or base-pair geometry might account for the different CD-spectra.

Complexes of α-bcd(T_{10}), and α-bcd(A_{10}) with Complementary DNA

The α-bicyclo-oligonucleotides α-bcd(A_{10}) and α-bcd(T_{10}) also form stable duplexes with their complementary natural β-deoxyoligonucleotide decamers as can be seen from the corresponding melting curves (Figure 6) with T_m's of 26.7° (α-bcd(T_{10}):β-d(A_{10})) and 32.9° (α-bcd(T_{10}):β-d(A_{10})) resp.

The latter of the two duplexes equals the purely β-anomeric, natural duplex β-d(T_{10}):β-d(A_{10}) which under the same conditions shows a T_m-value of 33°. In order to check the possibility of triplex formation, we recorded UV-mixing curves in the two pairing systems under discussion (Figure 4b). Under high salt conditions (1M NaCl) in neither of the two systems, triplex formation could be observed. The intersection points of both curves are in the range of 50 mol-% T, indicating only duplex formation.

CD-spectra of the two duplexes under discussion are displayed in Figure 5b. These spectra were recorded at 0° where only duplex and no single strands exist. In both cases, a characteristic strong negative cotton effect between 240 and 250 nm, and two positive maxima at 263nm and 282nm resp. were observed. The two duplexes mainly differ in the relative amplitude of the signal between 255nm -270nm. In general the spectra of both duplexes are surprisingly similar to that of the purely natural duplex β-d(T_{10}):β-d(A_{10}) but quite different from that of α-d(T_8):β-d(A_8) (13). This, again, indicates different structures for duplexes containing either α-deoxyribonucleosides or α-bicyclonucleosides. Since the preferred furanose

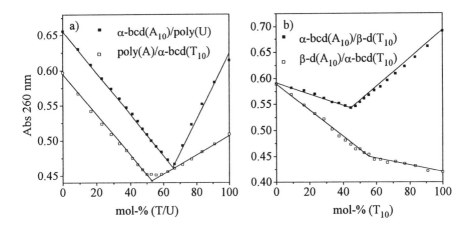

Figure 4. UV-mixing curves (260 nm) of a): bicyclo-DNA/RNA hybrids (c(tot. base) = 65 μM) and b): bicyclo-DNA/DNA hybrids (c(tot. strand) = 5.7 μM) in 10 mM NaH$_2$PO$_4$, 1M NaCl, pH 7.0.; T = 4°C).

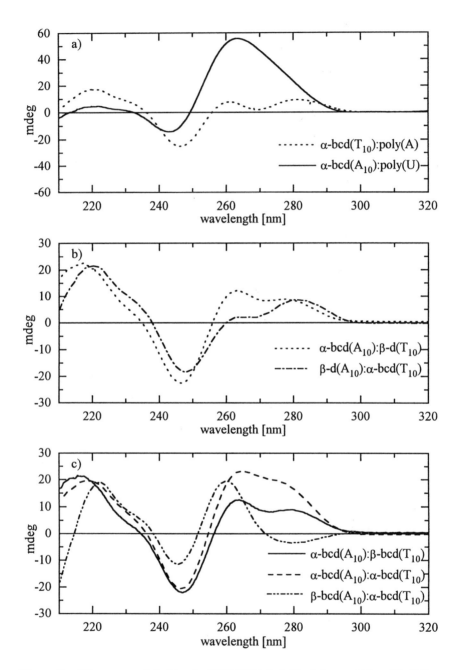

Figure 5. CD-spectra of a): bicyclo-DNA/RNA hybrids (c(base-pair) = 56 μM), b): bicyclo-DNA/DNA duplexes (c = 4.1 μM) and c): bicyclo-DNA/bicyclo-DNA duplexes (c = 4.3 μM); all in 10 mM NaH$_2$PO$_4$, 1M NaCl, pH 7.0; T = 0°C.

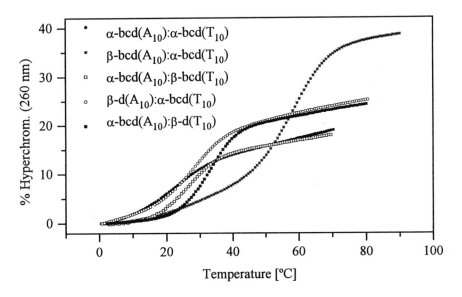

Figure 6.   Melting curves (260 nm) of selected duplexes containing α-bicyclo-DNA sequences. c = 5.2 - 6.7 µM in 10 mM NaH$_2$PO$_4$, 1M NaCl, pH 7.0.

conformations (2'-endo, 1'-exo) and the base orientation in the bicyclonucleosides are very similar to those of the α-deoxyribonucleosides, the structural differences most likely do not arise from differences in the furanose pucker within the two sugar systems. Whether torsional changes within the remaining part of the backbone, or different strand alignments (strand orientation and/or base-pair constitution) are responsible for the differences observed in the CD-spectra is not clear yet. An indication, that the former reason might be operative, comes from the x-ray structure of the parallel aligned, C-C$^+$-base-paired, miniduplex β-bcd(C$_2$) (24). In this duplex, the torsion angle γ is trans oriented whereas in natural DNA duplexes it generally adopts the gauche conformation. In bcd(C$_2$), the change of γ is compensated by a movement of the adjacent torsion angle β from trans to gauche resulting in an overall nucleotide conformation that retains most of the characteristics of B-DNA. Whether the structure of the α/β-hybrid-duplexes, described here, can be rationalized in this way, still needs to be proven.

We have calculated thermodynamic data for duplex formation for a series of duplexes (Table III) from $1/T_m$ vs. ln[c] plots (41) and have found remarkable differences in ΔH and ΔS values for duplex formation although the corresponding ΔG-values (and with this the melting temperatures) do not vary to a large extent.

Table III. Thermodynamic data and salt sensitivity of duplex formation

entry	ΔH [kcal/mol]	ΔS [cal/mol K]	$\Delta G(25^\circ C)$ [kcal/mol]	T_m [$^\circ$C]	$\delta T_m/\delta ln$ [NaCl]	Δn
α-bcd(A$_{10}$):α-bcd(T$_{10}$)	-41.4	-114.7	-7.2	22.0	9.6	4.9
α-bcd(A$_{10}$):β-bcd(T$_{10}$)	-60.5	-177.0	-7.7	25.2	6.8	4.7
β-bcd(A$_{10}$):α-bcd(T$_{10}$)	-48.8	-121.5	-12.6	58.5	11.0	4.9
β-d(A$_{10}$):α-bcd(T$_{10}$)	-45.0	-124.1	-8.0	26.7	7.8	3.9
α-bcd(A$_{10}$):β-d(T$_{10}$)	-64.6	-185.2	-9.4	32.9	5.4	3.7

The compensating effects of enthalpy and entropy balancing the free energy of duplex formation in DNA has already been referred to previously (42). The enthalpy of duplex formation is more favorable in the duplex with the bicyclic purine strand whereas the entropy of duplex formation is more favorable in the case where the pyrimidine strand is of bicyclic nature. In all duplexes investigated, the entropy term is numerically reduced to that of the natural duplex β-d(A$_{10}$):β-d(T$_{10}$).

Oligonucleotide duplex stability is strongly dependent of the amount of salt present in the medium. Increasing salt concentration stabilizes duplex formation because of the following two reasons: i) shielding of the negative charges by counter ions and ii) differential counter ion uptake upon duplexation. The latter effect, mainly entropic in nature, is dominant at monovalent salt concentrations below 1M. From plots of T_m vs. ln [NaCl] we evaluated the sensitivity of duplex formation to the concentration of NaCl (10mM-0.6M) in the systems mentioned. In both cases, as expected, a linear relationship between T_m and ln [NaCl] was observed (data not shown). Using the data for duplex formation enthalpy (ΔH) and the values $\delta T_m/\delta ln$[NaCl] (Table III) we have calculated the amount of cations that were taken up from the medium upon duplex formation (Δn) according to the polyelectrolyte theory (equation 1) (*43-44*):

$$\Delta n = -\frac{\partial Tm}{\partial \ln[NaCl]} \cdot \frac{2\Delta H^0_{cal}}{RT^2_m} \tag{1}$$

The $\delta T_m/\delta \ln[NaCl]$ values for the two duplexes are slightly different being larger in α-bcd(T$_{10}$):β-d(A$_{10}$) than in α-bcd(A$_{10}$):β-d(T$_{10}$). This relative difference only reflects the discrepancy between the pairing enthalpies in the two systems, since the extent of counter ion uptake for both systems is almost the same and in the range of the corresponding purely natural duplex β-d(T$_{10}$):β-d(A$_{10}$).

Duplex Formation Between Oligo-α- and β- Bicyclonucleotides.

From the four possible complexes in the row of the decamers of α- and β-bcd(A$_{10}$), and α- and β-bcd(T$_{10}$), the pairing behaviour of one of them (β-bcd(A$_{10}$):β-bcd(T$_{10}$)) has already been studied extensively (*26*). In the following we concentrate on the discussion of the remaining three duplexes. Melting curves of the three duplexes α-bcd(A$_{10}$):α-bcd(T$_{10}$), α-bcd(A$_{10}$):β-bcd(T$_{10}$) and β-bcd(A$_{10}$):α-bcd(T$_{10}$) provide a quite heterogeneous picture of the duplex stability in each system, characterizing the last duplex as the most exceptional one in terms of thermodynamic stability and structure.

From inspection of the data in Table III it becomes evident that the counter-ion uptake upon duplexation is similar within the purely bicyclic duplexes (Δn = 4.7-4.9) but is distinctly larger than in the mixed natural-bicyclic duplexes (Δn=3.7-3.9). This results in a higher sensitivity of the T$_m$ from the salt concentration and can be responsible for a considerable loss of binding energy at low salt concentration in the medium. If one assumes the spatial arrangement of phosphodiester groups in all duplex structures to be similar, being determined by the base-base stacking distance, then the additional counter-ion uptake in the purely bicyclic duplexes is due to intrastrand charge-condensation and is compatible with the view that the corresponding constituent single strands have a conformation with the phosphate groups being spatially further separated than in the deoxyribo-α- and -β- nucleotides. Bicyclo-DNA, therefore, is a prime example underlining the importance of electrostatic contributions in nucleic acid association.

The duplex β-bcd(A$_{10}$):α-bcd(T$_{10}$) is the thermodynamic most stable one in the series investigated here and similar to that of β-bcd(A$_{10}$):β-bcd(T$_{10}$), which under comparable conditions shows a T$_m$-value of 55°C. While the CD-spectra of the two duplexes containing α-bcd(A$_{10}$) (Figure 5c) are similar and essentially show the characteristics of that of natural β-d(A$_{10}$):β-d(T$_{10}$), again the duplex containing β-bcd(A$_{10}$) shows a CD spectrum, which differs significantly showing negative ellipticities below 215 nm and around 280 nm. Most interestingly, this CD-spectrum almost matches that of β-bcd(A$_{10}$):β-bcd(T$_{10}$) (*26*). It is not clear at present whether this similarity is the result of the conformation of the purine strand, dominating the CD-spectrum of these duplexes or whether it is due to a similar 3-dimensional arrangement of the strands (e.g. orientation, conformation, base-pair constitution) different from that of the other hybrid duplexes investigated, irrespective of the stereochemical difference at the anomeric center of the bicyclothymidylic acid strand.

Stability of α-bcd(A) and α-bcd(T)-Sequences Towards the Action of Phosphodiesterases:

We have shown previously that β-bicyclooligonucleotides are considerably more stable against nuclease S1 (S1) and calf spleen phosphodiesterase (CSP) (by factors of 10^2 and 10^3 resp.), and modestly more stable (by a factor of 3-6) against the activity of snake venom phosphodiesterase (SVP). It has also been shown that the change of configuration at the anomeric center of the deoxyribonucleosides leads to enhanced enzymatic stability of the corresponding α-DNA-sequences. We therefore tested the sequences α-bcd(A$_{10}$), α-bcd(T$_{10}$) and α-bcd(T$_{20}$) for resistance against phosphodiester hydrolysis catalyzed by SVP. Under comparable conditions half live times ($t_{1/2}$) of 156 min. (α-bcd(A$_{10}$)), 510 min. (α-bcd(T$_{10}$)) and 782 min. (α-bcd(T$_{20}$)) were measured. This indicates a more than 100-fold higher stability of α-bicyclo-DNA (with respect to natural DNA) against this highly potent enzyme. In order to compare the enzymatic resistance of α-bicyclo-DNA with that of α-DNA we synthesized the sequence α-d(T$_{20}$) and compared its enzymatic stability ($t_{1/2}$ against SVP = 806 min.) with that of α-bcd(T$_{20}$). We were surprised to find that the enzyme SVP hydrolyzes both sequences with almost identical rate. Essentially it is only the change of configuration at the anomeric center and not the structural differences within the carbohydrate residues that seem to determine the higher enzymatic stability against SVP. This also explains the relatively poor resistance of the β-bicyclo-DNA sequences to the action of this enzyme.

Antisense research has had in the past and will have in the future a great impact on the chemistry of DNA. This impact, however, is not only limited to the medicinal aspects of DNA chemistry, eventually leading to a powerful drug. In the design, synthesis and the study of DNA-analogues, especially such containing modified carbohydrate residues, lies also the key to an improved understanding of the driving forces directing selectivity and energy in DNA-recognition. In terms of supramolecular chemistry, DNA is a prime example of an 'instructed molecule' exhibiting many facettes of supramolecular assembly. It is with these type of molecules that we can learn more about non-covalent molecular interactions.

Acknowledgments

We gratefully acknowledge financial support from the Swiss National Science Foundation, from ETH and from Ciba-Geigy AG, Basel.

Literature Cited

1.   Zamecnik, P.C; Stephenson, M.L. *Proc. Natl. Acad. Sci.U.S.A.* **1978**, *75*, 280.
2.   Stephenson, M.L.; Zamecnik, P.C. *Proc. Natl. Acad. Sci. U.S.A.* **1978**, *75*, 285.
3.   Caruthers, M.H. *Acc. Chem. Res.* **1991**, *24*, 278.
4.   Uhlmann, E.; Peyman, A. *Chem. Rev.* **1990**, *90*, 543.
5.   Goodchild, J. *Bioconjugate Chem.* **1990**, *1*, 165.

6.   Englisch, U.; Gauss, D. *Angew. Chem. Int. Ed. Engl.* **1991**, *30*, 613.
7.   Séquin, U, *Experientia*, **1973**, *29*, 1059.
8.   Morvan, F.; Rayner, B.; Imbach, J-L.; Thenet, S.; Bertrand, J-R.; Paoletti, J.; Malvy, C.; Paoletti C. *Nucleic Acids Res.* **1987**, *15*, 3421.
9.   Morvan, F.; Rayner, B.; Imbach, J-L.; Chang, D-K.; Lown, J.W. *Nucleic Acids Res.* **1987**, *15*, 4241.
10. Gautier, C.; Morvan F., Rayner, B.; Huynh-Din, T.; Igolen, J.; Imbach, J-L.; Paoletti, C.; Paoletti, J. *Nucleic Acids Res.* **1987**, *15*, 6625.
11. Morvan, F.; Rayner, B.; Imbach, J-L.; Lee, M.; Hartley, J.A.; Chang, D-K.; Lown, J.W. *Nucleic Acids Res.* **1987**, *15*, 7027.
12. Morvan, F.; Rayner, B.; Leonetti, J-P.; Imbach, J-L. *Nucleic Acids Res.* **1988**, *16*, 833.
13. Durand, M.; Maurizot, J.C.; Hélène, C. *Nucleic Acids Res.* **1988**, *16*, 5039.
14. Praseuth, D.; Perrouault, L.; Le Doan, T.; Chassignol, M.; Thuong, N.; Hélène, C. *Proc. Natl. Acad. Sci. U.S.A.* **1988**, *85*, 1349.
15. Paoletti, J.; Bazile, D.; Morvan, F.; Imbach, J-L.; Paoletti, C. *Nucleic Acids Res.* **1989**, *17*, 2693.
16. Lancelot, G.; Guesnet, J-L.; Vovelle, F. *Biochemistry*, **1989**, *28*, 7871.
17. Morvan, F.; Rayner, B.; Imbach J-L. *Anti-Cancer Drug Design*, **1991**, *6*, 521.
18. Debart, F.; Rayner, B.; Imbach, J-L. *Tetrahedron Lett.* **1990**, *31*, 3537.
19. Debart, F.; Rayner, B.; Degols, G.; Imbach, J-L. *Nucleic Acids Res.* **1992**, *20*, 1193.
20. Rosemeyer, H.; Seela, F. *Helv. Chim. Acta*, **1991**, *74*, 748.
21. Rosemeyer, H.; Krecmerova, F.; Seela, F. *Helv. Chim. Acta*, **1991**, *74*, 2054.
22. Inoue, H.; Hayase, Y,; Imura, A.; Iwai, S.; Miura, K.; Ohtsuka, E.; *Nucleic Acids Res.* **1987**, *15*, 6131.
23. Tarköy, M.; Bolli, M.; Schweizer B.; Leumann, C. *Helv. Chim. Acta*, **1993**, *76*, 481.
24. Egli, M.; Lubini, P.; Bolli, M.; Dobler, M.; Leumann, C. *J. Am. Chem. Soc.* **1993**, *115*, 5855.
25. Tarköy, M.; Leumann, C. *Angew. Chem. Int. Ed. Engl.* **1993**, *32*, 1432.
26. Tarköy, M.; Bolli, M.; Leumann, C. *Helv. Chim. Acta*, **1994**, *77*, 716.
27. Cocu, F.G.; Posternak, T. *Helv. Chim. Acta* **1972**, *55*, 2838.
28. Dess, D.B.; Martin, J.C. *J. Org. Chem.* **1983**, *48*, 4155.
29. Vorbrüggen, H.; Krolikiewicz, K.; Bennua, B. *Chem. Ber.* **1981**, *114*, 1234.
30. Vorbrüggen, H.; Höfle, G. *Chem. Ber.* **1981**, *114*, 1256.
31. Vorbrüggen, H.; Bennua, B. *Chem. Ber.* **1981**, *114*, 1279.
32. Saenger, W. In Principles of Nucleic Acid Structure; Cantor, C.R., Ed.; Springer Advanced Texts in Chemistry, Springer-Verlag, New York, 1984.
33. Sundaralingam, M. *J. Am. Chem. Soc.* **1971**, *93*, 6644.
34. Davies, D.B. *Prog. NMR Spectrosc.* **1978**, *12*, 135.
35. McBride, L.J.; Caruthers, M.H. *Tetrahedron Lett.* **1983**, *24*, 245.
36. Jones, A.R., In Oligonucleotide Synthesis - A Practical Approach; Gait, M.J., Ed., IRL Press, Oxford, 1984, pp 23.
37. Hayakawa, Y.; Wakabayashi, S.; Kato, H.; Noyori, R. *J. Am. Chem. Soc.* **1990**, *112*, 1691.
38. Pieles, U.; Zürcher, W.; Schär, M.; Moser, H.E. *Nucleic Acids Res.* **1993**, *21*, 3139.
39.   Neidlein, U.; Leumann, C. *Tetrahedron Lett.* **1992**, *33*, 8057.
40. Job, P. *Anal. Chim. Acta* **1928**, *9*, 113.
41. Marky, L.A.; Breslauer, K.J. *Biopolymers*, **1986**, *26*, 1601.
42. Searle, M.S.; Williams, D.H. *Nucleic Acids Res.* **1993**, *21*, 2051.
43. Manning, G.S. *Quart. Rev. Biophys.* **1978**, *11*, 179.
44. Record Jr. M.T.; Anderson, C.F.; Lohman, T.M. *Quart. Rev. Biophys.* **1978**, *11*, 103.

RECEIVED July 6, 1994

Chapter 8

2',5'-Oligoadenylate Antisense Chimeras for Targeted Ablation of RNA

Paul F. Torrence[1], Wei Xiao[1], Guiying Li[1], Krystyna Lesiak[1,3],
Shahrzad Khamnei[1], Avudaiappan Maran[2], Ratan Maitra[2], Beihua Dong[2],
and Robert H. Silverman[2]

[1]Section on Biomedical Chemistry, Laboratory of Medicinal Chemistry,
National Institute of Diabetes and Digestive and Kidney Diseases,
National Institutes of Health, Bethesda, MD 20892
[2]Department of Cancer Biology, Cleveland Clinic Foundation,
Cleveland, OH 44195

The unique 2',5'-phosphodiester bond-linked oligonucleotide known as
2-5A ($p_n5'A2'(p5'A2')_mp5'A$) plays a key role in mediation of the anti-
encephalomyocarditis virus action of interferon. 2-5A acts as a potent
inhibitor of translation through the activation of a constituent latent
endonuclease, the 2-5A-dependent ribonuclease (RNase), which
degrades RNAs. Covalent linkage of the tetrameric
p5'A2'p5'A2'p5'A2'p5'A to an antisense deoxyribonucleotide provided
an adduct which was unimpaired in activation of the 2-5A-dependent
RNase and which annealed to the complementary sense DNA to give a
hybrid complex with a melting temperature similar to the unmodified
DNA antisense/sense duplex. Such 2-5A-antisense chimeras targeted
to a modified HIV mRNA or to the dsRNA-dependent protein kinase
(PKR) mRNA induced specific cleavage in their targets without
affecting non-targeted mRNA species. The unaided uptake of 2-5A-
antisense against the PKR mRNA in HeLa cells resulted in ablation of
the PKR mRNA, with no effect on β-actin mRNA. These findings
demonstrate that 2-5A-antisense chimeras are effective and versatile
reagents for the catalytic destruction of targeted RNA.

The developing field of antisense-oligonucleotide therapeutics holds great promise
(*1, 2*). For instance, products have been introduced into clinical trials for the *ex vivo*
treatment of chronic myelogenous leukemia, chemotherapy of human
immunodeficiency virus infection and AIDS, treatment of genital warts caused by
human papilloma virus, therapy of cytomegalovirus retinitis, and chemotherapy of
acute myelogenous leukemia. Oligonucleotides in the discovery or preclinical
phase include antisense compounds for the therapy of septic shock, restenosis,
respiratory syncytical virus caused bronchiolitis, psoriasis, Alzheimer's disease,
drug-resistant malaria, tumor   angiogenesis,   breast,   ovarian,   and   prostate

[3]Current Address: Center for Biologics Evaluation and Research, U.S. Food and Drug
Administration, Bethesda, MD 20872

cancers,   malignant melanoma, colorectal carcinoma, and a wide variety of viral diseases including herpes 1 and 2, and hepatitis viruses.

In addition, antisense oligonucleotides have been valuable as gene "knock out" reagents. For example, sequence-specific inhibition of mRNAs for the following gene products has been reported: c-myc, c-fos, c-kit, bcr-abl, retinoblastoma, erythropoeitin, IL-1, IL-1 receptor, myogenin, EGF, angiotensinogen (*3*). Developmental biology studies have used antisense oligomers   profitably to delineate the role of specific gene products. Heaseman and colleagues (*4, 5*) were able to generate a specific phenotype in the early Xenopus embryo by microinjection of oligonucleotides antisense to the adhesive protein EP-cadherin or the cytoskeletal protein cytokeratin.   Microinjection of the appropriate corresponding mRNA could "rescue" the observed phenotype. In essence, this antisense technique was able to bring about the equivalent of a maternal effect mutation in the embryo. Lough and collaborators (*6*) found a role for fibroblast growth factor-2 (FGF-2) in precardiac mesoderm proliferation during avian embryonic heart development through the use of oligonucleotides complementary to a sequence in FGF-2 mRNA. The transcription factor PAX-2 appears to be necessary for initiation of epithelial cell differentiation in the developing mammalian kidney, as determined with antisense oligonucleotides   (*7*). Sariola et al. (*8*) have employed antisense oligonucleotides to the low-affinity nerve growth factor receptor to establish a role for this receptor in kidney morphogenesis.

Such results as described above have stimulated efforts to discover even more potent and effective oligonucleotide antisense reagents to use as both experimental tools and as therapeutic agents.   One approach has been to synthesize antisense oligonucleotides with reactive functionalities which can covalently react with the complementary mRNA to inactivate it, sometimes by cleavage. Both chemical and photochemical approaches have been reported (*vide infra*).

The perceived need for chemical or photochemical crosslinking to RNA targets derives largely from the limitations associated with the two recognized mechanisms of antisense biological efficacy (*3, 9-12*). The first inhibitory mechanism has been termed passive hybridization. In this mode of action, the antisense oligomer anneals to a target sequence and therefore blocks it from functioning as template for protein synthesis, from interaction with proteins, from undergoing appropriate splicing reactions, or being transported from the nucleus.   Secondly, for a number of examined systems, the enzyme RNase H plays a vital role by effecting cleavage of the RNA strand of an *in situ* generated DNA antisense oligonucleotide-RNA hybrid (*13 -15*). This enzyme does not always appear to be essential for the inhibitory actions associated with antisense reagents since some modified oligomers (e. g., methylphosphonates) do not yield RNase H substrates upon complexation with RNA; nonetheless, such altered antisense oligonucleotides still exhibit inhibitory activities (*16*). On the other hand, other modifications of oligonucleotides have sought to retain or regain the potential for RNase H destruction of the targeted mRNA. Thus, one strategy has been to chemically modify antisense reagents to attack and destroy target sequences rather than relying on the relatively passive hybridization block and/or depending on the presence of RNase H activity to facilitate nucleic acid degradation.

Among the reactive oligonucleotides prepared to achieve this goal are the following:

a. oligonucleotides with alkylating functionalities which have the potential to react covalently with the targeted nucleic acid resulting in the formation of crosslinks (*17 -20*);

b. antisense oligonucleotides linked to chelating reagents which carry metal ions (Cu^{+2} or Fe^{+2}) capable of producing free radicals upon reaction with a reducing agent and oxygen to produce scission of the opposing strand ("oxidative nucleases") or one strand of a triple-helix (*21 -34*);

c. oligonucleotide antisense reagents bearing photoreactive moieties such as psoralen or porphyrin (*35 -39*).

These covalently reactive antisense reagents possess the capacity to irreversibly react with or cleave the targeted RNA (or DNA); however, some also can undergo non-specific reaction with untargeted species of nucleic acids. For instance, the reactivity of oligonucleotides with alkylating groups is not limited only to the final target species since such functional groups can react with any nucleophile (protein, nucleic acid, etc.) with which they come in contact. Possibly one of the greatest limitations of this chemoaffinity or photoaffinity approach to destruction of mRNA is that the approach is non-catalytic; i. e., it would require a minimum of a stoichiometric equivalent of antisense oligomer to target RNA to cause ablation of the message. Clearly, such a requirement dramatically increases the probability of non-specific, or in a therapeutic context, toxic side-reactions with untargeted mRNAs. Such a need for stoichiometric equivalency presumably may not be as great a problem for the so-called antigene approach wherein a specific gene sequence in the DNA is targeted in order to form a triplex. When just a single copy of a sequence must be eliminated, the amount of reagent applied to the cell would be considerably less than when many copies of the gene are embodied in the mRNA molecules needed for translation.

Such considerations have led researchers to the obvious promise of catalytic mechanisms to effect destruction of specific mRNAs. The ability of one mole of applied antisense agent to cause the turnover of an order of magnitude or more target RNAs would be a major step toward enhanced specificity (and reduced toxicity) of the antisense approach.

Even though, as mentioned above, the enzyme RNase H can be a significant player in mediating the biological effects of antisense oligonucleotides, at least in some conditions, the sometimes variable activity and relative lack of potency of simple DNA phosphodiester oligonucleotides has led to a search for other strategies which may afford more potent catalytic activity.

At least three different approaches can be summarized:

1. Ribozymes (*40*), or catalytically active RNA's can be linked to appropriate antisense oligonucleotides. Not only RNA ribozymes but also DNA-RNA ribozymes have been prepared by replacing the RNA flanking sequence with ribonuclease resistant DNA but retaining the catalytically active RNA ribozyme core. Such modified and unmodified catalytic RNA's have been shown to be capable of effecting cleavage of target RNA's in vitro and in cell culture, and can protect human lymphocytes against infection by HIV-1 (*41*).

2. A somewhat similar strategy relies upon the enzyme RNase P which normally functions to cleave tRNA precursors to yield the 5'-termini of mature tRNA's (*42 - 44*). RNase P is made up of protein and an RNA subunit, and the latter possesses the enzyme's catalytic activity. For the *Escherichia coli* enzyme, just one half-turn of the RNA helix and the 3'-proximal CCA sequence provide the minimum necessary recognition sites to behave as a substrate for RNase P. The 3'-terminal CCA sequence has been termed an "external guide sequence" (EGS) since it guides the RNase P to cleave the RNA at a specific site. Apparently, any RNA would be a candidate for cleavage by RNase P when it becomes hybridized to such an "external guide sequence" which has a partial complementary sequence and ends in the sequence ACCA. For human RNase P, however, a significantly more complex EGS is required due to the narrow range of substrate specificity of the human enzyme. Thus while a 25-mer EGS was able to direct scission by the *E. coli* enzyme, a 70-mer EGS was required to direct cleavage by the human RNase P. Application of this technique to RNA cleavage in intact cells has not been reported.

3. Another technique involves creation of hybrid nucleases by fusion of an oligonucleotide to a unique site on staphylococcal nuclease and RNase S (*45 - 49*). These hybrid nucleases have been shown to specifically hydrolyze both single-stranded RNA and single-stranded DNA as well as duplex DNA. This latter approach is based on the previously suggested concept of the introduction of new

binding domains into enzymes. Although developed for the purpose of generating a novel and general approach to DNA cleavage, this latter method also relied upon provision of a nuclease (class-IIS restriction) recognition site in the form of a hairpin structure covalently connected to a nucleotide sequence which was complementary to a specific region in the target nucleic acid. So in this instance, while the oligonucleotide fragment was not covalently linked to a nuclease, it supplied, before or after annealing, the appropriate recognition ligand for (in this instance) the restriction endonuclease. In a sense, this methodology can be considered distantly related to site-directed mutagenesis and chemical mutagenesis of enzyme active sites (*50*). This approach would presumably be of limited application to intact cells because of the need to assure uptake of the oligonucleotide-enzyme conjugate.

We have explored the potential of the 2-5A system as an approach to the targeted destruction of RNA (*51, 52*). The central molecule in this system, 2-5A (Figure 1), is an unique 2',5'-linked oligoadenylate which plays an important role in the interferon defense against virus infection.

There are three essential components of the 2-5A system (*53*) (Figure 2): 2-5A synthetase, the enzyme which, upon activation by double-stranded RNA, synthesizes 2-5A from ATP; the latent 2-5A-dependent ribonuclease (RNase L)

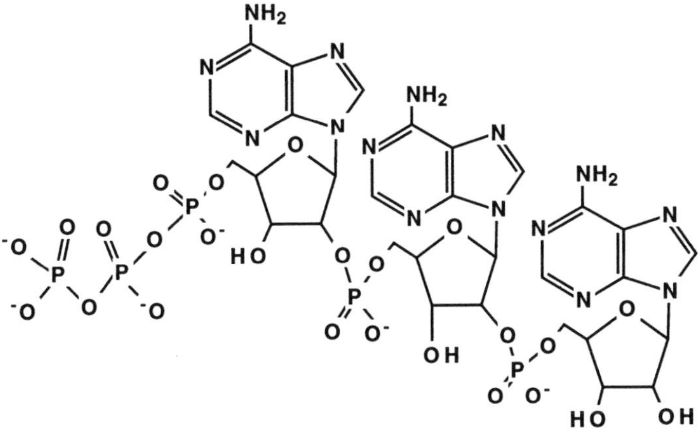

Figure 1. Structure of 2-5A Trimer Triphosphate: 5'-O-Triphosphoryl-Adenylyl(2'->5')Adenylyl(2'->5')-Adenosine.

which, upon activation by 2-5A, can degrade mRNA; the 2',5'-phosphodiesterase which degrades 2-5A to AMP and ATP, and presumably limits the cellular toxicity of 2-5A. These enzymes act in harmony as illustrated for the case of infection of an interferon-treated cell by encephalomyocarditis virus. Treatment of a cell with interferon induces enhanced levels of 2-5A synthetase. Then dsRNA, formed as an intermediate in viral replication, activates the synthetase to generate 2-5A from ATP. The 2-5A thus formed activates the latent 2-5A-dependent RNase which then degrades mRNA, thereby inhibiting translation. 2-5A is destroyed by the 2',5'-phosphodiesterase.

The enzyme responsible for 2-5A synthesis is called 2-5A synthetase, and it catalyzes the following reaction:

$$nATP \longrightarrow ppp5'A2'(p5'A)_{n-1} + (n-1)ppi \quad (n \geq 2).$$

The enzyme also can elongate by 2'-adenylation of nucleotides or oligonucleotides when these are supplied as primers (*55 - 57*).

$$C2'p5'A \quad + \quad nATP \longrightarrow \quad C2'p5'A2'(p5'A)_n \quad + n \, pp_i$$
$$A3'p5'A \quad + \quad nATP \longrightarrow \quad A3'p5'A2'(p5'A)_n \quad + n \, pp_i$$

The enzyme also displays a nucleotidyltransferase function through its ability to add a wide variety of nucleotides to the 2' hydroxyl terminus of a 2',5'-oligoriboadenylate (56) according to the following reaction:

$$ppp5'A2'(p5'A)_nA \quad + \quad TTP \longrightarrow \quad ppp5'A2'(p5'A)_nA2'pT \quad + \quad pp_i$$

The mechanism for chain elongation for 2-5A synthetase has been found to be non-processive, meaning that intermediates of polymerization dissociate from the enzyme before another nucleotide is added (54).

The 2-5A synthetases exist as a multienzyme family, and are all induced by interferon and all require dsRNA for activation, but they differ in specifics such as enzymatic properties, molecular weight, weights, their subcellular distribution, and their inducibility in different types of cells (58, 59). Human cells are capable of synthesizing at least three distinct classes of 2-5A synthetase isozymes which are antigenically unrelated (58).

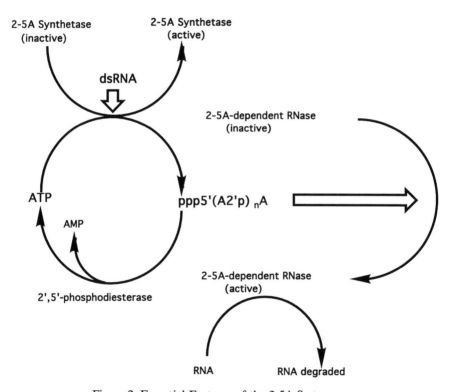

Figure 2. Essential Features of the 2-5A System.

Mouse cells also elaborate a set of 2-5A synthetase isozymes, including a 105 kDa form, synthetases in the 20-30 kDa range, and a cloned synthetase of 42 kDa which possesses 69% sequence homology with the human synthetase of the same molecular weight range (63). Deletion mutants of a murine 2-5A synthetase clone for a protein of 414 amino acids showed that a region between amino acid residues 104 and 158 was required for binding to dsRNA and another region between amino acids 320 and 344 was necessary for enzyme activity (*64*).

The major activity responsible for this degradation of 2-5A is a 2',5'-phosphodiesterase which hydrolyzes 2-5A to 5'AMP and 5'ATP. A phosphodiesterase has been partially purified from mouse L cells and from mouse reticulocytes (*65, 66*). This mouse enzyme had a molecular weight of 35,000-40,000 daltons, required magnesium ion, and could degrade both oligonucleotides with 2',5'-or 3',5'-linkages. A 2',5'-phosphodiesterase has also been purified from bovine spleen (*67*). This enzyme had a molecular weight of 65,000 daltons, and could cleave both 2',5'- and 3',5'-oligonucleotides.

The limited half-life of 2-5A in biological milieu (*53, 68, 69*) probably makes it imperative to stabilize 2-5A against such degradation if the biological activity of 2-5A is ever to be capitalized upon in a therapeutic sense.   The results of such stabilization can be dramatic. For instance, Defilippi et al. (*68*) microinjected 2-5A or the "tailed" analogue in which the 2'-terminal adenosine nucleotide was replaced with 2-(9-adenyl)-6-hydroxymethyl-4-hexylmorpholine.   The latter "tailed" derivative was 100-fold more active as an inhibitor of vesicular stomatitis virus replication as was unmodified 2-5A itself.

The murine 2-5A-dependent RNase has been purified to homogeneity (*70*), and both the human and mouse 2-5A-dependent RNase have been cloned and expressed (*71*). The full-length human 2-5A-dependent RNase, as produced in reticulocyte lysate, was determined to have a molecular weight of 83,539 Daltons, in excellent agreement with values obtained for RNase of natural sources (*70*). Although the 2-5A-dependent RNase is present in many species including reptiles, birds, and mammals, its basal level can be increased about 3-fold by interferon treatment of mouse cells. The murine and human enzymes contain a duplicated phosphate-binding loop motif which functions in the binding of 2-5A. The human 2-5A-dependent RNase gene (RNS 4 or RNase OA) has been mapped to chromosome 1q25 (*72*).

Other experiments have examined the RNA substrate preference of the 2-5A-dependent RNase (*73, 74*). The 2-5A activated enzyme cleaves single-stranded regions of RNA 3' of UpNp sequences, with a distinct preference for UpUp or UpAp. The homopolymer poly(U) was readily degraded by the pure recombinant human 2-5A-dependent RNase, but poly(C) was unaffected (*71*).

There is abundant evidence to implicate the 2-5A system and the 2-5A dependent endonuclease in the antiviral action of interferon.

1. Treatment of intact cells with 2-5A, using uptake-enhancement methodologies, results in inhibition of translation, increased RNA degradation, and inhibition of virus replication (e. g., encephalomyocarditis, mengo, vesicular stomatitis, and vaccinia viruses) (*75* , and reviewed in *53*).

2. The oligonucleotide 2-5A is detectable in interferon-treated cells infected with, for instance, encephalomyocarditis virus or reovirus. Similarly, such cells show a characteristic pattern of ribosomal RNA cleavage which is generated when ribosomes are treated in a 2-5A-dependent RNase-containing cell-free system with 2-5A (*76, 77*).

3. Addition of the double-stranded RNA, poly(I).poly(C), to interferon-treated HeLa cells, led to the formation of high concentrations of 2-5A (*78*).

4. Reovirus or encephalomyocarditis virus infection (which leads to production of viral double-stranded RNA) of interferon-treated cells leads to degradation of virus RNA (*79, 80*).

5. A 2-5A analogue-inhibitor (*81*) which prevents the action of the 2-5A-dependent RNase by competing with 2-5A for its binding site on the nuclease (*82*), provided an partial prevention of the anti-encephalomyocarditis virus activity of interferon.

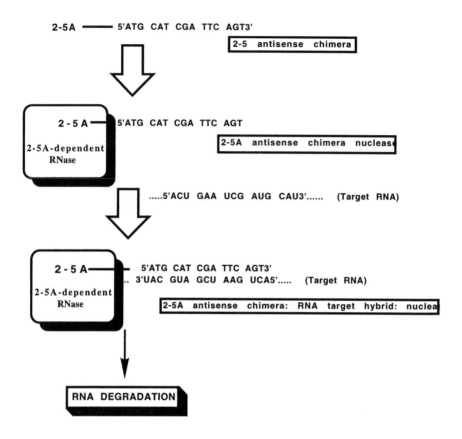

Figure 3. Postulated mechanism of selective destruction of a hypothetical RNA target by 2-5A-antisense chimeras.

6. Expression of the 2-5A synthetases, responsible for *in vivo* 2-5A synthesis, inhibited picornavirus replication to an extent directly related to the amount of 2-5A synthetase expressed by the specific clone (*83*).
7. Use of synthetase antibody showed that encephalomyocarditis virus RNA was associated with the activated 2-5A synthetase in interferon-treated virus-infected cells (*84*).
8. A dominant negative of the 2-5A-dependent RNase suppressed the antiviral and antiproliferative effects of interferon (*96*).
    We have applied the unique and potent 2-5A system to the specific degradation of mRNA through an examination of the biological activity of synthetic adducts between 2-5A and antisense oligonucleotides. Such composite nucleic acids could, through the antisense domain, target the chimera to a particular mRNA sequence which would then be targeted for destruction by the 2-5A component which would provide a localized activation of the latent 2-5A-dependent RNase (Figure 3). Presumably, this kind of chimeric molecule also could possess the mechanisms of action of classical antisense reagents, i. e., passive hybridization arrest and possibly RNase H degradation. Moreover, the antisense component of the chimera could provide a high degree of specificity normally missing from 2-5A-dependent RNase

cleavages. The antisense region of the composite molecule also might facilitate uptake of 2-5A since antisense oligonucleotides seem to be taken up by intact cells, perhaps by a specific mechanism. Finally, this approach would bring into play the potent and catalytic action of the latent 2-5A-dependent endonuclease, perhaps thereby substantially increasing the potency of the antisense approach.

The design of this molecule relied upon information gleaned from studies on the relationship between oligonucleotide structure and 2-5A-dependent RNase activation (*85*). The prototype 2-5A-antisense chimera consisted of an antisense domain made up of oligo(dT)$_{18}$ connected to 2-5A through a linker (51, 52). We chose the oligo(dT) to simplify initial synthesis and characterization of this novel molecule. We added the tetrameric 2-5A, p5'A2'p5'A2'p5'A2'p5'A to the 5'-terminus of the antisense oligomer. Only a 5'-monophosphate was required at the 5'-terminus of the 2-5A component since the 2-5A-dependent RNase of human CEM and Daudi lymphoblastoid cells can be activated by the monophosphate alone and does not require a 5'-di- or 5'-triphosphate. The 2-5A and antisense moieties were joined by two 1,4-butanediol molecules joined to each other and to 2-5A and antisense by phosphodiester bonds. Linkage to the 2-5A tetramer was at the 2'-terminal hydroxyl and linkage to the antisense oligonucleotide was at the 5'-terminal hydroxyl. It was considered necessary that the mode of linkage to the 2-5A component should be through the 2'-terminus of the oligomer since a free 5'-monophosphate was requisite for maximal 2-5A-dependent endonuclease activity. Earlier studies showed that if the 5'-terminal phosphate of 2-5A were incorporated into a phosphodiester grouping, a diminution of RNase binding resulted. Linker elements were used to join 2-5A and antisense rather than directly joining the terminal 2-5A adenosine nucleotide to the first nucleotide of the antisense sequence because of the possibility that the 2-5A-dependent RNase, once bound to the 2-5A component of the chimera, might disturb hybridization to target RNA, or conversely that the generated double helix from union of antisense oligo with sense RNA might interfere with maximum binding to the 2-5A-dependent RNase.

One expected advantage of this mode of conjugation of 2-5A to antisense was the resultant stabilization of the 2',5'-oligoadenylate to degradation by phosphodiesterases including that specific for 2',5'-linkages. Essentially, chimera construction as described above results in an antisense "tailing" of 2-5A and provides the DNA oligomer as an alternative substrate for phosphodiesterases.

Chimeric 2-5A-antisense congeners were undiminished in their ability to bind to the 2-5A-dependent RNase (*51,52*). In addition, the composite molecules did not adversely affect the ability of the antisense domain to hybridize to its target poly(A) (*52*). Finally, when the poly(A) target was pre-annealed with the chimera pA$_4$:dT$_{18}$, there resulted no significant decrease in 2-5A-dependent RNase binding ability (*52*). Thus, these novel conjugate molecules retained the ability of each partner in the construct to interact with the relevant biological receptors in a manner consistent with the proposed mechanism of action.

The 2-5A-antisense strategy for site-directed RNA cleavage first was evaluated in a cell-free system consisting of a postribosomal supernatant fraction of human lymphoblastoid Daudi cells, already known to contain basal levels of the 2-5A-dependent RNase. The target for the pA$_4$:dT$_{18}$ chimera was a engineered modified human immunodeficiency virus type 1 vif mRNA containing an internal 3',5'-oligoA tract; namely, TAR:A$_{25}$:vifRNA. This RNA was produced in vitro from a pSP64-derived plasmid containing a partial cDNA for the HIV vif protein interrupted with 25 adenylyl units. The resultant RNA was end-labeled with [^{32}P]-pCp to provide a tag to follow the fate of the RNA.

When the chimera, pA$_4$:dT$_{18}$, was added to the cell-free system containing labeled TAR:A$_{25}$:vifRNA, there was a nearly quantitative conversion of the target RNA to a specific cleavage product [*51*]. This cleavage could be induced by as little as 25 nM chimera. The exact sites of RNA cleavage were determined using primer-extension DNA synthesis.   The chimera, $pA_4:dT_{18}$, induced multiple cleavages within the oligo(A) tract of the $TAR:A_{25}$:vif RNA. No such specific cleavages were caused when target RNA was incubated without any added oligonucleotide or when pppA2'p5'A2'p5'A was added.

Various control experiments provided strong evidence that cleavage of the $TAR:A25$:vifRNA occurred by the postulated mechanism involving the 2-5A-dependent RNase (51).

a) When 2-5A itself was added alone with unlinked $(dT)_{20}$, no specific cleavage of the target RNA was observed.

b) In addition, RNA cleavage was dependent on a functionally active 2-5A derivative. As outlined above, activation of the human 2-5A-dependent RNase requires a 2',5'-oligoadenylate with at least one 5'-phosphate group.   Thus, consistent with this requirement, the 5'-unphosphorylated compound, $A_4:dT_{18}$, did not lead to detectable RNA cleavage.

c) The oligonucleotide, $(dT)_{20}$, should be able to block competively the action of the chimera $pA4:dT_{20}$ by annealing to the target A_{25} sequence in the RNA, thus preventing access of the chimera to the RNA. Indeed, when a tenfold molar excess of unlinked $(dT)_{20}$ was added to reaction mixtures containing chimeric $pA_4:dT_{18}$, specific cleavage was completely blocked.

d) The chimera $pA_4:dT_4$, which contains an antisense sequence too short to provide hybridization to the target A_{25} sequence under the cleavage conditions, failed to cause specific degradation of the modified HIV RNA.

e) If pA_4 were linked to antisense oligonucleotides which could not anneal to the A_{25} target because they contained non-complementary sequences, no specific cleavage resulted.

f) The composite oligonucleotide, $pA_4:dT_{18}$, was not capable of inducing any specific cleavages in a target RNA which lacked the A_{25} tract; namely, TAR:vifRNA.

g) Convincing proof for the vital role of the 2-5A-dependent RNase in obtaining specific cleavage of $TAR:A_{25}$:vifRNA obtained through the use of the 2-5A analogue-inhibitor of the 2-5A-dependent RNase, ppp5'I2'p5'A2'p5'A, an inosine-substituted congener of 2-5A (87). This analogue completely prevented the ability of chimeric $pA_4:dT_{18}$ to effect specific cleavage of TAR:A25:vifRNA, thereby establishing the 2-5A-dependent RNase as the enzyme responsible for RNA destruction.

In order to extend the 2-5A-antisense technology to a natural mRNA in intact cells, we targeted for cleavage the mRNA to "PKR" (88), the dsRNA-dependent protein kinase (p68 kinase, DAI, P1 kinase), in human HeLa cells.

PKR is an interferon-inducible protein which regulates protein synthesis (89) and which functions as a tumor suppressor factor in nude mice (90, 91). At the protein synthesis level, PKR phosphorylates the α subunit of protein synthesis initiation factor eIF-2, a component of the ternary complex, eIF-2:GTP:MET-tRNAᵢ. When the 80S ribosome is formed, the GTP in the ternary complex is hydrolyzed allowing release of eIF-2:GDP. The subsequent exchange of GTP for GDP on the eIF-2 is catalyzed by eIF-2B (GEF), a rate limiting factor in protein synthesis initiation (reviewed in 92). Phosphorylation of eIF2-α inhibits the eIF-2B catalyzed recycling reaction resulting in the cessation of protein synthesis due to an accumulation of inactive eIF-2:GDP complex. In addition, expression of PKR cDNA reduces replication of encephalomyocarditis virus providing convincing evidence that the kinase is involved in the antiviral activity of interferon (93). Furthermore, the tumor suppressor function of PKR has been demonstrated (90, 91).   Mutants of PKR apparently function as dominant negative inhibitors of endogenous PKR activity, leading to the formation of tumors in nude mice.

The PKR mRNA site chosen for the target of the 2-5A-antisense chimera was 60-

78 nucleotides from the start codon of the PKR mRNA in a region predicted by computer analysis to be single stranded. The following 2-5A-antisense chimera directed against the PKR mRNA thus was selected for synthesis.

p5'A2'p5'A2'p5'A2'p5'A2'p-C4-p-C4-p5'GTA CTA CTC CCT GCT TCT G3'

In this structure, the linkage between the 2',5'-oligoadenylate moiety and the 3',5'-deoxyribonucleotide antisense sequence was the same as for the chimeras targeted against the HIV TAR-A_{25}-vif RNA; namely, two 1,4-butanediol molecules linked by phosphodiester bonds and to the 2-5A moiety through a phosphodiester bond through the 2'-hydroxyl of the 2'-terminal adenosine and to the antisense sequence via the DNA's 5'-phosphate group.

To induce degradation of PKR mRNA, the oligonucleotides at 2 µM were added to HeLa cell cultures in microtiter wells. After four hours of incubation at 37°C the total RNA was isolated from the cells using RNazol reagent. To detect the PKR mRNA a reverse transcriptase-coupled PCR method was used. First the mRNA was reverse transcribed into cDNA using oligo(dT) as primer; then the cDNA was diluted three times at 1:4, and the PKR DNA in each dilution was amplified using primers to complementary sequences at opposite ends of the PKR coding sequence. To detect the amplified PKR DNA, it was blotted to Nytran membrane and then probed with radiolabeled PKR cDNA. In the absence of oligonucleotide treatment of the cells, the signal increased in proportion to the number of PCR cycles, and it decreased upon dilution of the cDNA . In contrast, there was no PKR mRNA detected after incubations of the cells with pA4-antiPKR. Therefore, this species of 2-5A-antisense caused the ablation of intact PKR mRNA in the HeLa cells. In contrast, the control oligonucleotides, antiPKR, pA4-antiHIV (directed against a splice-junction sequence in HIV RNA), and pA4-dA$_{18}$ had no effect on PKR mRNA levels. These results showed a requirement for the 2-5A moiety of 2-5A-antisense and indicated that the oligodeoxyribonucleotide part of the chimera must be complementary to the target mRNA for degradation to occur.

To determine if the 2-5A-antisense was specific in its mechanism of action, levels of β-actin mRNA were also measured by the same RT-PCR method. None of the oligonucleotides tested affected levels of the β-actin mRNA. Therefore, 2-5A-antisense did not damage mRNA which is not being targeted.

Several additional controls were performed to demonstrate the mechanism of action. For instance, addition to the cells of chimeras containing sense orientation sequence, pA4-sense PKR was without effect. Replacement of the 2-5A group with p(dA3'p)3dA, which cannot bind to or activate the 2-5A-dependent RNase, produced an inactive molecule when coupled to antisense to PKR (i. e., 3',5'-pdA4-antiPKR) . Removal of the 5'-phosphoryl group of 2-5A-antisense to give 2-5A core A4-antiPKR, inactivated the chimera as well. Also, pA$_2$-antiPKR was inactive. These latter two chimeric oligos possessed 2-5A congeners which either by virtue of the missing 5'-monophosphoryl moiety or the shortened (dimeric) length could not bind to or activate the 2-5A nuclease. The oligonucleotide m$_{10}$-pA4-antiPKR), in which ten mismatches were introduced into the antisense sequence, also was devoid of activity. Therefore, to induce RNA decay functional 2-5A must be present, and complementarity with the target sequence must be maintained.

Levels of the PKR protein *per se* were also greatly reduced in the HeLa cells treated with pA4-antiPKR. In these experiments, levels of PKR protein were determined on Western blots using monoclonal antibody to PKR and the enhanced chemiluminesence method. Treatment of HeLa cells for 16 hrs with pA4-antiPKR at 2 µM greatly reduced PKR levels.

The PKR protein, as outlined above, functions as a double-stranded RNA-activated protein kinase, and, when PKR is affinity adsorbed to a support of poly(I).poly(C) linked to cellulose, it can phosphorylate itself when supplied with γ -^{32}P-labelled ATP. This autophosphorylation can be followed conveniently by gel electrophoresis and autoradiography. The chimeric oligonucleotide, pA4-antiPKR, was added to HeLa cell cultures in 2 μM concentration at 0, 12, 24, 36, 48, 60, and 72 hours, and cells were harvested after a total of 84 hours of incubation. PKR in cell extracts was immobilized on the activating affinity matrix, poly(I)·poly(C)-cellulose, and then the washed matrix was incubated with[γ-^{32}P]-ATP. Polyacrylamide gel electrophoresis revealed that PKR activity was reduced to undectable levels. In contrast, neither the control chimeras, pA4-sense PKR nor the unphosphorylated chimera 2-5A core A4-antiPKR, had any effect on PKR activity. Thus, treatment of HeLa cells with 2-5A-antisense to the PKR message resulted in ablation of the PKR mRNA and loss of the PKR protein and its functional activity.

To determine if 2-5A-antisense was inhibitory to cell growth, we measured live cells cultured in the presence of 2-5A-antisense. Viable cell numbers were determined in the presence of trypan blue dye in a hemocytometer. Neither pA4-antiPKR or pA4-dA$_{18}$ both at 2 μM, were inhibitory to cell proliferation. AntiPKR also lacked any cytotoxicity in this assay. Therefore, 2-5A-antisense at 2 μM induced targeted ablation of mRNA in the absence of cytotoxicity.

Summary

The 2-5A-antisense approach to the targeted destruction of mRNA possesses several notable strong points:
1. it depends upon a novel mechanism for RNA destruction;
2. the key 2-5A-dependent RNase is constitutive in a wide variety of cells including those from reptiles, birds, and mammals (53);
3. the 2-5A-dependent RNase is latent until activated by nM concentrations of 2-5A;
4. essentially, a custom nuclease directed against a specific mRNA can be formed simply by an appropriate change in the antisense domain of the chimeric oligonucleotide;
5. these composite nucleic acids are as readily accessible as phosphodiester oligodeoxyribonucleotides (52);
6. modifications to the antisense region in order to enhance stability or other properties are not expected to adversely affect the activation of the 2-5A-dependent RNase, in contrast to the sensitivity of RNase H to certain alterations in chemical structure (13 -15);
7. 2-5A-antisense chimeras are of significantly lower molecular weight than either ribozyme constructs targeted to specific RNA sequences or oligonucleotides with external guide sequences for targeted cleavage by RNase P (41 - 44);
8. there exists the opportunity to modulate the effect of 2-5A-antisense oligoribonucleotides since levels of the 2-5A-dependent RNase can be changed in some systems by interferon treatment;
9. since 2-5A-antisense oligonucleotides achieve the destruction through an enzymic process, the mechanism is a catalytic one and one molecule of chimera can lead to the degradation of many molecules of mRNA.

At this time, there exist four possible considerations for adapting the 2-5A-antisense strategy to different systems:
1. while many cells have 2-5A-dependent RNase which needs only a 5'-monophosphate moiety for activation, 2-5A-dependent RNase activity from some

sources, chiefly mouse cell lines (*53*), require a polyphosphate group for optimal activation. If a cell's 2-5A-dependent RNase in fact requires a 5'-polyphosphate for effective nuclease activity, the greater lability of 5'-di- or 5'-triphosphates in the intra- of extracellular milieu may limit application;

2. the length of the 2-5A oligomer appended to the antisense domain may affect activity. For instance, 2-5A-dependent RNase from rabbit reticulocyte lysates, but not from other rabbit organs, required a tetramer rather than a trimer as the minimum oligomer which would activate the latent nuclease (*94*);

3. the 2-5A-dependent nuclease seems to be found in only reptiles, and avian and mammalian species. Only traces may be found in amphibia, and none has been detected in fish, insects, yeast, bacteria, or plants (*53*). Its application would be thus limited to such taxons with demonstrable nuclease;

4. levels of the 2-5A-dependent RNase can vary considerably during differentiation (*95*). This has been an area of very limited exploration, but one that should be considered if 2-5A-antisense is to be applied to developmental situations.

Many intriguing questions remain to be addressed concerning this burgeoning technique for the selective ablation of RNA species. For instance, where do the primary cleavages in the target RNA occur? How can that site be influenced by the structure and composition of the chimera? Can chimeric modifications be found to enhance biological stability and uptake? What is the range of cells to which this technique can be applied? What are the non-specific effects of such composite nucleic acids? Can 2-5A-antisense serve as the basis for therapeutic intervention in viral or neoplastic diseases?

Acknowledgments

This research was supported in part by the National Institutes of Health Intramural AIDS Targeted Antiviral Program (to P. F. T.) and in part by National Institute of Allegry and Infectious Disease Grant 5 R01AI28253 (Awarded to R. H. S.).

References

1. anonymous; *Drug & Market Dev.* **1993**, 4, 110.
2. anonymous; *Drug & Market Dev.* **1993**, 4, 179.
3. Stein, C. A.; Cheng, Y.-C.; *Science* **1993**, 261, 1004.
4. Heaseman, J.; Ginsburg, D.; Geiger, B.; Goldstone, K.; Pratt, T.; Yoshida-Noro, C.; Wylie, C. C.; *Development* **1994,** 120, 49.
5. Torpey, N.; Wylie, C. C.; Heaseman, J.; *Nature* **1992**, 357, 413.
6. Sugi, Y.; Sasse, J.; Lough, J.; *Dev. Biol.* **1993**, 157, 28.
7. Rothenpieier, U. W.; Dressler, G. R.; *Development* **1993**, 119, 711.
8. Sariola, H.; Saarma, M.; Saino, K.; Arumae, U.; Palgi, J.; Vaahtokari, A.; Thesleff, I.; Karavanov, A.; Science 1991, 254, 571.
9. Cohen, J., ed.; *Oligonucleotides. Antisense Inhibitors of Gene Expression* , CRCPress, Boca Raton, Fl,1989.
10. Mol, J. N. M. ; van der Krol, A. R. (eds); *Antisense Nucleic Acids and Proteins. Fundamentals and Applications***,** Marcel Dekker, Inc, N. Y., 1991.
11. Uhlmann, E.; Peyman, A.; *Chem. Rev.* **1990**, 90, 543.
12. Cook, P. D.; *Anti-Cancer Drug Des.* **1991**, 6, 585.
13. Walder, R. T.; Walder, J. A.; *Proc. Natl. Acad. Sci. USA* **1988**, 85, 5011.
14. Cazenave, C.; Stein, C. A.; Loreau, N.; Thuong, N. T.; Neckers, L. M.; Subasinghe, C.; Helene, C.; Cohen, J. S.; Toulme, J.-J.; *Nucl. Acids Res.* **1989**, **17**, 4255.
15. Cazenave, C.; Loreau, N.; Thuong, N. T.; Tuouleme, J. J.; Helene, C.; *Nucl. Acids. Res.* **1987**, 15, 4717.

16. Maher, L. J.; Dolnick, B. J.; *Nucl. Acids Res.* **1987**, **16**, 3341.
17. Knorre, D. G.; Vlassov, V. V.; Zarytova, V. F.; in Cohen, J., ed., *Oligonucleotides. Antisense Inhibitors of Gene Expression*, CRC Press, Boca Raton, Fl, pp. 173-196, 1989.
18. Vlassos, V. V.; Gaidamakov, S. A.; Zarytova, V. F.; Knorre, D. G.; Levina, A. S.; Nikonova, A. A.; Podust, L. M.; Fedorova, O. S.; , *Gene* **1988**, 72, 313.
19. Vlassov, V. V.; Zarytova, V. F.; Kutiavin, I. V.; Mamaev, S. V.; *FEBS Lett.* **1988**, 231, 352-354 .
20. Vlassov, V. V.; Zarytova, V. F.; Kutiavin, I. V.; Mamaev, S. V.; Podyminogin, M. A.; *Nucleic Acids Res.* **1986**, 14, 7661 .
21. Le Doan, T.; Perrouault, L.; Chassignol, M.; Thuomg, N. T.; and Helene, C.; *Nucl. Acids Res.* **1987**, 15, 8643.
22. Frolova, E. I.; Ivanova, E. M.; Zarytova, V. F.; Abramova, T. V.; Vlassov, V. V.; *FEBS Lett.* **1990**, 269, 101.
23. Chu, B. C. F.; Orgel, L. E.; *Proc. Natl. Acad. Sci. USA* **1985**, 82, 963.
24. Brosalina, A. S.; Vlassov, V. V.; Kazakov, S. A.; *Bioorg. Khim.* **1988**, 14, 125.
25. Le Doan, T.; Perrouault, L.; Helene, C.; Chassignol, M.; Thuong, N. T.; *Biochemistry* **1986**, 25, 6736.
26. Chen, C. B.; Sigman, D. S.; *Proc. Natl. Acad. Sci. USA* **1986**, 83, 7147.
27. Chen, C. B.; Sigman, D. S.; *J. Am. Chem. Soc.* **1988**, 110, 6570.
28. Francois, J.-C.; Saison-Behmoaras, T.; Chassignol, M.; Thuong, N. T.; Helene, C.; *J. Biol. Chem.* **1989**, 264, 5891..
29. Hertzberg, R. P.; Dervan, P. B.; *Biochemistry* **1984**, 23, 3934..
30. Moser, H. E.; Dervan, P. B.; *Science* **1987**, 238, 645.
31. Stroebel, S. A.; Moser, H. E.; Dervan, P. B.; *J. Am. Chem. Soc.* **1988**, 110, 7927.
32. Dreyer, G. B.; Dervan, P. B.; *Proc, Natl. Acad. Sci. USA* **1985,** 82, 968.
33. Strobel, S. A.; Dervan, P. B.; *Science* **1990**, 249, 73.
34. Praseuth, D.; Perrouault, L.; Le Doan, T.; Chassignol, M.; Thuong, N.; Helene, C.; *Proc. Natl. Acad. Sci. USA* **1988**, 85, 1349.
35. Pieles, U.; Englisch, U.; *Nucl. Acids Res.* **1989**, 17, 285.
36. Pieles, U.; Sproat, B. S.; Neuner, P.; Cramer, F.; *Nucl. Acids Res.* **1989**, 17, 8967.
37. Lee, B. L.; Blake, K. R.; Miller, P. S.; *Nucl. Acids Res.* **1988**, 16, 10681.
38. Lee, B. L.; Murakami, A.; Blake, K. R.; Lin, S.-B.; and Miller, P. S.; *Biochemistry* **1988**, 27, 3197.
39. Le Doan, T.; Praesuth, D.; Perrouault, L.; Chassignol, M.; Thuong, N. T. ; Helene, C.; *Bioconjugate Chem.* **1990**, 1, 108.
40. Cech, T. R.; Bass, B. L.; *Ann. Rev. Biochem.* **1986**, 55, 599.
41. Rossi, J. J.; Cantin, E. M.; Sarver, N.; and Chang, P. F.; *Pharmac. Ther.* **1991**, 50, 245.
42. Li, Y.; Guerrier-Takada, C.; Altman, S.; *Proc. Natl. Acad. Sci. USA* **1992**, 89, 3185.
43. Yuan, Y.; Hwang, E. ; Altman, S.; *Proc. Natl. Acad. Sci. USA* 1992, 89, 8006.
44. Yuan, Y.; Altman, S.; Science 1994, 263, 1269.
45. Corey, D. R.; Schultz, P. G.; *Science* **1987**, 238, 1401.
46. Zuckerman, R. N.; Corey, D. R.; Schultz, P. G.; *J. Am. Chem. Soc.* **1988**, 110, 1614.
47. Corey, D. R.; Pei, D.; Schultz, P. G.; *Biochemistry* **1989**, 28, 8277.
48. Zuckerman, R. N.; Schultz, P. G.; *Proc. Natl. Acad. Sci. USA* **1989**, 86, 1766.
49. Corey, D. R.; Pei, D.; Schultz, P. G.; *J. Am. Chem. Soc.* **1989**, 111, 8523.
50. Kaiser, E. T.; Lawrence, D. S.; *Science* **1984**, **226**, 505.
51. Torrence, P. F.; Maitra, R. K.; Lesiak, K.; Khamnei, S.; Zhou, A.; Silverman, R. H.; *Proc Natl. Acad. Sci. USA* **1993**, 90, 1300.
52. Lesiak, K.; Khamnei, S; Torrence, P. F.; *Bioconjugate Chem.* **1993**, 4, 467.

53. Johnston, M. I.; Torrence, P. F.; in *Interferon, Mechanisms of Production and Action,* v.3, ed. R. M. Friedman, Elsevier Science, Amsterdam, 1984, p. 189.
54. Justesen, J.; Ferbus, D.; Thang, M. N.; *Proc. Natl. Acad. Sci., USA* **1980**, 77, 4618.
55. Ferbus, D.; Justesen, J.; Besancon, F.; Thang, M. N.; *Biochem. Biophys. Res. Comm.*. **1981**, 100, 847.
56. Minks, M. A.; Benvin, S.; Baglioni, C.; *J. Biol. Chem.* **1980**, 255, 5031.
57. Ball, L. A.; *Ann, N. Y. Acad. Sci.* **1980**, 350, 486.
58. Hovanessian, A. G.; *J. Interferon Res.* **1991**, 11, 199.
59. Ilson, D. H.; Torrence, P. F.; Vilcek, J.; *J. Interferon Res.* **1986**, 6, 5.
60. Chebath, J.; Benech, P.; Hovanessian, A. G.; Galabru, J.; Revel. M.; *J. Biol. Chem.* **1987**, 262, 3852.
61. Willams, B. R. G.; Saunders, M. E.; Willard, H. F.; *Somat. Cell. Mol. Genet.* **1986**, 12, 403.
62. Saunders, M. E.; Gewert, D. R.; Tugwell, M. E.; McMahon, M.; Williams, B. R. G.; *EMBO J.* **1985**, 4, 1761.
63. Ichii, Y.; Fukunaga, R.; Shiojiri, s.; Sokawa, Y.; *Nucl. Acids. Res.* **1986**, 14, 10117.
64. Ghosh, S. K.; Kusari, J.; Bandyopadhyay, S. K.; Samanta, H.; Kumar, R.; Sen, G.; *J. Biol. Chem.* **1991**, 266, 15293.
65. Schmidt, A.; Zilberstein, A.; Shulman, L.; federman, P.; Berissi, H.; Revel, M.; *FEBS Lett.* **1978**, 95, 257.
66. Schmidt, A.; Chernajovsky, Y.; Shulman, L.; Federman, P.; Berissi, H.; Revel, M.; *Proc. Natl. Acad. Sci. USA* **1979**, 76, 4788.
67. Johnston, M. I.; Hearl, W. G.; *J. Biol. Chem.* **1987**, 262, 8377.
68. Defilippi, P.; Huez, G.; Verhaegen-Lewalle, M.; De Clercq, E., Imai, J.; Torrence, P. F.; Content, J.; *FEBS Lett.* **1986**,198, 326.
69. Imai, J.; Johnston, M. I.; Torrence, P. F.; *J. Biol. Chem.* **1982**, 257, 12739.
70. Silverman, R. H.; Jung, D. D.; Nolan-Sorden, N. L.; Dieffenbach, C. W.; Kedar, V. P.; SenGupta, D. N.; *J. Biol. Chem.* **1988**, 263, 7336.
71. Zhou, A.; Hassel, B. A.; Silverman, R. H.; *Cell* **1993**, 72, 753.
72. Squire, J.; Zhou, A.; Hassel, B. A.; Nie, H.; Silverman, R. H.; *Genomics* **1994**, 19, 174.
73. Wreschner, D. H.; McCauley, J. W.; Skehel, J. J.; Kerr, I. M.; *Nature* **1981**, 289, 414.
74. Floyd-Smith, G.; Slattery, E.; Lengyel, P.; *Science* **1981**, 212, 1020.
75. Williams, B. R. G.; Golgher, R. R.; Brown, R. E.; Gilbert, C. S.; Kerr, I. M.; *Nature* **1979**, 282, 582.
76. Wreschner, D. H.; James, T. C.; Silverman, R. H.; Kerr, I. M.; *Nucl. Acids Res.* **1981**, 9, 1571.
77. Silverman, R. H.; Skehel, J. J.; James, T. C.; Wreschner, D. H.; Kerr, I. M.; *J. Virol.* **46**, 1051.
78. Nilsen, T. W.; Maroney, P. A.; Baglioni, C.; *J. Biol. Chem.* **1981**, 256, 7806.
79. Nilsen, T. W.; Maroney, P. A.; Baglioni, C.; *J. Virol.* **1982**, 42, 1039.
80. Silverman, R. H.; Cayley, P. J.; Knight, M.; Gilgert, C. S.; Kerr. I. M.; *Eur. J. Biochem.* 1982, 124, 131.
81. Watling, D.; Serafinowska, H. T.; Reeses, C. B.; Kerr, I. M.; *EMBO J.* **1985**, 4, 431.
82. Torrence, P. F.; Imai, J.; Johnston, M. I; *Proc. Natl. Acad. Sci. USA* **1981**, 78, 5993.
83. Chebath, J.; Benech, P.; Revel, M.; Vigneron, M.; *Nature* **1987**, 330, 587.
84. Gribaudo, G.; Lembo, D.; Cavallo, G.; Landolfo, S.; Lengyel, P.; *J. Virol.* **1991**, 65, 1748.
85. Torrence, P. F.; Xiao, W.; Li, G.; Khamnei, S.; *Current Med. Chem.*, in press.
86.Torrence, P. F.; Imai, J.; Lesiak, K.; Jamoulle, J.-C.; and Sawai, H.; *J. Med. Chem.* **1984**, 27, 726.

87. Imai, J.; Lesiak, K.; Torrence, P. F.; *J. Biol. Chem.* **1985**, 260, 1390.
88. Clemens, M. J., Hershey, J. W. B., Hovanessian, A. C., Jacobs, B. C., Katz, M. G., Kaufman, R. J., Lengyel, P., Samuel, C. E., Sen, G., and Williams, B. R. G. (1993) J. Interferon Res. **13**, 241.
89. Hovanesssian, A. G.; *J. Interferon. Res.* **1989**, 9, 641.
90. Meurs, E. F.; Galabru, J.; Barber, G. N.; Katze, M. G.; Hovanessian, A.; *Proc. Natl. Acad. Sci. USA* **1993**, 90, 232.
91. Koromilas, A. E.; Roy, S.; Barber, G. N.; Katze, M. G.; Sonnenberg, N.; *Science* **1992**, 257, 1685.
92. Safer, B.; *Cell* **1983**, 33, 7.
93. Meurs, E. F.; Watanabe, Y.; Kaderiet, S.; Barber, G. N.; Katze, M. G.; Chong, K.; Williams, B. R. G.; Hovanessian, A.; *Virology* **1992**, 66, 5805.
94. Krause, D.; Silverman, R. H.; *J. Interferon Res.* **1993**,13,13.
95. Krause, D.; Silverman, R. H.; Jacobsen, H.; Leisy, S. A.; Dieffenbach, C. W.; Friedman, R. M.; Eur. J. Biochem. 1985, 146, 611.
96. Hassel, B. A.; Zhou, A.; Sotomayor, C.; Maran, A.; Silverman, R. H.; EMBO J. **1993**, 12, 3297

RECEIVED July 19, 1994

Chapter 9

Branched Nucleic Acids
Synthesis and Biological Applications

R. H. E. Hudson, K. Ganeshan, and M. J. Damha[1]

Department of Chemistry, University of Toronto, Mississauga,
Ontario L5L 1C6, Canada

A general method for the chemical synthesis of branched oligoribonucleic acids (bRNA) has been developed. The key features of the method are the use of the solid-phase "silyl-phosphoramidite" procedure to assemble linear oligoribonucleotide chains and the use of a nucleoside 2',3'-bis(O-phosphoramidite) reagent to join support-bound oligonucleotide chains. The synthesis of several "forked" RNA oligomers similar in structure to nuclear mRNA splicing intermediates is described. The method has also been adapted to the synthesis of novel biopolymer structures including a "dendritic" DNA molecule containing seven branch points and twelve termini at the periphery of the molecule. The usefulness of synthetic branched oligonucleotides as tools for studying small nuclear RNA:bRNA complexes that occur in the cell's nucleus, and as potential antisense agents for the control of gene expression is demonstrated.

Natural Occurrence

More than four decades ago Sir Alexander R. Todd and coworkers proposed that ribonucleic acids (RNA) contained "side chains" or "branches" and that these occurred at frequent intervals along the polynucleotide chain. They also speculated that the branches were short and were attached to the main chain at the C2' positions. These suggestions rested on the known chemical and nuclease lability of ribonucleic acids but no definite evidence was available to support them. In the concluding remarks of their 1952 publication (*1*) they noted:

[1]Corresponding author, current address: Department of Chemistry, McGill University, Montréal PQ H3A 2K6, Canada

0097–6156/94/0580–0133$08.00/0

"There can be no question of finality about any nucleic acid structure at the present time, since it is clear that there is no available method for determining the nucleotide sequence.   ...Nevertheless, we feel it desirable to indicate that (these branched) structures are capable of explaining a surprising number of the known facts about the hydrolytic behavior of ribonucleic acids, without invoking any linkages between nucleoside units other than simple phosphate groups between hydroxyl groups in the ribofuranose residues.   At present they represent a convenient working hypothesis which we hope to test in further synthetic studies now in progress."

Although it was later revealed that ribonucleic acids are not branched in structure, Todd was correct in proposing that some forms of RNA do possess this unusual architecture.   The first evidence was provided in 1983 by Wallace and Edmonds, who observed that nuclear RNA from HeLa cells contains branched molecules.   They speculated that these molecules were intermediates or products of messenger RNA splicing reactions (for a review, see 2).   The properties and structures of these bRNAs were subsequently studied by generating these molecules *in vitro* using HeLa cell extracts (3-6).

RNA Splicing.   In 1993 Phillip Sharp and Richard Roberts shared the Nobel prize in Physiology or Medicine for their research that revealed genes are often interrupted by non-coding sequences or introns.   The introns are transcribed as part of the precursor mRNA and subsequently deleted in a cleavage-ligation process termed "RNA splicing".   As indicated in Figure 1, nuclear pre-mRNA splicing is accomplished by two transesterification reactions (5, 7-9).   In the first step, the pre-mRNA is cleaved at the 5'-exon(E1)-intron(IVS) junction, and a 2',5'-phosphodiester bond is formed between the phosphate at the 5'-end of the intron and the ribose 2'-hydroxyl of an internal adenosine nucleoside.   In the second step, the RNA lariat intermediate (E1-IVS) is cleaved at the 3'-intron-exon junction, and the two exons are ligated to form the mature mRNA (E1-E2).   The internal sequence (IVS) is released in the unusual configuration of a circle with a tail containing a branch point (IVS-E2 or IVS), referred to as a "lariat" or "branched circle".   The entire cleavage-ligation process takes place within a multicomponent protein-RNA complex termed "the spliceosome", which consists of small nuclear ribonucleoprotein particles (snRNPs or "snurps").   The RNA components of these snRNPs are the small nuclear RNAs (snRNAs) U1, U2, U4, U5 and U6, all of which are directly involved in the splicing reaction (10).   U1, U2 and U5 bind to specific regions of the pre-mRNA (U1 binds to the 5'-splice site, U2 to the branch site, and U5 to the 3'-splice site) and U4 and U6 help assemble the spliceosome.   The snRNAs are believed to catalyze the cleavage/ligation steps whereas the protein components of the spliceosome serve as "scaffolding", *i.e.*, they bring the snRNAs and pre-mRNA into the correct conformation and proximity for the transesterification reactions to take place.

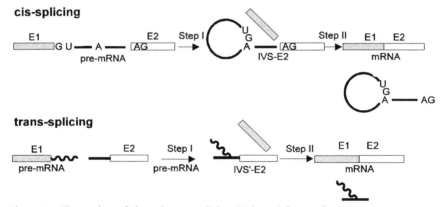

Figure 1: Illustration of cis and trans splicing (Adapted from refs. *2,12.*).

Group II introns, which are generally found in pre-mRNAs of mitochondria and chloroplasts, undergo splicing by the same lariat-formation mechanism as the nuclear introns. However, they are self-splicing, *i.e.*, splicing does not require the action of specialized RNA-protein complexes (*11*).

Nuclear intron "lariat" species have also been discovered in yeast cells (*3,4*). Non-circular branched structures such as a forked RNA branch or "Y"-structure have also been implicated in trans-splicing reactions of trypanosomal mRNA (Figure 1, for a review on trans-splicing see ref. *12*). The branch point in these species is also formed through a specific 2',5'-phosphodiester bond. An unusual satellite DNA called "multi-copy single-stranded DNA" (msDNA, Figure 2B) has also been discovered in *Myxococcus xanthus*, a Gram-negative bacterium living in soil (*13*). This unusual molecule consists of a 162-base single stranded DNA, the 5'-end of which is linked to a RNA of 77 bases by a 2,5'-phosphodiester linkage at the 2' position of an internal rG residue. Highly homologous msDNAs were subsequently discovered in other myxobacteria (*14*) and *E. coli* (*15*). Thus, a common structural feature of branched nucleic acids is the vicinal 2',5' and 3',5'-phosphodiester linkages at their branch point (Figure 2A). The branch core of RNA lariats is invariably ...5'...PuA(2'p5'G...)3'p5'Py...3'... where Pu stands for a purine and Py for a pyrimidine nucleoside.

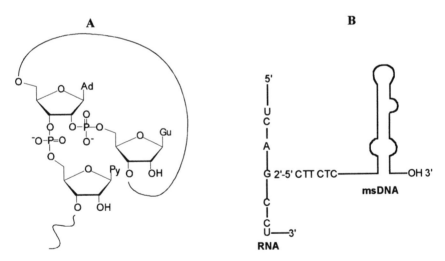

Figure 2: (A) The branched trinucleotide as part of the intron-lariat. (B) Branched core of msDNA, a chimera of RNA and DNA.

The discovery of bRNA has raised several important questions. What is the biological role of this unique nucleic acid? How does it interact with the spliceosomal components that recognize it? What are its conformational properties? Is there a unique three-dimensional structure at the branch point of RNA branches? Does this three-dimensional structure play an important role in pre-mRNA splicing? We believe that considerable insight can be obtained by chemically preparing bRNA fragments of defined structure and studying these molecules by spectroscopy and molecular biology techniques. The chemical synthesis of nucleic acids has led to several exciting discoveries including Z-DNA, G-tetraplexes and the development of the antisense and antigene technology. The synthesis, physical studies, and biological assays of bRNA's will lead to a better knowledge of the relationship between bRNA structure and biological activity. The main scope of this chapter is therefore to show how branched oligonucleotides have been synthesized in our laboratory, to demonstrate their potential as probes for studying RNA splicing, and to investigate whether their unique architecture provides a basis to design novel antisense oligomers of possible therapeutic value.

Chemical Synthesis of Branched RNA.

Several chemical routes to branched oligoribonucleotides have been explored recently by both J.B. Chattopadhyaya and coworkers (Uppsala University, Sweden) and our research group. Work from Chattopadhyaya's laboratory has yielded short forked ("Y"-like) ribonucleotides and small branched circles (*16,17*). Due to space limitations, the examples of synthetic procedures given below are limited to presentation of methods developed in the authors' laboratory (*18-21*). For other examples, we would refer the reader to three recent reviews on bRNA synthesis by Chattopadhyaya (*22*), Beaucage (*23*), and Damha and coworkers (*21*).

Solid-Phase Branched Oligoribonucleotide Synthesis. The introduction of 2'-O-silylated ribonucleoside 3'-O-phosphoramidite derivatives by Ogilvie and coworkers (*24*) has led to the routine synthesis of oligoribonucleotides (up to 80 bases) in many laboratories. The success of this procedure has been due to the development of a remarkably efficient coupling reaction (phosphite triester chemistry, *25,26*) and the use of the *t*-butyldimethylsilyl protecting group for the 2'-position (*27*). Furthermore, the "silyl-phosphoramidite" method is fully compatible, in terms of equipment and reagents, with RNA synthesis permitting the use of commercially available DNA synthesizers. Our solid-phase synthesis of branched RNA oligomers is similar to conventional procedures for preparing linear oligoribonucleotides, but with one key modification. Following oligonucleotide synthesis in the normal way, a ribonucleoside 2',3'-O-bisphosphoramidite synthon is used to join the 5'-ends of neighboring nucleotide chains, Figure 3.

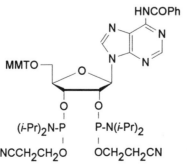

Figure 3: Key branching synthon: N^6-benzoyl-5'-(monomethoxytrityl)adenosine-2',3'-O-bis(2-cyanoethyl N,N-diisopropylphosphoramidite).

The standard tetrazole activation is used to carry out this step. The resulting vicinal 2',5' and 3',5'-phosphite-triester linkages are then oxidized with iodine/water to the desired phosphotriesters (Figure 4). In the next step, unreacted linear chains are capped by acylation with acetic anhydride. Further addition of building blocks to the vertex of the "V"-molecules generates forked (or "Y") molecules having similar structures to trans-splicing intermediates (Figure 4, c.f. Figure 1). Deprotection, purification and handling of these molecules is carried out in the conventional way (28). The ease of preparation of the bisphosphoramidite reagent (19) and the ready availability of silylated ribonucleoside phosphoramidite reagents from commercial sources allows anyone to synthesize a branched RNA sequence.

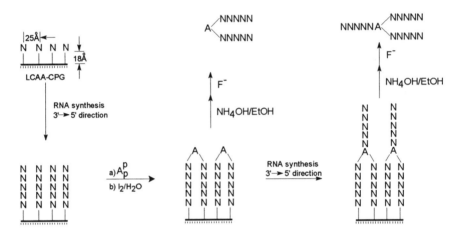

Figure 4: Synthetic methodology for preparing the "V" and "Y"-shaped molecules. A^P_P represents N^6-benzoyl-5'-(monomethoxytrityl)adenosine-2',3'-O-bis(2-cyanoethyl-N,N-diisopropylphosphoramidite).

As expected, the efficiency of the branching reaction is dependent on both the degree of derivatization of the solid support and the solution concentration of the bis(phosphoramidite) reagent. Highly substituted solid supports provide the best results since they ensure an appropriate distance between the reactive (5'-OH) end groups of the immobilized oligonucleotide chains. Conventional long-chain alkylamine controlled-pore glass (LCAA-CPG) is well suited for the branched RNA synthesis since it can be highly functionalized by procedures developed recently by Pon and coworkers (29) and our research group (30). Supports with a nucleoside "loading" of at least 20 μmol nucleoside/g of support should be used, 30-50 μmol/g giving the best results (commercial samples fall generally in the 20-40 μmol/g range). Low concentrations of the bisphosphoramidite reagent should be employed in the branching step. Typically, the bisphosphoramidite reagent is used as a 0.02-0.03 M acetonitrile solution which is directly attached to the fifth port of the synthesizer. If higher concentrations are used, yields of branched products are less, and mainly "extended" isomeric side products are generated, i.e., $A(2'p)3'p[(Np)_nN]$ and $A[2'p(Np)_nN]3'p$.

The standard 2'-silylated ribonucleoside 3'-phosphoramidite synthons are used as 0.10-0.15 M solutions.

The efficiency of the branching reaction can be assessed during synthesis by quantitation of the trityl cations released immediately before and after the branching reaction. Theoretically, 100% branch formation should give an "apparent coupling yield" of 50% since the trityl absorption (yield) following branching should be half of the previous absorption (during branching two nucleotide chains become joined to a single bisphosphoramidite synthon). Thus trityl yields significantly greater or lower than 50% indicate little branch formation (*e.g.*, 80% trityl yield indicates the formation of mainly the extended isomeric compounds $A(2'p)3'p[Np]_nN$ and $A[2'p(Np)_nN]3'p$). In a typical synthesis, *e.g.*, that of A(2'pUpUpUpUpU)3'pUpUpUpUpU, the branching step produces only 50% of the desired branched molecule. The regioisomeric hexamers 5'-MMT-A(2'p)3'pUpUpUpUpU and 5'-MMT-A(2'pUpUpUpUpU)3'p combined account for 25% (which also contribute to the trityl absorption), and the remaining 25% is the unreacted linear oligomer $5'-HO-(Np)_nN$ (which does not contribute). The overall effect of this is a 50% "coupling yield". Following deprotection, the desired branched molecule is easily identified as an oligonucleotide that migrates more slowly than the side-products by polyacrylamide gel electrophoresis. Overall yields of branched products after PAGE purification and a desalting step are generally very good. A typical 1 μmol synthesis of A(2'pUpUpUpUpU)3'pUpUpUpUpU yielded 54 A_{260} units of crude mixture, and 13 A_{260} units (25%) of the branched molecule after purification. This is sufficient for most biological and thermal denaturation analyses (*vide infra*). A large array of "V" and "Y" branched molecules of mixed base composition have been synthesized in this way, some related to mRNA splicing intermediates (*21*). More recently, hyperbranched (or "dendritic") molecules have been assembled by repeating the chain elongation and branching steps several times, Figure 5 (*31*). A number of chemical and enzymatic methods have been used to characterize our synthetic branched oligomers, including NMR (*19*), time-of-flight secondary ion mass spectrometry (*32,33*), and nuclease digestion (*21,34*).

Figure 5: A third generation "dendron" consisting of 87 nucleotide units with seven branch points.

Solid-Phase Synthesis of Regioisomeric Oligoribonucleotides. While the above synthetic strategy permits the rapid, high-yield preparation of branched sequences, it is clear that a major synthetic concern remains: the introduction of 2' and 3'-chains of different nucleotide composition at the branched adenosine cannot be accomplished with our present methodology. Early work clearly demonstrated that the stepwise assembly of the 2',5' and 3',5'-internucleotide linkages *via* phosphate triesters was not a straightforward matter since cyclic phosphates may be formed (*18,19,22*). This hurdle has been overcome by using phosphodiester intermediates as they are many orders of magnitude more stable to nucleophilic attack than the neutral derivatives of phosphotriesters (*22,35*). These syntheses however were carried out in solution and hence are only practical for the synthesis of short oligonucleotide sequences (<10 bases). Furthermore, they involve time-consuming chromatographic purification of charged phosphodiester products with a high consumption of reagents. By using a variation of our procedure, we have been able to assemble branched molecules containing different 2' and 3'-chain sequences. The key feature of this approach is the synthesis of two different chain sequences on the same support surface (Figure 6). This is followed by a branching step to generate four isomeric branched tridecaribonucleotides in very good yields (*ca.* 1:1:1:1 ratio). Following deblocking and deprotection, the oligomers were separated by capillary gel electrophoresis (*36*) and characterized by nucleoside and branched nucleotide composition analysis (Damha, M.J.; Ganeshan, K. *10th International Roundtable: Nucleosides and Nucleotides*, Park City, Utah, 09/16/1992). Although the method requires the separation of very similar branched molecules, it is the easiest method to prepare branched RNA with wild-type sequences.

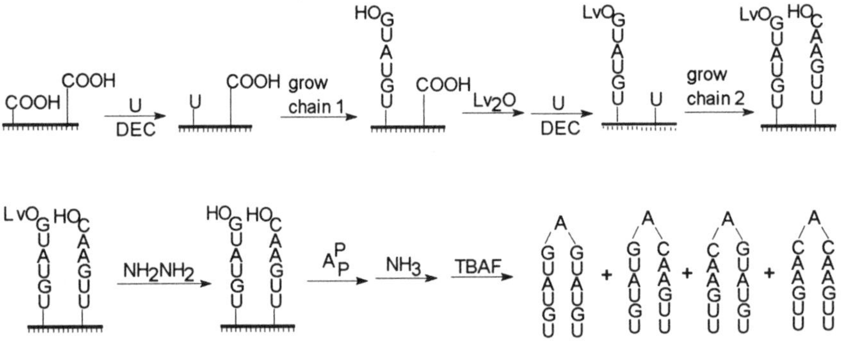

Figure 6: Synthetic methodology for the introduction of non-identical 2' and 3'-nucleotide sequences. Lv_2O, DEC and TBAF stand for levulinic anhydride, 1-(3-dimethylaminopropyl)-3-ethylcarbodiimide hydrochloride, and tetra(*n*-butyl) ammonium fluoride, respectively.

Applications of Branched Nucleic Acids.

Conformation of Branched RNA. A knowledge of the three dimensional structure of branched RNA oligomers is an essential requirement for the understanding of branched RNA recognition during mRNA splicing. NMR and circular dichroism have proven to be extremely useful in the investigation of short branched oligomers due to their sensitivity to small structural, conformational and environmental variations. Coupling constant data has provided information about the geometrical details of the phosphate and sugar rings. Furthermore, nuclear Overhauser effects and chemical shift data has been used to monitor intramolecular base stacking interactions (*37,38,39* and references therein). These studies have revealed that branched oligoribonucleotides adopt a unique structure in aqueous solution. The adenine ring of the branching nucleotide forms a base-base stack with the guanosine of the 2'-nucleotidyl unit (2'-5' stacking). This is true even for branched nucleotides containing a 2'-pyrimidine (C or U) rather than the natural 2'-guanine (*38,39*). The furanose ring of the branched adenosine nucleoside when linked to a guanosine via the 2',5'-linkage [*e.g.*, A(2'G)3'U] shows a lower preference (lower conformational purity) for the C2'-endo pucker conformation than an adenosine linked to a pyrimidine nucleoside through the same type of bond [*e.g.*, A(2'U)3'G]. This indicates a correlation between the sequence around the branch point and ribose conformational equilibrium (*38*).

In general, the residue attached to the branched adenosine residue via the 3',5'-linkage (3'-nucleotidyl residue) is "free" from the influence of the adenine and 2'-heterocyclic bases. The (2'-5') stacking is stronger for "V" oligomers than for "Y" oligomers, implying that the nucleotide immediately adjacent (5') to the adenosine residue promotes a "5'-conformational transmission effect" (*39*) along the 2'-5' arm in the branched molecules. This is in contrast to linear oligoribonucleotides since a "conformational transmission effect" can only occur along the 3'-5' axis. These studies suggest that adenine(2'-5')guanine stacking, if it exists in naturally occurring RNA lariats and forks, may serve as a conformational distortion point by altering the local RNA geometry of a normal-stacked structure (*39*). Such a local conformational change may itself serve as a recognition signal for late events in the splicing reaction (*i.e.*, snRNA binding/dissociation; 3'-exon cleavage and exon ligation).

Probing Small Nuclear RNA-Branched RNA Interactions. Genetic and biochemical analyses have established that the U2 snRNA binds to the branch region of yeast and mammalian pre-mRNAs (*40-43*). This Watson-Crick interaction displaces ("bulges") and presumably activates the adenosine that forms the branch point through the 2',5'-phosphodiester linkage, Figure 7. This nucleotide must be unpaired for proper function since insertion of a complementary base residue on U2 snRNA, which restores base-pairing, completely abolishes splicing (*44*). Thus the exquisite reactivity of the branch point adenosine residue can be explained by the particular conformation of the native snRNA U2:pre-mRNA complex.

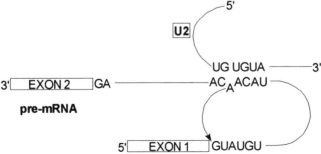

Figure 7: "Bulged" branch point nucleotide model during the interaction between U2 snRNA and pre-mRNA (Adapted from ref. *41*.).

Recently, Parker and Siliciano (45) proposed that the branch point adenosine remains "bulged out" following formation of the 2',5'-phosphodiester bond. One may be able to provide direct evidence for these interactions by studying synthetic bRNA-RNA complexes in solution. The first question we asked ourselves was,

Can bulged branch points stabilize bRNA/RNA complexes?

Preliminary thermal denaturation data obtained in our laboratory are not consistent with such a role. For example, we were unable to detect complex formation between **1** (X = A or C) and **2**. This, of course, may be due to the particular experimental conditions we chose. Other factors such as spliceosomal proteins, folding of the pre-mRNA, and non Watson-Crick base pairing interactions (4) may play important stabilizing roles *in vivo*. Interestingly, insertion of a complementary base residue on the linear target (*i.e.*, **3**), which allows for an additional base pair, results in duplex formation (X·Y = A·U or C·G, T_m = 48 °C). These complexes appear to be less stable than the corresponding linear complexes **4:5** (X·Y = A·U, T_m = 58 °C; X·Y = C·G, T_m = 60 °C) which may be due to helical distortions created by the dangling (2' or 3') tail. The interaction between the base of the branched residue (A or C) and its complementary base is confirmed by the observed lowering of the transition temperature (ΔT_m = 7-11 °C) in duplexes containing A/G and C/U mismatches (**1:3**, X·Y = C·U and A·G), see Figure 8. Clearly, the presence of a 2',5'-phosphodiester linkage destabilizes bRNA:RNA complexes but does not prevent Watson-Crick base pairing interactions around the branch point region. It would now be instructional to determine whether the adenine(2'-5')guanine stacking, so prevalent in "free" branched oligonucleotides, is maintained in bRNA:RNA complexes.

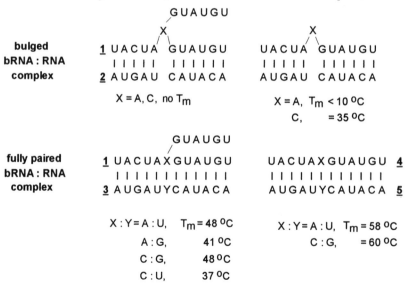

Figure 8: Modeling the pre-mRNA:U2-snRNA interaction. T_m's were determined by the maximum in the plot of the first derivative of absorbance *vs.* temperature. Values reported for T_m's are for the conditions of 1M NaCl, 10 mM $PO_4{}^{3-}$, and pH = 7.0.

Developing Antisense Strategies Based on Branched Nucleic Acids. The ability of single-stranded (ss) DNA or RNA to recognize and bind other single-stranded nucleic acids, in a sequence specific manner, is a fundamental biophysical characteristic of nucleic acids. Complementary nucleic acid sequences associate to form a duplex molecule. Double-stranded (ds) complexes can also be recognized by ssDNA or ssRNA wherein a third strand binds to the duplex through the major groove to form a triple-helix or triplex.

These types of interactions have become much more familiar to the chemist in recent years due to the developments in oligonucleotide synthesis (*vide supra*). Not only are "natural" oligonucleotides synthesized but a great variety of methods have also been developed for the preparation of many unnatural nucleic acid analogs (*46-48*). The retention of the fidelity of interaction of nucleic acids becomes an important consideration once the structures are chemically modified. Most frequently, the chemical constitution of the phosphate-carbohydrate backbone is the site for modification since the heterocyclic bases are key to the specificity of oligonucleotide interaction. Generally, these modified nucleic acids are linear polymers. Another form of modification involves changing the primary linear structure to that of a circular or branched polymer. "Comb"-like nucleic acids have been prepared by a combination of a linear synthesis and subsequent orthogonal synthesis involving branching off the nucleobase. These novel nucleic acids have found use as "amplification multimers" in bioassays targeting pathogens such as Hepatitis B (*49,50* and references therein). Branched geometric structures composed of catenated DNA or through the noncovalent association of several strands of DNA have also been synthesized (*51,52*).

The formation of triple helices have become an area of great interest to chemists for their possible role in the natural (*53-55*) and artificial regulation of gene expression (for reviews see: *56-62*), or for use in analytical, diagnostic or synthetic methods (for example: *63-66*). Much work has begun to focus on the physical and chemical requirements for triplex formation, yet the conditions required are still not fully elucidated (*67-71*).

Our interest in branched nucleic acids and recent investigations of nucleic acids as agents for control of gene expression, has led us to design branched oligonucleotides capable of forming triple helical complexes with single stranded nucleic acids. This strategy, which is an extension of the antisense approach (*46,56*), has so far been investigated with oligonucleotides that are linked at one end through a flexible linker (*72-75*) and with hairpin (*76*) and circular DNA (*77-79*). These analogues, including branched oligonucleotides, have an "entropic" advantage in recognizing substrates *via* triplex formation.

The Branched(poly-dT) Prototype System. The singly branched nucleic acids 6 and 7, represented below, were studied first due to their relative simplicity. Compound 6 is a "V"-shaped molecule, and can be regarded topologically as a linear yet kinked nucleic acid. Compound 7 has a well defined forked structure and is viewed as possessing a "Y" shape. The polarity of the T_{10} strands is identical being 5' → 3' from left to right, and is a consequence of the synthetic procedure.

$$\begin{array}{ll}
\text{2'-5'} & \text{2'-5'} \\
\quad\quad \nearrow \text{p-T-(p-T)}_8\text{-p-T-OH-3'} & \quad\quad \nearrow \text{p-T-(p-T)}_8\text{-p-T-OH-3'} \\
\text{5'-HO-}\underline{r}\text{A} & \text{5'-HO-T-(p-T)}_8\text{-p}\,\text{T-p-}\underline{r}\text{A} \\
\quad\quad \searrow \text{p-T-(p-T)}_8\text{-p-T-OH-3'} & \quad\quad \searrow \text{p-T-(p-T)}_8\text{-p-T-OH-3'} \\
\text{3'-5'} & \text{3'-5'}
\end{array}$$

6   **7**

Both compounds **6** and **7** bind linear complementary oligonucleotides as shown by cooperative transitions observed during UV-absorbance monitoring of thermal melting (melt curve or melt profile). The binding of complementary oligonucleotide by the branched DNA is at least as stable as that of linear oligodeoxynucleotides of similar composition and length as determined by the half dissociation temperatures (melting temperature; T_m), *vide infra*.

Association of 6 and dA$_{10}$. The melt curve for the complex formed between compound **6** and dA$_{10}$ is given in Figure 9A. The melting profile at 260 nm shows a monophasic, cooperative transition at 30 °C involving a hyperchromicity of 13%. This complex is of comparable stability to that of the duplex formed by T$_{10}$ and dA$_{10}$ (T_m = 29 °C) as inferred by the similar T_m's. However, the melting profile for **6**:dA$_{10}$ involves a significantly greater hyperchromic rise than for the melting of the T$_{10}$:dA$_{10}$ duplex. This may be attributed to the greater stacking interactions involved in the triple-stranded **6**:dA$_{10}$ complex as compared the the double stranded T$_{10}$:dA$_{10}$ complex.

Upon the addition of more than one mole-equivalent of linear complement, no further increase in hyperchromicity is observed. We attribute this to the ability of compound **6** to take up only one dA$_{10}$ molecule, leaving no other sites available for binding of additional complementary oligonucleotide. How could this occur, if the the branched DNA has two T$_{10}$ tails? It is possible that the branched molecule is sterically inhibited and simply cannot use both "arms" for hybridization. However, monitoring the same system at 284 nm (*80*) (Figure 9B) shows a cooperative transition of small but significant hyperchromicity which is consistent with triple-helix melting.

The stoichiometry of interaction of **6** with dA$_{10}$ has been independently verified by titrating the branched DNA with complement in the method of continuous variation (*81*), Figure 10. This method confirmed the stoichiometry of interaction to be 1:1, regardless of the direction of titration for compound **6** and dA$_{10}$.

Association of 7 and dA$_{10}$. The hybridization characteristics of compound **7** (Figure 11A) are quite similar to those exhibited by the "V"-shaped compound **6**. The dissociation of the complex formed by compound **7** and one equivalent of dA$_{10}$ shows a T_m of 39 °C with a hyperchromicity of 13%. The T_m is significantly greater than that of the duplex T$_{10}$:dA$_{10}$ (29 °C). The complex formed when dA$_{10}$:**7** are mixed in a 2:1 ratio, exhibits a T_m of 41°C and a hyperchromicity of 16%. When **7** is mixed with a greater molar excess of linear complement no further change in hyperchromicity or T_m is observed. This behavior is interpreted as compound **7** being saturated with two mole-equivalents of bound dA$_{10}$. These complexes also show a cooperative transition of significant hyperchromicity when monitored at 284 nm. The

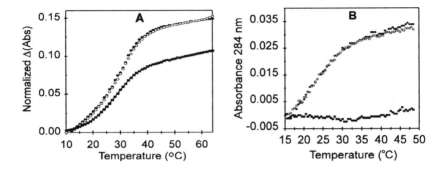

Figure 9: (A) Melting profile for the complex formed between 6 and dA_{10} (T_{10}:dA_{10}, 1:1, -■-; 6:dA_{10}, 1:1, -●-; 1:2, -o-) monitored at 260nm. Normalized change in absorbance was calculated by: $[(A_t - A_o)/(A_f - A_o)]$ where A_t = absorbance at any temperature; A_o = intial absorbance ; A_f = final absorbance. (B) Monitoring of melting at 284nm, symbols represent the same complexes as in A. Conditions for T_m determination are1M NaCl, 10mM PO_4^{3-}, pH = 7.0.

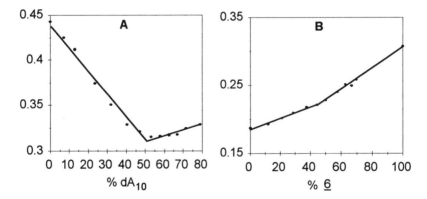

Figure 10: Determination of the stoichiometry of interaction for 6:dA_{10} by the method of continuous variation. The ordinate axis is in absorbance units at peak absorbance (*ca.* 260 nm). Conditions for stoichiometry determination are 50mM Mg^{2+}, 10mM PO_4^{3-}, pH = 7.0.

stoichiometry of association of $\underline{7}$ with dA_{10} was measured by the method of Job (81) and is shown in Figure 11B. The titration of compound $\underline{7}$ also indicated a 1:2 stoichiometry of complex formation with dA_{10}.

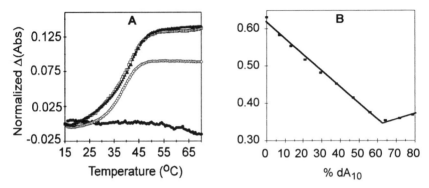

Figure 11: **(A)** Melt profiles for $\underline{7}$:dA_{10}: 1:0, -●-; 1:1, -○-; 1:2, -▲-; 1:3, -□-. Conditions for T_m determination are 1M NaCl, 10mM PO_4^{3-}, pH = 7.0. **(B)** Mixing curve for $\underline{7}$:dA_{10}, ordinate axis in absorbance units. Conditions for stoichiometry determination are 50mM Mg^{2+}, 10mM PO_4^{3-}, pH = 7.0.

Significance of Branched Poly(dT) Hybridization Patterns. The results of the branched(poly dT) hybridization with deoxyadenosine decamer can be rationalized in terms of triple helix formation. Similar systems of poly(dT)●poly(dA)∗poly(dT) have long been known to be able to exist in a triple helical configuration (82-84; in this representation "●" indicates Watson-Crick bonding and "∗" indicates Hoogsteen bonding, ref. 85).

Figure 12, below, illustrates the possible interactions of the "Y"-shaped molecule with dA_{10}. The modes of interaction account for the saturation of binding and also correlate with the evidence for triplex formation. Figure 13 details the hydrogen bonding scheme necessary for each triplex motif. Commonly, triplex dissociation occurs through a two step mechanism; dissociation of the triplex to a duplex and single strand followed by dissociation of the duplex to single strands. Each process may show a cooperative transition in the thermal melt profile when resolved by sufficiently differing T_m's. Alternatively, under certain circumstances, such as with the dissociation of the triplex formed between circular DNA and single stranded target (77-79), the dissociation is a one step process characterized by a single cooperative transition in the melt profile. Under the conditions used in this study, separate transitions for dissociative steps tsDNA $\rightarrow$ dsDNA $\rightarrow$ ssDNA were not observed. However, separate dissociations for tsDNA $\rightarrow$ dsDNA and dsDNA $\rightarrow$ ssDNA can observed for the complex formed between a 1:1 molar mixture of $\underline{7}$ and 5'-dA_{10}-T-dA_{10}-3' ($\underline{8}$) (data not shown). The melt profile displayed by this complex is biphasic, with the lower temperature transition nearly coincident with the transition given by the complex between compound $\underline{7}$ and dA_{10}. The higher temperature transition occurs at 54 °C, which is lower than the melt temperature for either the 1:1 complex between $\underline{8}$ and its complement (T_{10}-A-T_{10}; T_m = 65 °C) or for T_{19}:dA_{19} (T_m = 61 °C) (86). The two transitions displayed by the complex formed between $\underline{7}$ and $\underline{8}$ can be rationalized by initial duplex formation, followed by triplex formation with the third T_{10} arm of $\underline{7}$. The high temperature transition would correspond to the melting of the (imperfect) duplex, while the lower temperature transition (also seen when monitoring at 284 nm) indicates triplex melting. The two transitions are well separated, due in part to the difference in length of the duplex and triplex regions (*ca.* 21 *vs.* 10 base pairs). However, the potential length of the duplex and triplex involved in the complex of $\underline{7}$:dA_{10} are identical, so it is possible that the single transition observed in the melt profile represents a one step process from tsDNA → ssDNA.

Association of $\underline{6}$ and dA_{10}

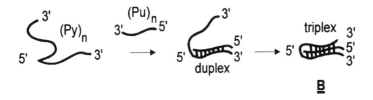

Association of $\underline{7}$ and dA_{10}

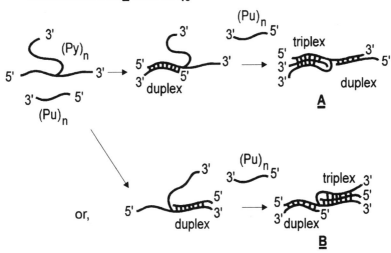

Figure 12: Possible interactions involved in the complexes conforming to the stoichiometry of 1:1, $\underline{6}$:dA_{10} (top) and 1:2, $\underline{7}$:dA_{10} (bottom). Note the different possibilities for third strand orientation are only possible in the "Y"-shaped molecule. Orientation **A** corresponds to the motif **A** in Figure 13, likewise orientation **B** corresponds to motif **B** (Adapted from ref. *88*).

A   B

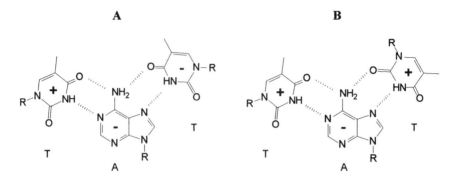

Figure 13: Configurations for the T-A-T triplex motifs. The (+) and (-) signs within the bases represent strand polarity (T = thymine; A = adenine; R = carbohydrate-phosphodiester backbone). The triplet on the left represents the more common motif wherein the third strand lies in the major groove parallel to the purine strand (Figure 12**A**). The representation on the right corresponds to Figure 12**B**, where the third strand of poly(T) lies antiparallel to the purine strand of the duplex (Adapted from ref. *87.*).

Expanding The Vocabulary. The versatility of branched nucleic acids to recognize single stranded target molecules may be expanded beyond the very specific example described in the preceding section. In the construction of the above molecules, an unmodified phosphodiester backbone was used. This arrangement could be elaborated for the possibility of recognizing more diverse substrates. Figure 14 shows a hypothetical branched nucleic acid, incorporating a flexible linker attached to the 2'-hydroxyl of a ribose moiety.

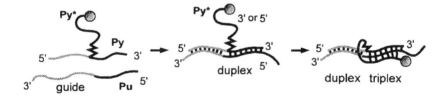

Figure 14: Design of second generation branched nucleic acid. The shaded segments represent a complementary "guide" sequence adjacent to a region targeted for triplex formation. The shaded circle represents conjugation to a nucleic acid modifying or cleavage agent.

The branched RNA molecules include three domains: a "guide" sequence containing any combination of the four naturally occurring bases, and two polypyrimidine sequences of opposite polarities (Py and Py*). The role of the "guide" and Py sequence is to capture the target sequence through specific hydrogen bonding (Watson-Crick), while the role of the Py* strand is to "fold" over and form the triplex. Also the triplex forming strand could be functionalized with intercalators to increase the stability of the complex or with modifying (or cleaving) agents to covalently alter the target nucleic acid (*75*). Because triple helix formation is also specified by base pairing (Hoogsteen-like), recognition of target (viral) RNA strand by branched sequences can be achieved with high affinity and without loss of specificity. Furthermore, in our system, the single-stranded target is recognized twice, once by the "guide" segment (duplex formation), and again by the third (branch) strand in triplex formation. Thus this strategy should maximize the binding affinity to the target, enabling lower dosing to achieve (antisense) inhibitory effects.

Thus this model system has produced intriguing preliminary results (*88*) and a rich and varied chemistry can be expected to follow.

Conclusions

A rapid, easy route to branched oligoribonucleotides has been developed. This method is based upon the solid-phase silyl-phosphoramidite method that is currently used worldwide for the assembly of normal RNA chains. Nucleic acids containing either single or multiple branch points have been prepared by this route, some with sequence similarity to the branch-region of naturally occurring "lariat" and "fork" RNA. A straightforward variation of the synthetic method allows for the synthesis of regioisomers differing in 2'-5' and 3'-5' nucleotide base composition. Triple helix forming branched poly(pyrimidine) oligonucleotides were also prepared.

The availability of branched RNA will allow the investigation of pre-mRNA splicing, lariat cleavage by the yeast debranching enzyme (Nam, K.; Hudson, R.H.E.; Chapman, K.B.; Ganeshan, K.; Damha, M.J.; Boeke, J.D. *J. Biol. Chem.*, **1994**, in press) and related phenomena. The capture of single-stranded target molecules *via* triple helix formation with branched nucleic acid analogues represents a new model for the antisense strategy, which we have termed the "branched" approach.

Further developments with this chemistry are anticipated to contribute much to our understanding of both the native role of branched RNA and the use of synthetic nucleic acids for the artifical manipulation of gene expression.

Acknowledgments

We thank the Natural Sciences and Engineering Research Council of Canada (NSERC), the Canadian Foundation for AIDS Reserach (CANFAR) and McGill Universtiy for generous support of this work. R.H.E.H. wishes to acknowledge NSERC and the University of Toronto for post-graduate scholarships. We also thank André Uddin and Ravinderjit Braich for critically reading this manuscript, and for helpful suggestions. Thanks to Tanya Tadey for capillary electrophoresis analysis of some of the branched nucleic acids that aided in the intrepetation of results presented in this chapter.

Literature Cited

(1)   Brown, D.M.; Todd, A.R. *J. Chem. Soc.* **1952**, 52.
(2)   Sharp, P.A. *Science* **1987**, *235*, 766.

(3)　Domdey, H.; Apostol, B.; Lin, R.; Newman, A.; Brody, E.; Abelson,J. *Cell*
　　　1984, *39*, 611.
(4)　Rodriguez, J.R.; Pikielny, C.W.; Rosbash, M. *Cell* **1984**, *39*, 603.
(5)　Ruskin,B.; Krainer, R.R.; Maniatis,T.; Green, M.R. *Cell* **1984**, *38*, 317.
(6)　Padgett, R.A., Grawoski, P.J.; Konarska, M.M.; Seiler, S.; Sharp, P.A. *Ann.*
　　　Rev. Biochem. **1986**, *55*, 1119.
(7)　Padgett, R.A.; Konarska, M.M.; Grabowski, P.J.; Hardy, S.F.; Sharp, P.A.
　　　Science **1984**, *225*, 898.
(8)　Konarska, M.M.; Grabowski, P.J.; Padgett, R.A.; Sharp, P.A. *Nature* **1985**, *313*,
　　　552.
(9)　Maschhoff, K.L.; Padgett, R.A. *Nucleic Acids Res.* **1993**, *21*, 5456.
(10)　Guthrie, C. *Science* **1991**, *253*, 157.
(11)　Perlman, P.S.; Peebles, C.L.; Daniels, C. In *Intervening Sequences in Evolution*
　　　and Development; Editors, Stone, E.M.; Schwartz, R.J., Oxford University
　　　Press, New York, New York, 1990, pp. 112-161.
(12)　Laird, P.W. *Trends Genet.* **1989**, *5*, 204.
(13)　Yee, T.; Furuichi, T.; Inouye, S. *Cell*, **1984**, *38*, 203.
(14)　Inouye, S.; Furuichi,T.; Dhundale, A; Inouye, M. In *Molecular Biology of RNA:*
　　　New Perspectives, Editors Inouye, M; Dudock, B.S., Academic Press, Inc. San
　　　Diego, California, 1987, pp. 271-284.
(15)　Lampson, B.C.; Sun, J.; Hsu, M.-Y.; Vallejo-Ramirez, J.; Inouye, S., Inouye, M.
　　　Science **1989**, *243*, 1033.
(16)　Zhou, X.X.; Remaud, G.; Chattopadhyaya, J. *Tetrahedron* **1988**, *44*, 6471.
(17)　Sund, C.; Agback, P.; Chattopadhyaya, J. *Tetrahedron* **1991**, *47*, 9659.
(18)　Damha, M.J.; Pon, R.T.; Ogilvie, K.K. *Tetrahedron Lett.* **1985**, *26*, 4839.
(19)　Damha, M.J.; Ogilvie, K.K. *J. Org. Chem.* **1988**, *53*, 3710.
(20)　Damha, M.J.; Zabarylo, S.V. *Tetrahedron Lett.* **1989**, *46*, 6295.
(21)　Damha, M.J.; Ganeshan, K., Hudson, R.H.E.; Zabarylo, S.V. *Nucleic Acids Res.*
　　　1992, *20*, 6565.
(22)　Sund, C.; Foldesi, A.; Yamakage, S.-I.; Chattopadhyaya, J. *Nucleic Acids Res.*
　　　Symp. Series **1991**, *24*, 9.
(23)　Beaucage, S.L.; Iyers, I.P. *Tetrahedron* **1993**, *49*, 10441.
(24)　Ogilvie, K.K.; Usman, N.; Nicoghosian, K; Cedergren, R.J. *Proc. Natl. Acad.*
　　　Sci. U.S.A. **1988**, *85*, 5764.
(25)　Letsinger, R.L.; Lunsford, W.B. *J. Am. Chem. Soc.* **1976**, *98*, 3655.
(26)　Beaucage, S.L.; Caruthers, M.H. *Tetrahedron Lett.* **1981**, *22*, 1859.
(27)　Ogilvie, K.K.; Sandana, K.L.; Thompson, E.A.; Quilliam, M.A.; Westmore, J.B.
　　　Tetrahedron Lett. **1974**, *15*, 2861.
(28)　Damha, M.J.; Ogilvie, K.K. In *Protocols for Oligonucleotides and Analogs -*
　　　Synthesis and Properties, Editor, S. Agrawal, *Methods in Molecular Biology*,
　　　The Humana Press, Inc. Totowa, N.J., 1993, *20*, 81-114.
(29)　Pon, R.T.; Usman, N.; Ogilvie, K.K. *BioTechniques* **1988**, *6*, 768.
(30)　Damha, M.J.; Giannaris, P.G.; Zabarylo, S.V. *Nucleic Acids Res.* **1990**, *13*,
　　　3813.
(31)　Hudson, R.H.E.; Damha, M.J. *J. Amer. Chem. Soc.* **1993**, *115*, 2119.

(32) Lafortune, F.; Standing, K.G.; Westmore, J.B.; Damha, M.J.; Ogilvie, K.K. *Org. Mass Spectrom.* **1988**, *23*, 228.

(33) Lafortune, F.; Damha, M.J.; Tang, X.; Standing, K.G.; Westmore, J.B.; Ogilvie, K.K. *Nucleosides and Nucleotides* **1990**, *9*, 445.

(34) Reilly, J.D.; Wallace, J.C.; Melhem, R.F.; Koop, D.W. Edmonds, M. In *RNA Processing, Part A, General Methods*, J.E. Dahlberg; J.N. Abelson, Ed.s; Methods in Enzymology; Academic Press, Inc., San Diego, California, 1989, pp 177.

(35) Kierzek, R.; Kopp, D.W.; Edmons, M.; Caruthers, M.H. *Nucleic Acids Res.* **1986**, *14*, 4751.

(36) Tadey, T.; Purdy, W.C. *J. Chromatogr.*, in press.

(37) Lee, M.; Huss, S.; Gosselin, G.; Imbach, J.-L.; Hartley, J.A.; Lown, J.W. *J. Biomol. Struct. Dyn.* **1987**, *5*, 651.

(38) Damha, M.J.; Ogilvie, K.K. *Biochemistry* **1988**, *27*, 6403.

(39) Sund, C.; Agback, P.; Koole, L.H.; Sandström. A; Chattopadhyaya, J. *Tetrahedron* **1992**, *48*, 695.

(40) Parker, R.; Siliciano, P.G.; Guthrie, C. *Cell* **1987**, *49*, 229.

(41) Guthrie, C. *Amer. Zool.* **1989**, *29*, 557.

(42) Zhuang,, Y.; Weiner, A.M. *Genes Dev.* **1989**, *3*, 1545.

(43) Wu, J.; Manley, J.L. *Genes Dev.* **1989**, *3*, 1553.

(44) Schmelzer, C.; Schwayen, R. *Cell* **1986**, *46*, 557.

(45) Parker R.; Siliciano, P.G. *Nature* **1993**, *361*, 660.

(46) Uhlmann, E.; Peyman, A. *Chem. Rev.* **1990**, *90*, 543.

(47) Englisch, U.; Gauss, D.H. *Angew. Chem. Int. Ed. Eng.* **1991**, *30*, 613.

(48) Goodchild, J. *Bioconjugate Chem.* **1990**, *1*,165.

(49) Horn, T.; Urdea, M.S. *Nucleic Acids Res.* **1989**, *17*, 6959.

(50) Chang, C-A.; Horn, T.; Ahle, D.; Urdea, M.S., *Nucleosides and Nucleotides* **1991**, *10*, 389.

(51) Seeman, N.C. *DNA and Cell Biol.* **1991**, *10*, 475.

(52) Chen, J.; Seeman, N.C. *Nature* **1991**, *350*, 631.

(53) Lyamichev, V.I; Mirkin, S.M.; Frank-Kamenetskii, M.D. *J. Biomol. Struct. Dyn.* **1986**, *3*, 667.

(54) Lyamichev, V.I; Mirkin, S.M.; Frank-Kamenetskii, M.D. *J. Biomol. Struct. Dyn.* **1987**, *5*, 275.

(55) Mirkin, S.M.; Lyamichev, V.I; Drushlyak, K.N.; Dobrynin, V.N.; Filippov, S.A.; Frank-Kamenetskii, M.D. *Nature* **1987**, *330*, 495.

(56) Thuong, N.T.; Hélène, C. *Angew. Chem. Int. Ed. Eng.* **1993**, *32*, 666.

(57) Hélène, C. *Anti-Cancer Drug Des.* **1991**, *6*, 569.

(58) Hélène, C.; Toulmé, J.-J. *Biochem. Biophys. Acta* **1990**, *1049*, 99.

(59) Maher III, L.J.; Dervan, P.B.; Wold, B. In *Prospects for Antisense Nucleic Acid Therapy of Cancer and AIDS*; E. Wickstrom, Ed., Wiley-LISS, Inc., New York, New York, 1991, 227-242.

(60) Gee, J.E.; Miller, D.M. *Med. Sci.* **1992**, *304*, 366.

(61) Maher III, L.J. *BioEssays*, **1992**, *14*, 807.

(62) Crooke, S.T. *Annu. Rev. Pharmacol. Toxicol*, **1992**, *32*, 329.

(63) Cooney, M.; Czernuszewicz, G.; Postel, E.H.; Flint, S.J.; Hogan, M.E. *Science,* **1988,** *241,* 456.
(64) Skoog, J.U; Maher III, L.J. *Nucleic Acids Res.* **1993,** *21,* 2131.
(65) Huamin, J.; Smith, L.M. *Anal. Chem.* **1993,** *65,* 1323.
(66) Luebke, K.J.; Dervan, P.B. *J. Amer. Chem. Soc.* **1989,** *111,* 8733.
(67) Roberts, R.W.; Crothers, D.M. *Science* **1992,** *258,* 1463.
(68) Wilson, W.D.; Hopkins, H.P.; Mizan, S.; Hamilton, D.D.; Zon, G. *J. Amer. Chem. Soc.* **1994,** *116,* 3607.
(69) Frossella, J.A.; Kin, Y.J.; Shih, H.; Richards, E.G.; Fresco, J.R. *Nucleic Acids Res.* **1993,** *21,* 4511.
(70) Cheng, A.-J.; Van Dyke, M.W *Nucleic Acids Res.* **1993,** *21,* 5630.
(71) Han, H.; Dervan, P.B. *Proc. Natl. Acad. Sci. U.S.A.* **1993,** *90,* 3806.
(72) Letsinger, R.L.; Chaturvedi, S.K.; Farooqui, F.; Salunkhe, M. *J. Amer. Chem. Soc.* **1993,** *115,* 7535.
(73) Salunkhe, M.; Wu, T.; Letsinger, R.L. *J. Amer. Chem. Soc.* **1992,** *114,* 8768.
(74) Giovannangeli, C.; Thuong, N.T.; Hélène, C. *Proc. Natl. Acad. Sci. U.S.A.* **1993,** *90,* 10013.
(75) Giovannangeli, C.; Montenay-Garestier, T.; Rougee, M.; Thuong, N.T.; Hélène, C. *J. Amer. Chem. Soc.* **1991,** *113,* 7775.
(76) Brossalina, E.; Pascolo, E.; Toulmé, J.-J. *Nucleic Acids Res.* **1993,** *21,* 5616.
(77) Prakash, G.; Kool, E.T. *J. Chem. Soc. Chem. Commun.* **1991,** 1161.
(78) Kool, E.T. *J. Amer. Chem. Soc.* **1991,** *113,* 6265.
(79) D'Souza, D.J.; Kool, E.T. *J. Biomol. Struct. Dyn.* **1992,** *10,* 141.
(80) Pilch, D.S.; Levenson, C.; Shafer, R.H. *Proc. Natl. Acad. Sci. U.S.A.* **1990,** *87,* 1942.
(81) Job, P. *Ann. Chim. (Paris)* **1928,** *9,* 113.
(82) Felsenfeld, G.; Davis, D.R.; Rich, A. *J. Amer. Chem. Soc.* **1957,** *79,* 2023.
(83) Arnott, S.; Selsing, E. *J. Mol. Biol.* **1974,** *88,* 509.
(84) Riley, M.; Maling B.; Chamberlin, M.J. *J. Mol. Biol.* **1966,** *20,* 359.
(85) Hoogsteen, K. *Acta Crystallogr.* **1963,** *16,* 907.
(86) Kibler-Herzog, L.; Zon, G.; Whittier, G.; Shaikh, M.; Wilson, W.D. *Anti-Cancer Drug Des.* **1993,** *8,* 65.
(87) Beal, P.A.; Dervan, P.B. *Science,* **1991,** *251,* 1360.
(88) Hudson, R.H.E.; Damha, M.D. *Nucleic Acids Res. Symp. Ser.* **1993,** *29,* 97.

RECEIVED June 16, 1994

MODIFICATIONS OF PHOSPHODIESTER LINKAGE

Chapter 10

Anti-Human Immunodeficiency Virus Activity and Mechanisms of Unmodified and Modified Antisense Oligonucleotides

T. Hatta, S.-G. Kim, S. Suzuki, K. Takaki, and H. Takaku

Department of Industrial Chemistry, Chiba Institute of Technology, Tsudanuma, Narashino, Chiba 275, Japan

We demonstrated that unmodified and modified (phosphorothioate) oligonucleotides prevent cDNA synthesis by AMV or HIV reverse transcriptases. Antisense oligonucleotide/RNA hybrids specifically arrest primer extension. The blockage involves the degradation of the RNA fragment bound to the antisense oligonucleotide by the reverse-transcriptase-associated RNase H activity. However, the phosphorothioate oligomer inhibited polymerization by binding to the AMV RT rather than to the template RNA; whereas, there was no competitive binding of the phosphorothioate oligomer on the HIV RT during reverse transcription. We also describe the anti-HIV activities of phosphorothioate oligonucleotides and point out some of the problems that still need to be solved.

Antisense RNA or antisense oligodeoxyribonucleotides within cells targeted toward the RNA transcript of a specific gene can inhibit the expression or promote the degradation of the transcript, resulting in suppression of the function coded for by the gene. The addition of chemically modified antisense oligomers to culture medium and their uptake by cells has been used to inhibit the expression of specific target genes (1-4). These compounds have been used as antisense inhibitors of gene expressions in various culture systems and are considered to be potential therapeutic agents (5,6).

Antisense oligonucleotides complementary to viral RNA inhibit viral replication in cells cultured with Rous sarcoma virus (7), Human immunodeficiency virus (8-11), Vesicular stomatitis virus (12,13), Herpes simplex virus (12-14), and Influenza virus (15,16).

However, the mechanism by which the antisense oligonucleotide inhibited retroviral protein synthesis, syncytia formation, and reverse transcriptase activity has not been fully elucidated. Recently, Toulmé et al. have reported that unmodified oligonucleotides indeed arrested cDNA synthesis by AMV and MMLV RTs, which have RNase H activity, but that α-oligonucleotide analogues did not (17,18).

On the other hand, Matsukura et al. reported that the inhibition of de novo infection by S-ODNs is both composition- and length-dependent. For example, homo-oligo S-dC28 is a better inhibitor than S-dC20 or S-ODNs (20-mer, coding exon I of art/trs gene in HIV) (9). However, S-dC28 did not inhibit the p24 gag expression in chronically infected T cells; whereas, S-ODNs complementary to the initiation sequence of HIV-rev inhibited the production of several viral proteins in chronically

HIV-infected T cells (*19*). On the other hand, Lisziewicz et al. (*20*) have reported that chemotherapy based on specifically targeted antisense oligonucleotide phosphorothioates is an effective means of reducing the viral burden in HIV-1 infected individuals at clinically achievable oligonucleotide concentrations. However, despite advances in AIDS therapy, there still remains the issue of how to select an effective target sequence and length for the phosphorothioate oligonucleotides. To define the dependence on both the target sequences and length of the phosphorothioate oligonucleotides of optimal anti-HIV activity, we synthesized phosphorothioate oligonucleotide analogues.

We present here a detailed analysis of the effect of unmodified and modified (phosphorothioate) oligodeoxyribonucleotides (*21*) on cDNA synthesis by AMV and HIV (*22*) RTs. We also describe the synthesis of one normal chain and the chain of either the whole-PS group or that capped with PS at both 3' and 5'-ends in sense, random, homo-oligomeric or antisense sequences with five target sites (gag, pol, rev, tat, and tar) within the HIV gene (Figure 1). The phosphorothioate oligonucleotides that are nuclease-resistant analogues of oligodeoxyribonucleotides can be used to prevent reverse transcription.

Mechanisms of the Inhibition of Reverse Transcription by Unmodified and Modified Antisense DNA

Materials. The unmodified oligonucleotide derivatives 5'-d[TTGTGTCAAAAGCAA -GT] [17 cap (n)], 5'-d[CACCAACTTCTTCCACA] [17 sc (n)], and 5'-d[TGCCCA-GGGCCTCAC] [15 sc (n)], and modified (phosphorothioate) oligodeoxyribonucleotide derivatives 5'-d[TsTsGsTsGsTsCsAsAsAsAsGsCsAsAsGsT] [17 cap (s)] and 5'd[Cs-AsCsCsAsAsCsTsTsCsTsTsCsCsAsCsA] [17 sc (s)] were synthesized on a Biosearch synthesizer. The oligonucleotide derivatives were purified by reverse phase HPLC on an oligo-DNA column. Purified oligomers were evaluated by resolving ^{32}P-labeled by electrophoresis samples on 20% polyacrylamide/7 M urea gels.

Methods. Rabbit ß-Globin RNA (50 ng, containing 0.3 pmol of intact ß-globin), primer (50 pmol), and the desired amount of antisense oligonucleotides were pre-incubated for 30 minutes at 39° C. After adding 1 ml of 10 x RT buffer (1M Tris/HCl, pH 8.3, 720 mM KCl, 100 mM MgCl$_2$, 100 mM dithiothreitol) containing 8 units of RNasin, 2 pmol of [α-^{32}P]dCTP (3000Ci/mmol; Ci= 37GBq; NEN), 5 nmol of the three dNTPs, and 2.5 nmol of dCTP, the volume of the mixture was adjusted to 10 μl with sterile water. AMV RT (1-10 units-i.e., 0.13-1.3 pmol) was then added. The reaction with HIV RT was allowed to proceed with 1 unit, incubated for 1 hour at 39°C. The cDNA was chloroform-extracted according to standard procedures and loaded on a 10% PAGE. The results obtained for cDNA synthesis were corrected for the number of labels incorporated into each fragment.

The cDNA Synthesis by AMV-RT Using Unmodified and Modified Antisense Oligonucleotides. Reverse transcription of rabbit ß-globin mRNA by AMV RT was primed with 17 sc, complementary to oligonucleotide 113-129 (Figure 2), giving rise to the predicted cDNA fragment of about 130 nucleotides (Figure 3a). In contrast, when the polymerization was performed in the presence of 17 cap (n) (0.05-2 μM), an oligonucleotide targeted to the cap region of the mRNA, a shortened DNA fragment, was synthesized at the expense of the full-length product. The size of the cDNA fragment corresponded to the distance between the primer and binding site of the antisense oligonucleotides (Figure 2). Therefore, the hybridization of this antisense oligonucleotide with the complementary sequence of the b-globin mRNA prevents transcription of this region. The inhibitory efficiency was dependent on the 17 cap (n) concentration. At concentrations as low as 1.0 μM, 17 cap (n), 96% inhibition of

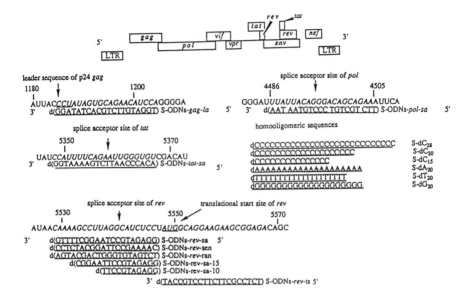

Figure 1. DNA Sequences of the Leader Site of Gag, the Splice Acceptor Site of Pol, Rev, or Tat, and the Translation Start Site of Rev Tested. The random sequence has the extract base content as antisense rev-sa but has <70% homology with any portion of the HIV sequence as antisense or sense. Homo-oligomers of the four bases (A, C, G, and T) were synthesized in four lengths (10, 15, 20, and 28 mer). The phosphorothioate derivatives are denoted by "S".

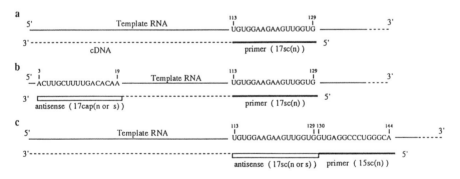

Figure 2. Inhibition of Reverse Transcription by Complementary Oligonucleotides. The full length cDNA products derived from 17 sc (n) or 15 sc (n) (a) could be blocked by the antisense oligomers 17 cap (n or s) and 17 sc (n or s) bound to a template sequence (b) or adjacent (c) to the primer-binding site.

reverse transcription was observed. This resulted from competition between the 17 sc (n) primer and the 17 cap (n) antisense oligomer.

Next, we examined the cDNA synthesis by AMV reverse transcriptase using theÅ@17 cap (s) phosphorothioate oligomer instead of 17 cap (n). Figure 3b shows that 17 cap (s) reduced the synthesis of the 110 nucleotide transcript; at concentrations of 1, 2, and 4 μM, the percentages of inhibition were, respectively, 55%, 74%, and 82%. However, these values are lower than that of the unmodified oligonucleotide at 1 μM. This suggested that part of the phosphorothioate oligomer binds to the RT enzyme. As a result, the synthesis of full length cDNA cannot be completely blocked by reduced binding of the phosphorothioate oligomer to mRNA. This process was essentially sequence independent and was the result of the preferential binding of the modified oligomers to the RT enzyme compared with unmodified oligomers.

To characterize the inhibitory process of the phosphorothioate oligomers, we incubated antisense oligomers 17 cap (n) or 17 cap (s) (2 μM) and the 17 sc (n) primer (5 μM) with AMV RT at concentrations of 1 or 10 units under the same conditions as described above. Smaller amounts of AMV RT were used because the polymerase activity aborts synthesis. The synthesis of cDNA with 17 cap (n) was inhibited by 1 unit of AMV RT. However, cDNA synthesis was significantly decreased when the modified oligomer 17 cap (s) was used (Figure 4). It should be noted that the cDNA could not be synthesized by a fall of ability of the polymerization due to the competitive binding of phosphorothioate oligomers on the RT enzyme. On the other hand, when the AMV RT concentration was 10 units, the short 123 nucleotide transcript was detected in all cases (Figure 4). This suggests a relationship between the RT enzyme and phosphorothioate oligomer and its inhibitory efficiency (9,23). It should be pointed out that the antisense phosphorothioate oligomer did not compete directly with the mRNA as the RT enzyme.

The cDNA Synthesis by HIV-RT Using Unmodified and Modified Antisense Oligonucleotides. To further characterize the inhibitory process of the phospho-rothioate oligomers, we examined how to prevent cDNA elongation by HIV RT instead of AMV RT under the same conditions as described above. Figure 5 shows that the antisense oligomers 17 cap (n) and 17 cap (s) induced the characteristic shortened cDNA fragments when the reaction proceeded with HIV RT. The 17 cap (s) at a concentration as low as 1.0 μM resulted in a shortened DNA fragment at the expense of the full-length product. This indicated that the phosphorothioate oligonucleotide inhibited the production of cDNA by being a retroviral polymerase bound to the RNA downstream from the primer. However, the inhibitory efficiency of a phosphorothioate oligonucleotide is influenced by its relationship with RT.

We speculated that an oligomer adjacent to the primer (Figure 2) could function by a different mechanism: the primer-antisense tandem may be viewed as a single complementary sequence by the priming RT molecule. To test this mechanism, we synthesized the unmodified [17 sc (n)] and the modified [17 sc (s)] oligonucleotides complementary to nucleotides 113-129 of rabbit ß-globin mRNA and unmodified 15 sc (n) complementary to nucleotides 130-144. Figure 6 shows that 17 sc (n) or 17 sc (s) reduced the synthesis of the 144 nucleotide long transcript; complete inhibition was achieved at <2 μM. This suggests that the oligomer complementary to a site immediately adjacent to the primer binding site could then act as a primer, which would be lengthened by polymerase. The inhibition involved degradation of the template by RNase H. However, the efficiency of an oligomer adjacent to the primer binding site was lower than that of an antisense oligomer bound to the 5'-end of mRNA. Furthermore, increased concentrations of the 17 sc (s) phosphorothioate oligomer in the reverse transcription mixture containing AMV RT (1 unit) led to the disappearance of the cDNA transcript. This result also suggested that the antisense

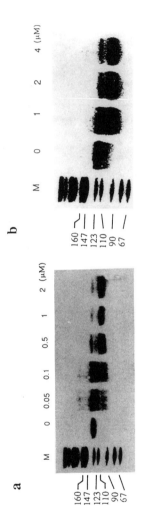

Figure 3. Effect of the Unmodified and Modified Antisense Oligonucleotides on DNA Synthesis. (a) Reverse transcription proceeded as indicated using 10 units of AMV RT primed by 5 μM oligomer 17 sc (n) without (0) or with various amounts (mM) of oligomer 17 cap (n). (b) cDNA analysis of the 17 cap (s) used in place of the 17 cap (n) in a. First lanes on left (M) in a and b correspond to DNA size markers.

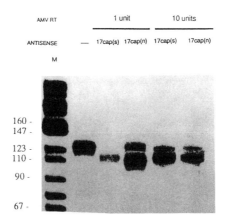

Figure 4. The Interaction Between the Antisense Oligomers and AMV RT on the Reverse Transcript. Analysis of cDNA fragments synthesized by 1 unit (left) or 10 units (right) of AMV RT. The reaction primed by a17 [17 sc (n) (5 µM)] complementary to the 113-129 oligonucleotide without (-) or with 2 µM antisense oligonucleotides, 17 cap (n) and 17 cap (s). DNA size markers were run in lane M.

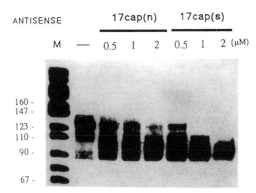

Figure 5. Effect of Antisense Oligonucleotides on cDNA Synthesis Using HIV RT. Reverse transcription proceeded as indicated using 1 unit of HIV RT primed by a 17 mer [17 sc (n) (5 µM)] complementary to the 113-129 oligonucleotide without (-) or with various amounts (µM) of antisense oligonucleotides, 17 cap (n) (left) and 17 cap (s) (right). DNA size markers were run in lane M.

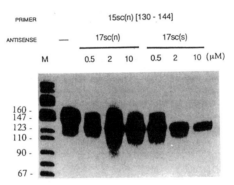

Figure 6. Effect of the Relative Location of Antisense and Primer-Binding Sites. Analysis of cDNA fragments synthesized by 1 unit of AMV RT. The reaction primed by a 15 complementary to the 130-144 oligonucleotide (5 mM) without (-) or with the antisense oligonucleotides 17 sc (n) or 17 sc (s) at the concentrations (mM) indicated above the lanes. DNA size markers were run in lane M.

phosphorothioate oligonucleotide inhibited polymerization by binding to the enzyme RT rather than to the template RNA. It is notable that the antisense phosphorothioate oligonucleotide participates through a mechanism of interaction different from those of the RT enzymes in adherence and structural differences. However, the antisense oligonucleotides can be used with intact cells, and they prevent the development of retrovirus in culture.

Antiviral Effect of Oligonucleotide Phosphorothioates Complementary to HIV

Materials. The phosphorothioate oligonucleotide analogues (S-ODNs or SO-ODNs) were prepared on a synthesizer using our new phosphate approach and purified by HPLC according to published procedures (*24*). We used five genes [gag, pol, rev, tat, and tar] as targets for antisense interruption of viral gene expression. To clarify the sequence specificity, we tested the phosphorothioate oligomers containing the rev sense sequence (S-ODNs/sen), six different gene antisense sequences (S-ODNs-gag-leader sequence (ls), pol-splice acceptor (sa), rev-sa, rev-translation start site (ts), tat-sa, tar), a random sequence with the same base composition as rev-sa (S-ODNs/ran), and 15-28 mer homo-oligomers [(dC)28, (dC)20, (dA)20, (dT)20, (dG)20, and (dC15; (S-dC)28, (S-dC)20, S-(dA)20, S-(dT)20, S-(dG)20, and S-(dC)15; Figure 1].

Methods. The anti-HIV activity of test compounds in fresh cell-free Å@HIV infection was determined as protection against HIV-induced cytopathic effects (CPE). Briefly, MT-4 cells were infected with HTLV-IIIB at a multiplicity of infection (MOI) of 0.01. HIV-infected or mock-infected MT-4 cells (1.5 x 105/mL, 200 mL) were placed into 96 well microtiter plates and incubated in the presence of various concentrations of test compounds (dilution 2 to 5; 9 concentrations of each compound were examined in triplicate). After a 5-day incubation at 37°C in a CO_2 incubator, the cell viability was quantified by a calorimetric assay that monitored the ability of the viable cells to reduce 3-(4,5-dimethylthiazol-2-yl)-2,5-diphenyl-tetrazolium bromide (MTT) to a blue formazan product. The absorbance was read in a microcomputer-controlled photometer (Titertek MultiskanR; Labsystem Oy, Helsinki, Finland) at two wave lengths (540 and 690 nm). The absorbance measured at 690 nm was automatically subtracted from that at 540 nm to eliminate the effects of non-specific absorption. All data represent the mean values of triplicate wells. These values were then translated into percentage cytotoxicity and percentage antiviral protection, from which the 50% cytotoxic concentration (CC50), the 50% effective concentration (EC50), and the selectivity indexes (SI) were calculated (*25*).

Sequence-Specific Inhibition of Viral Expression in HIV-1 Infected Cells. To examine the selective antisense inhibition of HIV with phosphorothioate oligomer, we selected targets that were involved in the viral recognition step (Figure 1). These include the AUG initiator codons and splice acceptor sites involved in processing RNA. Splice sites are good targets for HIV and also for herpes simplex virus (*26*). These sites involved in translation and processing of RNA should have some homology with host sequences, and oligonucleotides should bind the corresponding viral sequences.

All the antisense oligomers inhibited HIV without toxicity in cell culture (Figure 7); moreover, antisense S-ODNs of targeted internal splice sites (rev and tat splice acceptor sites) and the initiator codon (AUG) were very active, causing more than 95% inhibition at 0.02-0.5 μM. However, only the pol-sa gene oligomer was less active than other antisense oligomers (S-ODNs-gag-ls, rev-sa, rev-ts, tat-sa).

On the other hand, we could not detect any inhibitory effects of the 3',5'-capped phosphorothioate substituted oligomer (SO-ODNs-rev-sa) at a concentration of 0.5 μM (Figure 7). The weaker anti-HIV activity of SO-ODNs oligomer than that of S-ODNs

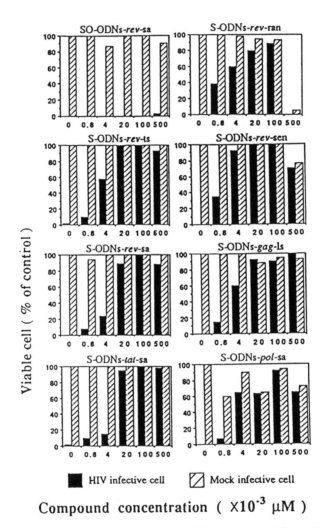

Compound concentration (×10⁻³ µM)

Figure 7. Sequence-specific inhibition of HIV-1 by modified oligodeoxyribo-nucleotides including replacement of either the whole PS (S-ODNs) or end capped with PS (SO-ODNs) at both 3'-and 5'-ends.

oligomer was supported by the uptake studies (27-29). Figure 8 shows the effect of time on the uptake of the 3'-[^{32}P] labeled SO-ODNs-tat-sa and S-ODNs-tat-sa in MOLT-4 cells. The uptake was found to plateau after 1 hour, and the SO-ODNs oligomer yielded about 10 times less cell-associated counts than the S-ODNs oligomer. It should be noted that the capped oligonucleotides used in this study contain 4 phosphorothioate linkages but retain the phosphorodiester linkages. It is conceivable that oligonucleotides with all phosphorothioate linkages (S-ODNs) might enter cells as intact oligonucleotides under tissue culture conditions, while the 3',5'-capped phosphorothioate substituted oligomers (SO-ODNs) were degraded. The 3',5'-capped phosphorothioate substituted oligomer (SO-ODNs) was weakly taken up in cells, which decreased the antiviral efficiency of the phosphorothioate group in our assay. Thus, in a defined concentration, the 3',5'-capped phosphorothioate substituted oligomer seemed to interact directly with the cell surface but not with HIV, thus inhibiting HIV replication. In addition, whole phosphorothioate oligomers (S-ODNs-rev or tat) showed anti-HIV activity, which was probably due mainly to the relative resistance of S-ODNs to nucleases. However, the 3',5'-capped P-S oligomers did not show any anti-HIV activity in our assay. This finding suggests that they are digested by the enzyme in an endonucleolytic manner rather than by exonucleolytic cleavage. The anti-HIV activity of antisense oligomers is influenced by the resistance of oligonucleotides to nucleases rather than the stability of RNA-DNA duplexes.

We also tested a sequence complementary to the double helix site (5'-GGUCUCUC-UGGUUAGACCAG-3', 1-20, tar-1) and the loop site (5'-GGCCAGAUCUGAGCC-UGGGA-3', 15-34, tar-2) of the tar RNA required for the viral tat protein to bind (Figure 9). We found that the S-ODNs-tar oligomer was equally active as the S-ODNs-pol-sa gene oligomer (Figures 7 and 10). This suggests a marked decrease in the ability of the antisense oligomers S-ODNs-tar-1 and 2 to bind to the tar RNA because of the spanning sequences on both the three bulges (GGUCUCU, 1-7; UUAGACCAG, 12-20; AGAUCUGAG, 19-27) and the very short loop sequence (CUGGGA, 29-35). The former binds to tat protein and the latter to a 68 kD a cellular protein. Secondary or tertiary RNA structures would not readily be susceptible to the antisense hybridization perhaps at the initiation of transcription or at mRNA processing. In pol-sa, this target sequence also has a splice acceptor site lying near the vif within HIV gene, and antisense agents complementary to some sequences close to a frame shift region would not exhibit potent antiviral activities.

Included in this series were two different phosphorothioate derivatives containing the rev sense sequence (S-ODNs/sen) and a random sequence (S-ODNs/ran). These compounds were as active as the antisense phosphorothioates at 0.1 μM but at 0.5 μM were relatively toxic. This finding differs from those of Matsukura et al. (9). However, only the antisense sequence of the phosphorothioate oligomers inhibited viral expression without toxicity, supporting the notion of the sequence-specific regulation of viral expression of HIV in chronically infected cells.

Furthermore, the phosphorothioate oligonucleotides were studied, and a target site on HIV RNA was used as with a splice acceptor site of the HIV rev gene (Figure 9). The phosphorothioate oligonucleotide, a chain length of 15 (S-ODNs-rev-sa-15), was as active as the 20-mer phosphorothioate oligomer (S-ODNs-rev-sa-20), but the phosphorothioate 10-mer (S-ODNs-rev-sa-10) had no activity. Of particular interest was S-ODNs-rev-sa-15, which had activity in the same order as the phosphorothioate 20-mer (S-ODNs-rev-sa-20). However, oligomer chain length had a greater effect on antiviral activity, since longer compounds were more potent. Further, specifically targeted antisense-phosphorothioate oligonucleotides such as rev or tat may be useful for treating HIV-infected patients.

Non-sequence-Specific Inhibition of Viral Expression in HIV-1 Infected Cells. The adenosine, thymidine, and guanosine 20 mers and cytidine 15, 20, and 28 mers were

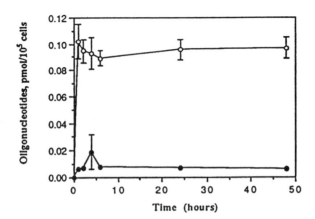

Figure 8. Cellular Uptake of the S-ODNs-rev-ts (5'-^{32}pTsCsTsCsCsGsCsTsTsCs-
TsTsCsCsTsGsCsCsAsT-3', O) and the SO-ODNs-rev-ts (5'-^{32}pTsCsTCCGCT-
TCTTCCTGCCsAsT-3', ●) in Molt-4 Cells. Radioactive counts associated with cell
lysates are expressed as pmol/105 cells over time. Vertical bars represent standard
deviation.

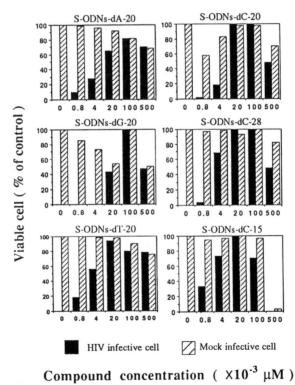

Figure 9. Effect of the Chain Lengths of Oligonucleotides on the Ability of an
Antisense Phosphorothioate Oligonucleotide (S-ODNs-rev-sa) to Inhibit HIV-1
Replication.

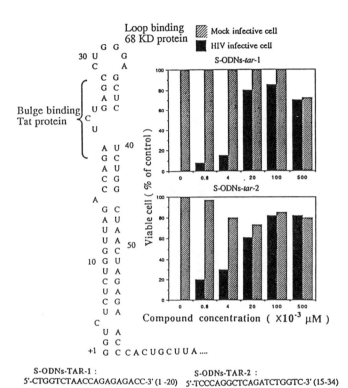

S-ODNs-TAR-1 :
5'-CTGGTCTAACCAGAGAGACC-3' (1 -20)

S-ODNs-TAR-2 :
5'-TCCCAGGCTCAGATCTGGTC-3' (15-34)

Figure 10. The Secondary Structure and Anti-HIV Activity (S-ODNs-tar-1 and 2) of Tar RNA.

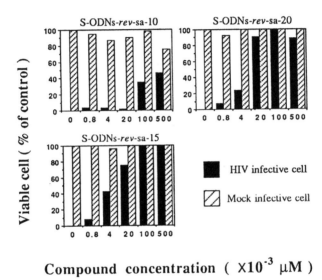

Compound concentration ($\times 10^{-3}$ μM)

Figure 11. Inhibitory Effects of the Chain Lengths and Non-Sequence-Specific effects of HIV-1 by Homo-Oligonucleotide Phosphorothioates.

tested for their comparative anti-HIV activity in HIV infected MOLT-4 cells (Figure 11). In this series, the homo-oligonucleotide phosphorothioates (S-dC28 and S-dC20) had activity of the same order as the antisense phosphorothioate 20-mer (S-ODNs-rev-sa-20) at a concentration of 0.1 μM. However, at 0.5 μM it had showed low activity and some toxicity. On the other hand, S-dC15 was less active than 20-mers (S-dC28 and S-dC20). This observation is somewhat different from a previous description of homo-oligonucleotide phosphorothioates (9). Furthermore, the anti-HIV activity of the homo-oligomers S-dA20 and S-dT20 was also tested in infected cells. S-dA20 and S-dT20 were both somewhat more active at all concentrations, and they inhibited up to 75% at concentrations from 0.1 to 0.5 μM without toxicity. These results did not indicate any correlation of inhibitory activity with the G+C content.

Thus, homo-oligomers appear to behave differently from hetero-oligomers and may inhibit HIV by some mechanism other than by antisense competitive hybridization. Homopolymers are better substrates than heteropolymers for reverse transcriptase (30,32). Consequently, the phosphorothioate homo-oligomers might be competitive inhibitors of reverse transcriptase or DNA polymerase. While the antiviral properties of the homo-oligomers are worth further consideration, they showed to be less selective in anti-HIV activity and potentially more toxic than antisense oligonucleotides.

The sense- and random-oligomers or homo-oligomers inhibited polymeriation by binding to the reverse transcriptase or DNA polymerase rather than to RNA at low concentrations. However, at high concentrations, the excess amount of non-sequence-specific oligonucleotide phosphorothioates may degrade without binding to mRNAs under tissue culture conditions (28, 32) and may be due to a cytotoxic effect (20).

Antisense oligonucleotides enter MOLT-4 cells, hybridize with complementary segments of RNA or DNA, and selectively inhibit viral replication and expression. In cells infected with HIV, perfectly matched phosphorothioate oligomers were more effective inhibitors than sense- and random-oligomers or homo-oligomers. A chain length of 15-20 mer units was optimal under our tissue culture conditions. They, thus, reinforce the concept that specific base pairing is a crucial feature of the oligonucleotide inhibition of the human immunodeficiency virus.

Acknowledgments

This work was supported by a grant-in-aid for scientific research on priority areas no. 05265220 and no. 05262102 from the Ministry of Education, Science and Culture Japan, and a research grant from the Human Science Foundation.

Literature Cited

1. Jaskulski, D.; DeRiel, J.K.; Mercer, W.E.; Calabretta, B.; Baserga, R. *Science* **1989**, *240*, 1544-1546.
2. Lewin, B. *Cell* **1990**, *61*, 743-752.
3. Travali, S.; Reiss, K.; Ferber, A.; Petralia, S.; Mercer, W.E.; Calabretta, B.; Baserga, R. *Mol. Cell. Biol.* **1991**, *11*, 731-736.
4. Calabretta, B. *Cancer Res.* **1991**, *51*, 4505-4510.
5. Stein, C.A.; Cohen, J.S. *Cancer Res.* **1988**, *48*, 2659-2668.
6. Toulmé, J.J.; Hélène, C. *Gene* **1988**, *72*, 51-58.
7. Zamecnik, P.C.; Stephenson, M.L. *Proc. Natl. Acad. Sci. USA.* **1978**, *75*, 280-284.
8. Zamecnik, P.C.; Goodchild, J.; Taguchi, Y.; Sarin, P.S. *Proc. Natl. Acad. Sci. USA.* **1986**, *83*, 4143-4146.
9. Matsukura, M.; Shinozuka, K.; Zon, G.; Mitsuya, H.; Reit, M.; Cohen, J.C.; Broder, S. *Proc. Nalt. Acad. Sci. USA.* **1987**, *84*, 7706-7710.

10. Kim, S.-G.; Suzuki, Y.; Nakashima, H.; Yamamoto, N.; Takaku, H. *Biochem. Biophys. Res. Commun.* **1991**, *179*, 1614-1619.
11. Kim, S.-G.; Kanbara, K.; Nakashima, H.; Yamamoto, N.; Murakami, K.; Takaku, H. *Bioorg. Med. Chem. Lett.* **1993**, *3*, 1223-1228.
12. Miller, P.S.; Agris, C.H.; Aurelian, L.; Blake, K.R.; Murakami, A.; Reddy, M.P.; Spitz, S.A.; Ts'o, P.O.P. *Biochimie* **1985**, *67*, 769-776.
13. Lemaitre, M.; Bayard, B.; Lebleu, B. *Proc. Natl. Acad. Sci. USA.* **1987**, *84*, 648-652.
14. Smith, C.C.; Aurelian, L.; Reddy, M.P.; Miller, P.S.; Ts'o, P.O.P. *Proc. Natl. Acad. Sci. USA.* **1986**, *83*, 2787-2791.
16. Zerial, A.; Thuong, N.T.; Hélène, C. *Nucleic Acids Res.* **1987**, *15*, 9909-9919.
15. Leiter, J.M.E.; Agrawal, S.; Palese, P.; Zamecnik, P.C. *Proc. Natl. Acad. Sci. USA.* **1990**, *87*, 3430-3434.
17. Loreau, N.; Boiziau, C.; Verspieren, P.; Shire, D.; Toulmé, J.J. *FEBS Lett.* **1990**, 52, 53-56.
18. Boiziau, C.; Thuong, N.T.; Toulmé, J.J. *Proc. Natl. Acad. Sci. USA.* **1992**, *89*, 768-772.
19. Matsukura, M.; Zon. G.; Shinozuka, K.; Robert-Guroff, M.; Shimada, T.; Stein, C.A.; Mitsuya, H.; Wong-Staal, F.; Cohen, J.S.; Broder, S. *Proc. Natl. Acad. Sci. USA.* **1989**, *86*, 4244-4248.
20. Lisziewicz, J.; Sun, D.; Klotman, M.; Agrawal, S.; Zamecnik, P.; Gallo, R. *Proc. Natl. Acad. Sci. USA.* **1992**, *89*,11209-11213.
21. Stec, W.; Zon, G.; Egan, W.; Stec, B. *J. Am. Chem. Soc.* **1984**, *106*, 6077-6079.
22. Starnes, M.C.; Gao, W.Y.; Ting, R.Y.; Cheng, Y.-C. *J. Biol. Chem.* **1988**, *263*, 5132-5134.
23. Majumdar, C.; Stein, C.A.; Cohen, J.S.; Broder, S.; Wilson, S.H.; *Biocemistry* **1989**, *28*, 1340-1346.
24. Hosaka, H.; Suzuki, S.; Sato, H.; Kim, S.-G.; Takaku, H. *Nucleic Acids Res.* **1991**, *19*, 2935-2940.
25. Nakashima, H.; Pauwels, R.; Baba, M.; Schols, D.; Desnyrer, J.; De Clercq, E. *J. Virol. Method* **1989**, *26*, 319-330.
26. Smith, C.; Aurelian, L.; Reddy, M.P.; Miller, P.S.; Ts'o, P.O.P. *Biochemistry* **1986**, *83*, 2187-2191.
27. Stein, C.A.; Mori, K.; Loke, S.L.; Subasinghe, C.; Shinozuka, K.; Cohen, J.S.; Necker, L.M. *Gene* **1988**, *72*, 333-341.
28. Shoji, Y.; Akhtar, S.; Periasamy, A.; Herman, B.; Juliano, R.L. *Nucleic Acids Res.* **1991**, *19*, 5543-5550.
29. Krieg, A.M.; Tonkinson, J.; Matson, S.; Zhao, Q.; Saxon, M.; Zhang, L.M.; Bhanja, U.; Yakubov, L.; Stein, C.A. *Proc. Nalt. Acad. Sci. USA.* **1993**, *90*, 1048-1052.
30. Allaudeen, H.S. In *Inhibition of RNA and DNA Polymerase*; Sarin, P.S.; Gallo, R.C., Eds.; New York: Pergamon, 1980, pp. 1-26.
31. Hatta, T.; Kim, S.-G.; Nakashima, H.; Yamamoto, N.; Sakamoto, K.; Yokoyama, S.; Takaku, H. *FEBS Lett.* **1993**, *330*, 161-164.
32. Agrawal, S.; Temsamani, J.; Tang, J.-Y. *Proc. Nalt. Acad. Sci. USA.* **1991**, *90*, 7595-7599.

RECEIVED June 16, 1994

Chapter 11

Carboranyl Oligonucleotides for Antisense Technology and Boron Neutron Capture Therapy of Cancers

Raymond F. Schinazi[1,2], Zbigniew J. Lesnikowski[1,2],
Géraldine Fulcrand-El Kattan[1,2], and David W. Wilson[3]

[1]Laboratory of Biochemical Pharmacology, Department of Pediatrics,
Emory University School of Medicine, Atlanta, GA 30322
[2]Atlanta Veterans Affairs Medical Center, Decatur, GA 30033
[3]Department of Chemistry, Georgia State University, Atlanta, GA 30303

The methodology for the synthesis of two novel types of oligonucleotide analogues, bearing o-carboranyl cage, within the internucleotide linkage (o-carboranylmethylphosphonate oligonucleotide analogue), and at 5-position of uracil residue (CDU-oligonucleotides) is described. The increased lipophilicity of these new compounds and high boron content make them candidates for boron neutron capture therapy (BNCT) of cancers, and also for antisense/antigene oligonucleotide therapy (AOT). The syntheses, physicochemical properties and biological studies of some of these new promising oligonucleotides are discussed.

The use of boron-containing compounds in the treatment of malignancies is based on the property of non-radioactive boron-10 nuclei to absorb low energy neutrons. When this stable isotope is irradiated with thermal neutron, an alpha (α) particle (helium) and lithium-7 nuclei are released through the nuclear reaction producing about 100 million times more energy than that which was initially used. The generated radiation, destroys the target tumor cells (1). This combined modality, originally proposed by Locher in 1936 (2), is known as BNCT. Selectivity of boron trailers toward tumor cells is one of major obstacles of BNCT (3). We hypothesize that o-carboranyl modified oligonucleotides could solve these important problems (4, 5). These novel compounds can potentially target cancer cells by interacting with overexpressed or unique genes found in these cells (6). The o-carboranyl residue is lipophilic, and should facilitate boron-modified oligonucleotide transport through cellular membranes (7, 8). The oligonucleotides can bear several o-carboranyl residues, each containing 10 boron atoms, which effectively increase the boron content per carrier molecule. An additional potential advantage of boron-modified oligonucleotides is the finding that oligonucleotides accumulate in the nucleus (9-11) which may further decrease the boron levels in the tumor cell essential for BNCT (12).

0097–6156/94/0580–0169$08.00/0

Molecular Basis for BNCT

One of most lasting effect of radiation on living cells is damage to the genetic material. In this respect the critical damage is that to the DNA component of the chromosome (13-17). In tumor therapy selective killing of malignant cells is required. Cellular repair mechanisms efficiently repair the bulk of DNA. Therefore, it is crucial for effective tumor treatment to reach the level of DNA damage overwhelming the cellular repair capacity. To reach this goal, in the case of BNCT, a sufficient number of α-particles should be generated inside tumor cells.

Decomposition of water is of fundamental importance for DNA damage caused by ionizing radiation since the changes in DNA leading to cell death are a result of chemical reactions between DNA components (sugar moieties, nucleic bases and to lesser extend internucleotide phosphodiester groups) and products of water radiolysis (18, 19). Primary products of water radiolysis react among themselves and with surrounding water generating a cascade of reactive species such as hydroxyl radicals (OH·), hydrogen radicals (hydrogen atom, H·), superoxide anion radical (O·⁻₂), perhydroxyl radical (HO₂·) and hydrated electron (e⁻aq). These species diffuse towards DNA causing chemical damages followed by structural changes.

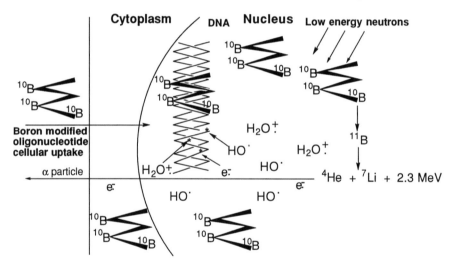

Fig. 1. Hypothetical mechanism of action of boronated oligonucleotides for BNCT. $H_2O^{+\cdot}$ (water radical cation), HO· (hydroxyl radical), e: (electron). Star depicts DNA damages such as single-strand break (ssb), double-strand break (dsb), and base modification.

The radiation DNA damages mediated by products of water radiolysis can be classified, in general, as (a) single-strand break (ssb) or double-strand break (dsb) resulting from breakage of phosphodiester bonds, (b) nucleic base modifications leading to mutation of genetic information, and (c) DNA-protein cross-links (20) (Figure 1). All these damages lead to disruption of genetic information and may cause cell death. The long lasting goal of radiation therapy of tumors is to achieve selective killing of tumor cells while sparing the healthy ones. BNCT seems ideally suited to perform such a goal if tumor selective boron carriers can be synthesized. This is mainly due to the properties of α-particles, generated during the micronuclear reaction produced by the boron neutron capture. The high α-particle linear energy transfer (high LET) (21), together with short ionization track about 0.01 mm long, which is the equivalent of approximately one cell diameter, limits damage primarily to the cell containing boron. Because the cell nucleus is known to be a "critical site" for radiation action it is highly desirable that boron carriers accumulate in these organelles. This is of special importance since it was found that the average number of α-particle traversals through nucleus, producing a single lethal lesion is greater than one (22). The location of boron directly in the "critical site" increases the probability of this desired event. It was observed that oligonucleotides injected into the cytoplasm accumulate in the nucleus (9-12). We believe that this will also be the case for carboranyl-containing oligonucleotides, a new class of boron trailers for BNCT. We hypothesize that the affinity of suitably designed carboranyl oligonucleotides towards malignant cells may be increased by their antigene (or antisense) type interaction with overexpressed or unique tumor genes (8, 23).

Rationalizing the Synthesis of Boronated Compounds

It is generally accepted that 30 ppm of ^{10}B in tumor tissue is adequate for effective neutron capture reaction *in vivo* (24). This concentration is based on the assumption that 2 to 3 α-particles trajectories through the nucleus lead to cell death. To generate these heavy particles, a determinate number of boron fission must occur which in turn depends on a minimum amount of ^{10}B atoms present in the cells. Calculations were performed to rationalize the content of ^{10}B atoms in a carrier. These calculations were based on the following assumptions: 1) 10-30 ppm of ^{10}B is needed; 2) uniform distribution of the compound occurs in the tumor tissue; and 3) the volume of 1 g of tumor tissue is equivalent to 1 mL. Figure 2 depicts the variation of the intracellular concentration of the carrier *versus* the variation of the ^{10}B atoms number per trailer molecule. The equations for the curves are $Y = a / X$, where $a = 3,000$ μM when $X = 1$ for the cytoplasm; and $a = 1,000$ μM when $X = 1$ for the nucleus. This indicates that if the carrier contains only one ^{10}B atom per molecule, a 3,000 μM concentration of the carrier must be reached intracellularly for cell death. If the carrier contains one carboranyl residue (10 ^{10}B atoms) per molecule, a 10-fold decreased concentration is needed. In this regard, it is apparent that a plateau is quickly reached when the number of carboranyl moieties per trailer molecule is 2 or more for nuclear delivery (Figure 2, inset).

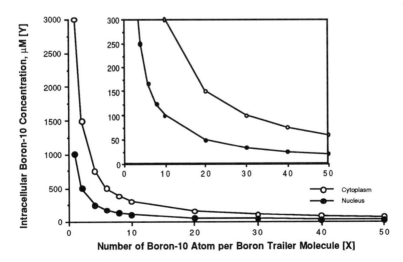

Fig. 2. Estimate of ^{10}B concentration required to produce a lethal boron neutron capture reaction.

However, this calculation does not reflect the effect of compartmentalization of the compound in the cells. If the boron carrier binds to the surface membrane, penetrates into the cytoplasm, or localizes in the nucleus, one should expect different results. For example, in the case of boron trailer-antibody conjugates, it was estimated that about 1000 [10]B atoms per molecule is necessary to reach the minimum level of boron required (28). This substantial requirement of boron results from the limited number of cellular antibody-receptors available, thereby restricting the optimum concentration of the boronated-bioconjugate that can be delivered into a cell. Moreover, if the [10]B atoms are confined to the cell nucleus, the potency of the boron neutron capture reaction can be increased by a factor of 2 to 5 (25, 26).

These data indicate that there is a limit to the number of [10]B atoms that are needed to be incorporated in a carrier since beyond 2 to 4 carboranyl moieties no apparent benefit will result assuming unrestricted cellular uptake. Therefore, it is essential to focus on the ability of the boron carrier to localize, and ideally to accumulate in the nucleus. The latter is a critical site for the reaction to take place. Based on these considerations, it appears that carefully designed carboranyl containing oligonucleotides may be valuable BNCT sensitizers since it has been demonstrated that most oligonucleotides once incorporated into cells accumulate in the nucleus (10).

Carboranyl Oligonucleotides

Although attempts to develop tumor-selective boron carriers date back to the early 1960s (27), the problem of selective delivery of boron to tumor cells remains. Among the many compounds synthesized for BNCT are boron-containing amino acids, porphyrins, and low-density lipoproteins (3, 28). However, the clinical applications of boron derivatives are limited to few compounds such as L-4-(dihydroxyboryl)-phenylalanine (BPA) (29) and the sodium salt of thioborane anion ($Na_2B_{12}H_{11}SH$, BSH) (30). Another promising class of boron carriers are nucleic acid bases and nucleosides developed by the groups of Schinazi, Soloway, Spielvogel, and Yamamoto (31). A new generation of radiosensitizers for BNCT described only recently are polymers and biopolymers bearing one or more carboranyl residues. This class of boron trailers includes carboranyl peptides (32), carboranyl oligophosphates (33, 34) and oligonucleotides (1, 4, 5, 35-37, Fulcrand-El Kattan, G.; Lesnikowski, Z.J.; Yao, S.; Tanious, F.; Wilson, D.W.; Schinazi, R.F. J. Am. Chem. Soc., in press). In this paper we focus on novel chemistry and biophysical properties of two novel oligonucleotide modifications: neutral o-carboranylmethylphosphonate analogue bearing modifications within internucleotide linkage, and oligomers containing the modified nucleoside, 5-o-carboranyl-2'-deoxyuridine (CDU).

Synthesis of Carboranyl Oligonucleotides

Di(thymidine o-carboranylmethylphosphonate) (3) (4). 3',5'-Protected dinucleotide 3 was prepared using "in solution" triester method of oligonucleotide synthesis (38). Thus, the monomer, 5'-O-monomethoxytritylthymidine 3'-O-(o-carboranylmethylphosphonate (1) was first activated with triisopropylbenzene-sulfonylchloride then treated with nucleoside component, 3'-O-acetylthymidine (2) (Figure 3). The reaction was performed in anhydrous acetonitrile or tetrahydrofuran as a solvent and in the presence of 2,4,6-collidine and 1-methylimidazole as the nucleophilic catalyst. The yield of 3',5'-protected 3 after chromatographic purification was 30% [δ ^{31}P NMR ($CDCl_3$) 21.16 and 22.95 ppm]. Removing of 3'-acetyl group was performed using conc. NH_3OH/CH_3OH (39) and the 5'-O-monomethoxytrityl group was removed by treatment with 80% CH_3COOH (40). Crude dinucleotide 3 was purified by silica gel column chromatography.

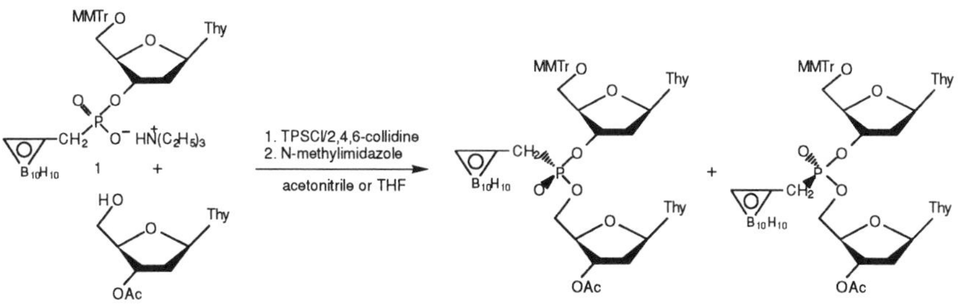

Fig. 3. Synthesis of 5'-O-monomethoxytritylthymidine(3',5')3'-O-acetyl-thymidine o-carboran-1-ylmethylphosphonate.

The key intermediate in the synthesis of monomer **1** is the borophosphonylating agent O-methyl-o-carboranylmethylphosphonate. This new compound was obtained in a 3-step procedure. In the first step, propargyl bromide was reacted with trimethyl-phosphite, using the Michaelis-Arbuzov type reaction (*41*), yielding O,O-dimethyl-propargylphosphonate in 45% yield. As a major by-product, O,O-dimethylmethyl-phosphonate was isolated. In the second step, O,O-dimethylpropargylphosphonate was reacted with decaborane in acetonitrile, according to the general reaction of decaborane and acetylene derivatives described by Heying et al. (*42*). The desired O,O-dimethyl-o-carboranylmethylphosphonate was obtained in 40% yield [δ ^{31}P NMR (CDCl$_3$) 20.70 ppm]. O-Methyl-o-carboranylmethylphosphonate was obtained in 79% yield [δ ^{31}P NMR (CDCl$_3$) 14.80 ppm] when O,O-dimethyl-o-carboranyl-methylphosphonate was treated with a mixture of thiophenol/triethylamine/dioxane (*43*).

Fully protected monomer 5'-O-monomethoxytritylthymidine 3'-O-(O-methyl-o-carboranylmethylphosphonate) was synthesized in the reaction of 5'-O-mono-methoxytritylthymidine and O-methyl-o-carboranylmethylphosphonate in the presence of triisopropylbenzenesulfonylchloride and 2,4,6-collidine and 1-methyl-imidazole as described for coupling reaction leading to dinucleotide **3**. 5'-O-Mono-methoxytritylthymidine 3'-O-(O-methyl-o-carboranylmethylphosphonate) was obtai-ned in 45% yield after standard work up and chromatographic purification [δ ^{31}P NMR (CDCl$_3$) 33.00 ppm]. Selective demethylation of fully protected monomer with thiophenol/triethylamine/dioxane was performed as described for O-methyl-o-carboranylmethylphosphonate synthesis. After $\approx$ 5 min, the triethylammonium salt of **1** was isolated by precipitation with hexanes in 80% yield [δ ^{31}P NMR (CDCl$_3$) 12.85 ppm]. Crude **1** was sufficiently pure by TLC analysis to be used for the synthesis of **3**.

Dinucleotides such as **3** can serve, after selective deprotection and phosphitylation of the 3'-end with 2-cyanoethyl *N,N*-diisopropylchlorophosphora-midite, as building blocks for the synthesis of longer oligonucleotides bearing one or more alternating o-carboranylmethylphosphonate linkages by automatic synthesis on solid support. This work was recently published by our group (*4*).

5-o-Carboranyl-2'-deoxyuridine containing oligonucleotides (5). (CDU-oligonucleotides).

CDU was synthesized from 5-iodo-2'-deoxyuridine in a five-step procedure, as described previously (*44, 45*). 5-(o-Carboranyl)-5'-O-dimethoxytrityl-2'-deoxyuridine-3'-[*N,N*-diisopropyl-2-cyanoethylphosphoramidite] (**4**) [δ ^{31}P NMR (CDCl$_3$) 149.61 and 149.90 ppm] was obtained in a two-step procedure. First, the 5'-hydroxy function of CDU was protected with dimethoxytrityl group then the corresponding partially protected CDU was reacted with 2-cyanoethyl *N,N*-diisopropylchlorophosphoramidite yielding the phosphoramidite monomer (*46*).

Dodecathymidylic acid $d(T)_{12}$ analogues bearing one or two 5-(o-carboranyl)-uracil residues at the 1st, 2nd, 7th, 11th, and both 10th and 11th and 1st and 11th (**5**) locations of the 12-mer $d(T)_{12}$ were obtained by solid phase automated synthesis using standard 2-cyanoethyl cycles (47) (Figure 4).

After removal of the 5'-dimethoxytrityl group, the oligonucleotides were then cleaved from the support by incubation with concentrated NH_4OH at room temperature for 1 h. We have observed that NH_4OH treatment leads to partial transformation of $closo$-[$closo$-1,2-$C_2B_{10}H_{11}$] form of carboranyl cage to its $nido$-[$nido$-7,8-$C_2B_9H_{11}$] form. The deprotected oligonucleotides were purified by HPLC and, for selected cases, separated into $nido$- and $closo$-form (48).

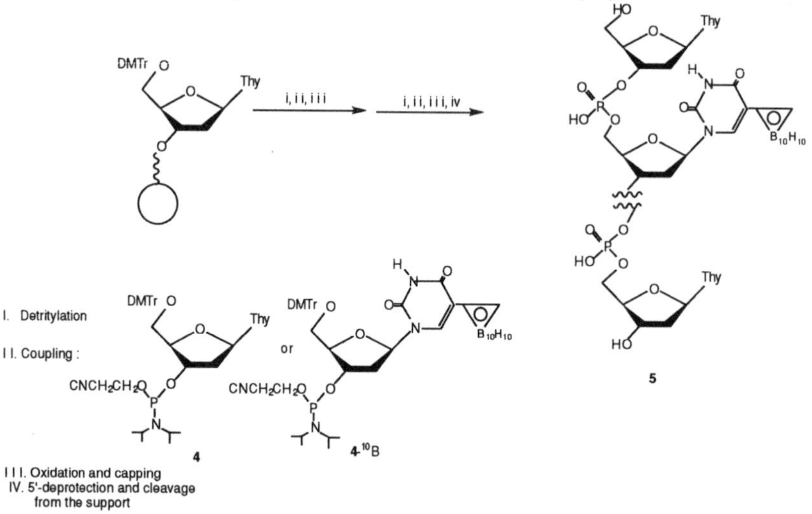

Fig. 4. Automated synthesis of CDU containing dodecathymidylates.

The yield for the overall synthesis of CDU-containing oligonucleotides was comparable to unmodified $d(T)_{12}$ as determined by quantitation of trityl release during the automated synthetic procedure. The exception was the 12-mer containing two CDU units at 10th and 11th locations where we observed ca. 25% drop of overall oligonucleotide yield due to the low efficiency of the two first coupling reactions. Further studies are necessary to optimize the synthesis of oligonucleotides containing adjacent CDU monomers.

Structure of Carboranyl Cage and its Effect on CDU-Oligonucleotide Diastereoisomerism

Three peaks were identified in the HPLC chromatograms of oligonucleotides containing one CDU residue and up to nine for oligonucleotides containing two CDU residues for compounds 5'-CDUd(T)$_{11}$, 5'-d(T)$_6$CDUd(T)$_5$, 5'-d(T)$_{10}$CDUd(T), 5'-d(T)$_9$CDU$_2$d(T), and 5'-CDUd(T)$_9$CDUd(T). The multiplicity of the peaks displayed on the HPLC chromatograms is due to the existence of CDU carboranyl cage in three distinctive forms, one $closo$ and two $nido$. The two $nido$ forms result from subtraction of boron number 3 or 6 from $closo$-cage which leads to two $nido$-carboranyl enantiomers ($nido$-1 and $nido$-2) (49) (Figure 5).

Since the nucleoside units of oligonucleotide chain are chiral, the CDU-oligonucleotide exists as one species for the $closo$-form of the carboranyl cage and as two diastereoisomers for the $nido$-form of the carboranyl residue. In the case of oligonucleotides containing two CDU residues, the number of possible oligonucleotide components is defined by the formula m^n, where m is a number of forms carboranyl cage can exist (*closo* and two *nido*) and n is the number of carboranyl cages (CDU residues) per oligonucleotide chain. Since some of the diastereoisomers may overlap, the number of peaks detected may be lower than calculated. A similar phenomenon is observed for P-chiral oligonucleotide analogues. The number of diastereoisomeric oligomers is defined by the formula 2^n, where 2 is the number of isomers per P-chiral internucleotide linkage, and n is the number of P-chiral linkages (*50*).

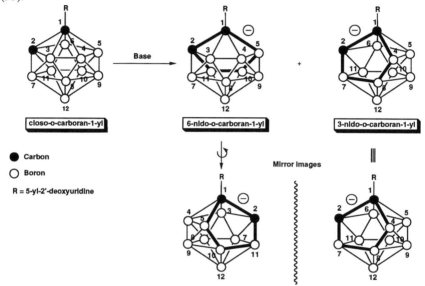

Fig. 5. Stereochemistry of *closo* to *nido* transformation of carboranyl cluster.

Comparison of HPLC traces for 5'-CDUd(T)$_9$CDUd(T) before and after pyrrolidine treatment (*17*), leading to selective transformation of *closo*- to *nido*-form of the carboranyl cage, is insightful. The HPLC trace for 5'-CDUd(T)$_9$CDUd(T) before pyrrolidine treatment shows nine peaks: one CDU residue can exist as *nido*-1, *nido*-2 or *closo*-form of the carboranyl cage, therefore, $m = 3$; for two CDU units, $n = 2$ one can expect 3^2 or 9 compounds, which appears as nine peaks. After pyrrolidine treatment the *closo*-form of the carboranyl cage is transformed into *nido*-1 and *nido*-2 enantiomers, therefore $m = 2$, and the number of expected isomers decreases to 2^2 or 4. Because it seems that the retention time (R$_t$) for two of isomers is the same, under the eluting condition used, the two peaks overlap generating three signals with integration area ratio of 1:2:1, instead of four peaks with ratio 1:1:1:1. Mass spectrometry analyses of the separated CDU oligonucleotide isomers confirmed the presence of *nido*-carboranyl cage in two enantiomeric forms (Table 1).

Physicochemical Properties of Carboranyl Oligonucleotides

The lipophilicity of CDU-oligonucleotides 5'-CDUd(T)$_{11}$, 5'-d(T)$_6$CDUd(T)$_5$ and 5'-d(T)$_{10}$CDUd(T) (mixture of *nido*-1, *nido*-2, and *closo*) was compared with d(T)$_{12}$ by coinjection experiment using a reverse phase column. HPLC analysis indicated not only different affinity (lipophilicity) for the *closo/nido* carboranyl forms to C$_{18}$ resin, but also a correlation between the CDU location within the oligonucleotide chain and oligonucleotide R$_t$. The oligomers lipophilicity as measured by R$_t$ increased in the order *nido*-1 < *nido*-2 < *closo*- and with respect to CDU location, in the order d(T)$_{12}$ < central CDU < 5'-terminal CDU < pseudo 3'-terminal CDU (Table 1).

Circular dichroism (CD) studies and melting temperature (T_m) measurements of CDU-oligonucleotides (5) were performed with the carboranyl oligonucleotides and compared to the corresponding natural oligomer. Inspection of CD spectra of single-stranded $d(T)_{12}$ and CDU modified $d(T)_{12}$ (5), recorded under analogous conditions, showed them to be almost identical in term of their shapes and molecular ellipticity values. This suggests a lack of effect of CDU on single-stranded oligonucleotide conformation in solution, at least for oligomers containing CDU at certain positions.

Table 1. HPLC and MS characteristics of CDU-containing dodecathymidylic acids

Oligonucleotide	Carboranyl cage	R_t [min][a]	T_m [°C][b]	T_m [°C][c]	Molecular formula	MS
$d(T)_{12}$		15.5	29.0	30.0	$C_{120}H_{157}N_{24}O_{82}P_{11}$	-
5'-CDUd(T)$_{11}$	nido-1	21.0	28.8	28.0	$B_9C_{121}H_{165}N_{24}O_{82}P_{11}$	3706[d]
	nido-2	21.5	28.4		$B_9C_{121}H_{165}N_{24}O_{82}P_{11}$	3706[d]
	closo-	25.8	28.0		$B_{10}C_{121}H_{165}N_{24}O_{82}P_{11}$	3716[e]
5'-d(T)$_6$CDUd(T)$_5$	nido-1	19.9	15.2	15.2	$B_9C_{121}H_{165}N_{24}O_{82}P_{11}$	3704[f]
	nido-2	20.3	15.3		$B_9C_{121}H_{165}N_{24}O_{82}P_{11}$	3704[f]
	closo-	23.1	15.3		$B_{10}C_{121}H_{165}N_{24}O_{82}P_{11}$	3716[e]
5'-d(T)$_{10}$CDUd(T)	nido-1	25.0	20.5	18.5	$B_9C_{121}H_{165}N_{24}O_{82}P_{11}$	3705[e]
	nido-2	25.5	20.4		$B_9C_{121}H_{165}N_{24}O_{82}P_{11}$	3704[f]
	closo-	32.8	20.9		$B_{10}C_{121}H_{165}N_{24}O_{82}P_{11}$	3716[e]

[a]HPLC analysis was performed on a Hewlett-Packard 1050 system, using a Whatman Partisphere C_{18} 5 μm, 4.7 x 235 mm column. All analyses were performed at room temperature. The following gradient of CH_3CN in 0.05 M triethylammonium acetate (TEAA) buffer was used: from 5% to 20% (20 min), 20% (5 min), from 20% to 30% (3 min), and 30% (3 min); T_m in PIPES buffer at 20 mM NaCl; linear plots of T_m *versus* log(sodium ion activity) gave slopes of 15 ± 1°C. [b] $d(A)_{12}$ in base molar ratio 1:1 was used as complementary strand for duplex formation; [c] poly r(A) in base molar ratio 1:1 was used as target. [d][M+1]; [e][M]; [f][M-1].

Interestingly, the CD spectra of duplexes formed between CDU-modified oligothymidylates (5) or unmodified $d(T)_{12}$ and $d(A)_{12}$ as complementary strand showed a reduction in the magnitude of molecular ellipticity at 246 nm, which correlated with increased thermal stability of the duplexes. Thus, the least stable duplex formed by the oligonucleotide modified with CDU at the 7th position [5, 5'-$d(T)_6$CDUd(T)$_5$] was characterized by the highest value of molecular ellipticity of the negative CD band at 246 nm, and consequently the most stable unmodified and 5'-end modified dodecathymidylic acid [5, 5'-CDUd(T)$_{11}$] were characterized by the lowest value of molecular ellipticity of the negative CD band. Both also showed a distinctive peak at 257 nm. The higher Cotton effect observed between 240 and 260 nm for stable duplexes suggests less disturbance of base stacking which indicates a stable, well arranged double helix structure.

Mixing curves for unmodified $d(T)_{12}$ and $CDUd(T)_{11}$ (*nido*-1) were obtained by titrating of $d(T)_{12}$ (50 mM) or $CDUd(T)_{11}$ (40 mM) with $d(A)_{12}$ (50 mM) at 20°C in PIPES buffer [1,4-piperazinebis(ethanesulfonic acid), 10 mM], pH 7.0, containing 1 mM EDTA, and 100 mM NaCl. Absorption change at 260 nm was followed as a function of $d(T)_{12}$ or $CDUd(T)_{11}$ (*nido*-1)/$d(A)_{12}$ ratio. With both oligomers a break in the mixing curve plot occurred at the ratio of 1:1. No evidence for formation of a 1A:2T triplex was observed under these conditions.

T_m measurements of the duplexes between CDU-modified $d(T)_{12}$ (**5**), and either $d(A)_{12}$ or polyr(A) as a complementary sequence, were compared with those formed between unmodified $d(T)_{12}$ and template. Significant effects on T_m were noted, depending on location of the modification within the oligonucleotide chain (Table 1). First, oligonucleotides modified with CDU at either the 1st [**5**, 5'-$CDUd(T)_{11}$], 7th [**5**, 5'-$d(T)_6CDUd(T)_5$], or 11th [**5**, 5'-$d(T)_{10}CDUd(T)$] position were separated into *closo*- and *nido*-derivatives by HPLC, and their T_m values were determined as duplexes with $d(A)_{12}$. The T_m for the *closo*- and *nido*-forms were almost identical within the same type of oligonucleotide modification (pseudo-3'-, 5'-end, or middle position). This may be due the high lipophilicity of the boron cage and low density of the negative charge in the *nido* compounds. However, the effect of the CDU location was striking. 5'-Modifications at the 1st and 2nd position [respectively: **5**, 5'-$CDUd(T)_{11}$, T_m for the *closo*-form was 28°C and **5**, 5'-$d(T)CDUd(T)_{10}$, T_m for the *closo-/nido*-form mixture was 27.2°C] did not influence markedly the stability of the duplex compared to unmodified $d(T)_{12}$ (T_m = 29°C). In contrast, modification in the central position of the oligonucleotide chain [**5**, 7th position, 5'-$d(T)_6CDUd(T)_5$] resulted in a marked decrease of duplex stability as noted by a low T_m value of 15.2°C (*nido*-forms). A less pronounced effect was generated by the presence of CDU at the 3'-end of the oligonucleotide (11th position). The T_m of the duplex formed by this oligonucleotide [**5**, 5'-$d(T)_{10}CDUd(T)$] decreased to 20.5°C. Inserting a second CDU nucleoside at the 3'-end further destabilized the duplex, decreasing the T_m value to 15.3°C [**5**, 5'-$d(T)_9CDU_2d(T)$, mixture of *closo-/nido*-form]. The diverse consequences of 3'- and 5'-end modifications upon the duplex stability were evident from the above data. Insertion of the CDU nucleoside at the 5'-end has a much more favorable effect than at the 3'-end. This is well illustrated by comparing T_m values between oligonucleotides where the CDU nucleotide is located at the second position from the 5'-end [**5**, 5'-$d(T)CDUd(T)_{10}$] and from the 3'-end [**5**, 5'-$d(T)_{10}CDUd(T)$], respectively. The difference in T_m is 7-8°C, which is significant and reveals the importance of the carboranyl cluster interaction with adjacent bases.

These results were in agreement with T_m measurements of duplexes formed between CDU-modified oligomers **5** and polyr(A). In effect, the location of CDU within the oligonucleotide chain induced a greater destabilization, which seemed more pronounced when the modification was closer to the 3'-end. Although, in the case of polyr(A), the length of template was much longer than the dodecamer $d(A)_{12}$, this unfavorable effect along with the CDU position was further amplified compared to the DNA/DNA duplex. To our knowledge, this work represents the first systematic studies on the relationship between the location of pyrimidine nucleoside modified at position five with bulky cluster within an oligonucleotide chain and duplex stability. However, it is of interest that a similar effect of location was reported within the oligonucleotide chain of base modified nucleoside bearing a biotynylated linker (*51*). The potential steric interactions of carboranyl cluster with other components of double helix is further discussed below.

Biological Properties of Carboranyl Oligonucleotides

The stability of the oligonucleotides towards nucleases is an important factor for their future applications as potential therapeutics. It is generally accepted that 3'-exonuclease activity is responsible for most of the unmodified antisense oligonucleotide degradation in serum (*7*). For example, effective ways to secure resistance against nucleases is modification of internucleotide linkages. Replacement of all natural phosphodiester linkages within oligonucleotide chain by methylphosphonates ensures complete stability of the oligonucleotide towards exo- and endonucleases (52). However, there are several literature reports on increased enzymatic stability of oligonucleotides modified only at 3'- or 3',5'-flanking positions (53, 54).

Di(thymidine o-carboranylmethylphosphonate) (3) contains methylphosphonate type of internucleotide linkage which is known to be completely resistant to enzymatic digestion. Indeed, results of enzymatic assays indicated that 3 was, as anticipated (55), resistant to cleavage by calf spleen phosphodiesterase and snake venom phosphodiesterase (SVPD). To test resistance of CDU-oligonucleotides towards 3'-exonucleolytic activity, SVPD from *Crotalus durissus terrificus* was used. Oligomers (0.25 optical density unit) were treated with enzyme (3.8 x 10^{-3} unit) for 4 h at 37°C. The presence of CDU at the 3'-end of oligonucleotides effectively improved their stability towards SVPD. As expected, the oligonucleotide resistance towards nucleolytic activity increased in the order unmodified $d(T)_{12}$ ($t_{1/2}$ = 0.5 min) < $d(T)_{12}$ modified with CDU at the 11th position [5'-$d(T)_{10}$CDUd(T), *nido*-forms] ($t_{1/2}$ = 6.7 min) ~ $d(T)_{12}$ modified with two CDU residues at the both 1st and 11th position [5'-CDUd(T)$_9$CDUd(T), mixture of *closo-/nido*-forms] ($t_{1/2}$ = 5.1 min) << $d(T)_{12}$ modified with two CDU residues at the 10th and 11th position [5'-$d(T)_9$CDU$_2$d(T), mixture of *closo/nido*-forms] ($t_{1/2}$ = 76.2 min). The substantial change in half-life caused by CDU modification (150-fold increase for oligonucleotide bearing two CDU molecules at the 3'-end) compared to unmodified oligomer may have practical implications.

It seems that the presence of a bulky lipophilic carboranyl substituent at the 5 position of pyrimidine nucleoside prevented effective interaction of the modified oligonucleotide fragment with the active center of the enzyme. This reasoning is supported by the finding that the stability of oligonucleotide with two adjacent CDU residues is notably higher than the oligonucleotide with one modification, which suggests further disturbance of oligonucleotide-enzyme interaction. It should be noted that due to the method of modified oligonucleotide synthesis used, the CDU residue is located at the penultimate position of the oligomer, and a thymidine is the actual 3'-terminal nucleoside. We assume that 3'-thymidylate is rapidly removed in the first stage of enzymatic digestion and then the resultant undecamer terminated with CDU is slowly digested by the enzyme.

Studies are underway to determine the resistance of CDU-modified oligonucleotides against other nucleolytic enzymes and in various sera. 5'-Modified oligonucleotide [5, 5'-CDUd(T)$_{11}$] was completely resistant towards 5'-exonucleolytic activity of phosphodiesterase from calf spleen. It is also of practical importance to note that all the oligonucleotides described are 5'-end phosphorylated by polynucleotide kinase which allow their 5'-end [^{32}P] labeling. Preliminary results suggest that some of the CDU-modified oligonucleotides evaluated serve as primers for various polymerases, including human immunodeficiency virus type 1 reverse transcriptase (Schinazi, R.F.; Lesnikowski, Z.J.; Fulcrand-El Kattan, G.; Lloyd, R. M., Jr. - unpublished data).

Stereochemical Considerations

CDU-oligonucleotides (5). To evaluate the observed differences in duplex stability for different CDU-modified oligonucleotides, molecular mechanics approaches were used. For this purpose, the AMBER all-atom force field method (56) and equations were applied. Initial models of $d(T)_{12} \cdot d(A)_{12}$ DNA duplexes constructed with a carborane cage at different positions were energy minimized along with an unsubstituted control. There were significant unfavorable interactions of the carborane substituent with adjacent bases, and the interactions were asymmetric in orientation due to the right-handed twist of DNA.

The refined models clearly demonstrate that the carborane interacts strongly, as a consequence of steric clash, with the base on the 5'-side of the CDU, but has essentially no interaction with the base on the 3'-side of the carborane substituted base. In agreement with these observations, energy minimization results with d(T)$_{12}$·d(A)$_{12}$ showed that substitution on the 5'-end of the d(T)$_{12}$ strand gave overall helix energies that were significantly lower than for the duplex substituted on the 3'-side of d(T)$_{12}$. Substitution of an interior base of the duplex caused even larger helix destabilization. With the 5'-substituted duplex, there is little interaction of the carborane with bases and the overall helix geometry is similar to the unsubstituted duplex. To relieve some of the strain from the steric interaction at the 3'-end of the duplex, the bases were distorted due to end effects and freedom of motion of base pairs at the end of the double helix. CDU substituted bases in the center of the hybrid have similar steric interaction, but they do not have the flexibility of base pairs located at the 3'-end of the helix.

Thus, duplexes with central CDU substitutions exhibited the highest energy, indicating that this was the most unfavorable position for CDU modification. These observations were in complete agreement with the T$_m$ results, where CDU substitution at the 5'-end of the d(T)$_{12}$ strand caused little change in T$_m$ relative to the unsubstituted duplex. Substitution at the 3'-end destabilized the helix structure, while central substitution caused the largest decrease in T$_m$.

Di(thymidine o-carboranylmethylphosphonate) (3). Replacing one of the anionic oxygen atoms by the o-carboranylmethyl moiety generates a new center of chirality at the phosphorus of dinucleotide 3. Due to this modification and the non-stereo-selectivity of the coupling reaction used, the o-carboranylmethylphosphonate oligonucleotide is obtained as a mixture of diastereoisomers (Figure 3). The effect of absolute configuration at phosphorus of P-chiral antisense oligonucleotides on their physicochemical and biochemical properties has been previously studied (57). We anticipated that duplex formation by the oligonucleotide containing several o-carboranylmethylphosphonate modifications would be severely affected by the absolute configuration (R$_P$ or S$_P$) at the phosphorus atom of internucleotide o-carboranylmethylphosphonate moiety. If the complex between carboranylmethyl-phosphonate modified oligonucleotide and complementary nucleic acid have B-type geometry, the o-carboranylmethyl group [1,2-C$_2$B$_{10}$H$_{11}$CH$_2$] of the S$_P$-isomer is oriented "inward", towards the DNA double-helix, near the hydrophobic base-stacking region of the complex, and that of the R$_P$-isomer is oriented "outward" away from the DNA double-helix (into the solvent). It has been shown for P-homochiral methylphosphonate oligonucleotides, that DNA duplexes with substituent directed outward into the solvent (R$_P$-isomer) are markedly more stable than those whose methyl groups project into the wide groove (S$_P$-isomer). In fact, duplexes formed by R$_P$-oligomethylphosphonates are more stable than those formed by their unmodified counterparts (58). Since the steric effect of the substituent at the phosphorus depends strongly on its bulkiness, one can expect that the destabilizing influence of the S$_P$-carboranylmethylphosphonate modification will be more pronounced than that of the methylphosphonate group (50). Therefore, it seems that synthesis of o-carboranyl-methylphosphonate modified oligonucleotides containing R$_P$ o-carboranylmethyl-phosphonate moieties should resolve the potential problems caused by the steric hindrance of bulky carboranyl group. Future work will take advantage of recent progress in the synthesis of P-chiral oligonucleotide analogues to obtain longer, P-stereodefined oligomers for biological evaluation (45). We have demonstrated that the potentially lower stability of duplexes formed by carboranyl oligonucleotides and complementary nucleic acids due to the bulkiness of carboranyl substituent can be reduced in the case of CDU-oligonucleotides by proper location of the carboranyl modified nucleoside within the oligonucleotide chain, and in the case of carboranylmethylphosphonate oligonucleotides by synthesizing of oligonucleotide with the R$_P$ configuration at the phosphorus of o-carboranylmethylphosphonate moiety. It should be pointed out that the chemistry we have developed readily permits the synthesis of suitable carboranyl oligonucleotides with carboranyl residue linked to a phosphorus atom or 5-carbon on the pyrimidine base through the proper spacer which should reduce unfavorable steric interactions. Multiple carboranyl modifications can be strategically placed on oligomer to favor strong interaction with complementary strand and to resist degradation.

Prospects

BNCT is binary system for treatment of malignancies which combines two, separately nonlethal constituents, a boron-10 radiosensitizer with nonionizing neutron irradiation. An advantage of such a system is that each component can be manipulated independently to maximize selectivity and efficiency. The carboranyl oligonucleotides proposed as new boron carrier warheads for BNCT add a new dimension to AOT due to the information encoded in the base sequence of the oligonucleotide part of the carrier. We hypothesize that this may allow more selective targeting of tumor cells by the sequence specific interaction of carboranyl oligonucleotides with overexpressed or unique tumor genes. The carboranyl oligonucleotides could be also potentially used independently as therapeutic antisense agents.

An added advantage of carboranyl oligonucleotides is the known tendency of oligonucleotides to accumulate in nucleus. The recent development of ion microscopy (IM) technique (59-61) provides a powerful tool which permits determination of intracellular location of carboranyl oligonucleotides. Studies on the cellular uptake of these compounds in relevant cells will provide insights which are critical for successful BNCT and AOT. Carboranyl oligonucleotides applied as molecular probes is another important future practical application.

Acknowledgments.

Supported by NIH grant CA 53892, and the Department of Veterans Affairs. Note that decaborane is extremely toxic and present an explosion hazard. It should be handled with proper care as it readily sublimes into air.

References.

1. *Advances in Neutron Capture Therapy;* Soloway, A.H.; Barth, R.F., Carpenter, D.E., Eds. Plenum Press: New York, NY, 1993.
2. Locher, G.L. *Am. J. Roentgenol. Radium Ther.* **1936**, *36,* 1.
3. Hawthorne, M. F. *Angew. Chem. Int. Ed. Engl.* **1993**, *32,* 950.
4. Lesnikowski, Z.J.; Schinazi, R.F. *J. Org. Chem.* **1993**, *58,* 6531.
5. Shaw, B.R.; Madison, J.; Sood, A.; Spielvogel, B.F.; *Protocols for Oligonucleotides and Analogs*; Agrawal, S., Ed.; Methods in Molecular Biology; Humana Press Inc.: Totowa, NJ, 1993, Vol. *20*; 225.
6. Zon, G. *Pharm. Res.* **1988**, *5,* 539.
7. Vlassov, V.V.; Yakubov, L.A. *Prospects for Antisense Nucleic Acid Therapy of Cancer and AIDS*, Wickstrom, E., Ed.; Wiley-Liss Inc.: New York, NY, 1991, 243.
8. Milligan, J.F.; Matteucci, M.D.; Martin, J.C. *J. Med. Chem.* **1993**, *36,*1923.
9. Bennett, C.F.; Chiang, M-Y.; Chan, H.; Shoemaker, J.E.E.; Mirabelli, C.K. *Mol. Pharm.* **1992**, *41,* 1023.
10. Leonetti, J.P.; Mechti, N.; Degols, G.; Gagnor, G.; Lebleu, G. *Proc. Natl. Acad. Sci. U.S.A.* **1991**, *88,* 2702.
11. Fisher, T.L.; Terhorst, T.; Cao, X.; Wagner, R.W. *Nucl. Acids Res.* **1993**, *21,* 3857.
12. Kobayashi, T.; Kanda, K. *Boron Neutron capture Therapy for Tumors*, Hatanaka, H., Ed.; Nishimura Co. Ltd.: Nigata, Japan, 1986, 293.
13. Chatterjee, A. *Radiation Chemistry, Principles and Applications*, Farhatziz and Rodgers, M.A.J., Eds.; VCH Publishers, New York, NY,1987, 1.
14. Sonntag von, C. *The Chemical Basis of Radiation Biology*; Taylor & Francis; Philadelphia, PA, 1987.

15. Lett, J.T. *Prog. Nucleic Acid Res. Mol. Biol.* **1990**, *39*, 308.
16. Becker, D.; Sevilla, M.D. *Advances in Radiation Biology*, Lett, J.T.; Sinclair W.K., Eds.; Academic Press, Inc.: San Diego, CA, 1993, 121.
17. Allen, A.O. *The Radiation Chemistry of Water and Aqueous Solutions*, D. Van Nostrand Co. Inc., Princeton, NJ, 1961.
18. Mee, L.K. *Radiation Chemistry, Principles and Applications*, Farhatziz and Rodgers, M.A.J., Eds.; VCH Publishers, New York, NY, 1987, 477.
19. Spinks, J.W.T.; Woods, R.J. *An Introduction to Radiation Chemistry*, John Wiley & Sons, Inc., New York, NY, 1990.
20. Huttermann, J.; Kohnlein, W.; Teoule, R.; Bertinchamps, A.J. *Effect of Ionizing Radiation on DNA. Physical, Chemical and Biological Aspects*;; Molecular Biology, Biochemistry and Biophysics; Springer-Verlag, Berlin, Germany, 1978, Vol. 27.
21. The greater biological damage caused by high linear energy transfer (high LET) radiation (α, heavy ions) than low LET radiation (β, γ, X) results from a high energy deposition per track length unit than from radiation with low deposition rate for a given total energy loss. This can be illustrated by the comparison of water decomposition efficacy of 40 keV electron (LET 0.08 eV/A), 18 MeV deuteron (LET 0.5 eV/A) and 5.5 MeV α-particle (LET 9.0 eV/A). Assuming 20 eV per water molecule decomposition needed, the energy deposited on 250 Å track corresponds to 1, 6 and 112 water molecules decomposed for electron, deuteron and α-particle of above energies, respectively (*17*).
22. Raju, M.R.; Eisen, Y.; Carpenter, S.; Inkret, W.C. *Radiat. Res.* **1991**, *128*, 204.
23. Neckers, L.; Whitesell, *L. Am. J. Physiol. (Lung Cell Mol. Physiol. 9)*, **1993**, *265*, L1.
24. Fairchild, R.G.; Bond, V.P. *Int. J. Radiat. Oncol. Biol. Phys.* **1985**, *11*, 831.
25. Kobayashi, T.; Kanda, K. *Radiat. Res.* **1982**, *91*, 77.
26. Gabel, D; Foster, S.; Fairchild, R.G. *Radiat. Res.* **1987**, *111*, 14.
27. Soloway, A.H.; Wright, R.L.; Messer, J.R. *J. Pharmacol. Exp. Ther.* **1961**, *134*, 117.
28. Barth, R.F; Soloway, A.H. *Cancer*, **1992**, *70*, 2995.
29. Ichihashi, M.; Nakanishi, T.; Mishima, Y. *J. Invest. Dermatol.* **1982**, *78*, 215.
30. Hatanaka, H.; Kamano, S.; Amano, K.; Hojo, S.; Sano. K.; Egawa, S.; Yasukochi, H. *Boron Neutron capture Therapy for Tumors*, Hatanaka, H., Ed.; Nishimura Co. Ltd.: Nigata, Japan, 1986, 349.
31. Goudgaon, N.M.; Fulcrand-El Kattan, G.; Schinazi, R.F. *Nucleosides Nucleotides* **1994**, *13*, 849.
32. Kane, R.R.; Pak, R.H.; Hawthorne, M.F. *J. Org. Chem.* **1993**, *58*, 991.
33. Kane, R.R.; Lee, C.S.; Drechsel, K.; Hawthorne, M.F. *J. Org. Chem.* **1993**, *58*, 227.
34. Kane, R.R.; Drechsel, K.; Hawthorne, M.F. *J. Am. Chem. Soc.* **1993**, *115*, 8853.
35. Powell, W.J.; Spielvogel, B.F.; Sood, A.; Shaw, B.R. Imeboron VIII, Knoxville, Tennessee, 1993, 125.
36. Shaw, B.R.; Sood, A.; Powell, W.; Tomasz, J.; Porter, K.; Spielvogel, B.F. International Conference on Nucleic Acid Medical Applications, Cancun, Mexico, 1993.
37. Sood, A.; Spielvogel, B.F. ; Tomasz, J.; Porter, K. W.; Shaw, B.R. Boron USA Third Workshop, Washington, Washington DC, 1992.
38. Reese, C.B. *Tetrahedron*, **1978**, *34*, 3143.
39. Lohrman, R.; Khorana, H.G. *J. Am. Chem. Soc.* **1964**, *86*, 4188.
40. Weber, H.; Khorana, H.G. *J. Mol. Biol.* **1972**, *72*, 219.
41. Arbuzov, B.A. *Pure and Applied Chemistry*, **1964**, *9*, 307.
42. Heying, T.L.; Ager, J.W. Jr.; Clark, S.L.; Mangold, D.J.; Goldstein, H.L.; Hillman, M.; Polak, R.J.; Szymanski, J.W. *Inorg. Chem.* **1963**, 1089.
43. Daub, G.W.; Van Tamelen, E.E. *J. Am. Chem. Soc.* **1977**, *99*, 3526.
44. Yamamoto, Y.; Seko, T.; Rong, F.G.; Nemoto, H. *Tetrahedron Lett.*, **1989**, *30*, 7191.

182   CARBOHYDRATE MODIFICATIONS IN ANTISENSE RESEARCH

45. Schinazi, R.F.; Goudgaon, N.M.; Fulcrand, G.; El Kattan, Y.; Lesnikowski, Z.; Ullas, G.V.; Moravek, J.; Liotta, D.C. *Intl. J. Radiation Oncol. Biol. Phys.*, **1994**, *28* , 1113.
46. Beaucage, S.L. *Protocols for Oligonucleotides and Analogs*; Agrawal, S., Ed.; Methods in Molecular Biology; Humana Press Inc.: Totowa, NJ, 1993, Vol. *20*; 33.
47. *Applied Biosystems USER Bulletin* No. 43, **1987**, Applied Biosystems, Foster City, CA.
48. Kane, R.R.; Lee, C.S.; Drechsel, K.; Hawthorne, M.F. *J. Org. Chem.*, **1993**, *58*, 3227.
49. Hawthorne, M.F.; Young, D.C.; Garret, P.M.; Owen, D.A.; Rchwerin, S.G.; Tebbe, F.N.; Wegner, P.A. *J. Am. Chem. Soc.* **1968**, *90*, 862.
50. Lesnikowski, Z.J. *Bioorg. Chem.* **1993**, *21*, 127.
51. Le Brun, S.; Duchange, N.; Namane, A.; Zakin, M.M.; Huyng-Dinh, T.; Igolen, J. *Biochimie*, **1989**, *71*, 319.
52. Miller, P.S. *Bio/Technology*, **1991**, *9*, 358.
53. Sarin, P.S.; Agrawal, S.; Civeira, M.P.; Goodchild, J.; Ikeuchi, T.; Zamecnik, P.C. *Proc. Natl. Acad. Sci. U.S.A.*, **1988**, *85*, 7448.
54. Vespieren, P.; Cornelissen, A.W.C.A.; Thuong, N.T.; Hélene, C.; Toulmé, J.J. *Gene*, **1987**, *61*, 307.
55. Agrawal, S.; Goodchild, J. *Tetrahedron Lett.*, **1987**, 3539.
56. Singh, U.C.; Weiner, P.K.; Caldwell, J.; Lollman, P.A. AMBER 3.0, **1986**, University of California: San Francisco, CA.
57. Uhlman, E.; Peyman, A. *Chem. Rev.* **1990**, *90*, 544.
58. Lesnikowski, Z.J.; Jaworska, M.; Stec, W.J. *Nucl. Acids Res.*, **1990**, *18*, 2109.
59. Castaing, R.; Slodzian, G. *J. Microscopie (Paris)*, **1962**, *1*, 395.
60. Bennett, B.D.; Zha, X.; Gay, I.; Morrison, G.H. *Biol. Cell.*, **1992**, *74*, 105.
61. Zha, X.; Ausserer, W. A.; Morrison, G.H. *Cancer Res.*, **1992**, *52*, 5219.

RECEIVED July 18, 1994

RNA AND RNA ANALOGUES

Chapter 12

Advances in Automated Chemical Synthesis of Oligoribonucleotides

Nanda D. Sinha and Stephen Fry

Millipore Corporation, 75A Wiggins Avenue, Bedford, MA 01730

Synthesis of oligodeoxynucleotides (DNA) by chemical methods is now a standard process due to automation and optimization in synthetic methodologies. Synthetic strategies for oligoribonucleotides (RNA) are essentially the same as for DNA; however, progress in this field is somewhat limited due to the presence of 2′-hydroxyl group. Proper selection of its protection during chain assembly and complete removal at the end of synthesis are essential steps in RNA synthesis. Various approaches have been explored to optimize the chemical synthesis of RNA. In addition, several modifications and improvements have been introduced to facilitate post synthesis workup. Presently, it is possible to synthesize oligoribonucleotide chains in sufficient quantities to study catalytic activities and X-ray crystallographic structures of synthetic RNA using t-BPA as a labile exocyclic amine and t-BDM silyl for 2′-hydroxyl protections.

Approaches for the chemical syntheses of DNA and RNA were simultaneously developed in the 1960's by Khorana et al., which ultimately resulted in the synthesis of the alanine t-RNA gene (1) and the codons for amino acids, i.e., 64 ribonucleotide triplets (2). The progress and refinements in the synthetic chemistry for internucleotide bond formations [phosphate triester (3) - phosphite (4) - phosphoramidite (5)/H-phosphonate (6)] have made the chemical synthesis of DNA a standard and routine process. These successful refinements have also been explored in the synthesis of oligoribonucleotides (7) with limited success, due to the presence of the 2′-hydroxyl function. In the synthesis of RNA chains, three hydroxyl groups and the exocyclic amine functions of the ribonucleoside have to be dealt with. The 2′-hydroxyl group has to be protected during chain assembly and unprotected at the end of the synthesis.

The selection of 2′-OH protecting group is such that it should be compatible with the chemistry and reagents used in chain extension, and at the end of the synthesis easily and quantitatively removed without any significant side reactions. In recent years, the 2′-hydroxyl groups of ribonucleosides have been protected with acetal (8), benzyl (9), or silyl (10) derivatives and have been used in oligoribonucleotide synthesis following phosphotriester and phosphoramidite chemistries. Two of these protecting group classes, acetals [Mthp (11); Fpmp (12)] and silyls [t-BDM silyl (13) and TIPS (14)], are presently used for 2′-hydroxyl protection in conjunction with the use of dimethoxytrityl (DMT) for 5′-hydroxyl protection. Other combinations such as

0097–6156/94/0580–0184$08.00/0

(i) acid labile Mthp for 2′-OH and base labile levulinyl for 5′-OH and (ii) acid labile Mthp for 2′-OH and base labile fluorenylmethoxy carbonyl FMOC for 5′-OH have also been used with some success for RNA synthesis (*15, 16*).

Quantitative internucleotide bond formation at the chain extension step is also a significant challenge in RNA synthesis. Two approaches have been utilized to address this goal: using activators other than standard tetrazole (*17*) or replacing standard N,N-diisopropylamino phosphoramidite with more reactive N,N-diethylamino/N,N-ethylmethylamino phosphoramidites (*18, 19*). Similarly, to insure high quality synthetic RNA using silyl derivative for 2-hydroxyl protection, more base labile exocyclic amine protecting groups (*20, 21*) have been introduced. The milder de-acylation of nucleoside exocyclic amines at room temperature and shorter exposure time to basic conditions minimizes the de-silylation (*21*). This in turn minimizes RNA chain degradation due to the exposure to bases.

Since the discovery of self-splicing activity of RNA chains, tremendous efforts have been directed towards the utilization of synthetic RNA for therapeutic purposes based on ribozyme technology. The quantity as well as the quality of synthetic RNA necessary for therapeutic purposes and for understanding protein-RNA binding activities is important. Various protocols have been reported for the complete removal of protecting groups and removal of excess reagents used in carrying out successful deprotections. Additionally, to synthesize high quality RNA, the quality of monomers is important. The monomers should be free from regioisomeric phosphoramidites such as 5′- or 2′-*O*-phosphoramidite.

This chapter describes the improvements carried out in our laboratory to fulfill some of the requirements (scale and quality) to make synthetic RNA as a viable therapeutic agent and to understand the interaction of protein-RNA binding by their crystal structure. The strategy of our synthesis follows standard phosphoramidite chemistry for synthesis of RNA on solid supports, i.e., (i) 5′-hydroxyl group protected with acid labile dimethoxytrityl (DMT), (ii) 2′-hydroxyl protected as silylether using ṭ-BDM silyl, (iii) exocyclic amines of nucleosides protected with labile ṭ-butylphenoxyacetyl, (iv) 3′-OH functionalized with either N,N-diethylamino phosphoramidite or N,N-diisopropylamino phosphoramidite, (v) activation of phosphoramidite for internucleotide bond formation with alkyl-thio tetrazole for larger scales, (vi) removal of exocyclic amine protection with conc. NH4OH/EtOH mixture at R.T., (vii) desilylation with tetrabutylammonium fluoride, (viii) removal of excess TBAF by quaternary methyl ammonium membrane (QMA-MemSep) cartridge, (ix) purification by two step chromatography (reverse-phase and ion exchange).

We have recently introduced ṭ-butylphenoxyacetyl as exocyclic amino group protection in DNA synthesis (*21*). Although labile to milder basic conditions, this is stable towards most of the reagents used in DNA synthesis and minimizes depurination. Complete removal of this group is achieved by short exposure to conc. NH4OH (2 hr vs. 48 hr at R.T. or 15 min. vs. 16 hr at 55° C) or conc. NH4OH/EtOH mixture (3:1) 2 hr at 55° C. This faster removal of the exocyclic amino protecting group should benefit the chemical synthesis of RNA using 2′-*O*-silyl protected monomers.

The synthesis approach for monomeric building blocks is outlined in Scheme 1, and some of the steps were carried out by following literature procedures (*18, 21*).

(a)   N-ṭ-Butylphenoxyacetyl-ribonucleosides: The protection of exocyclic amine was carried out following transient protection strategy (*22*) using ṭ-butylphenoxyacetic acid activated with carbonyl diimidazole in pyridine/acetonitrile mixture for A, C, and ṭ-butylphenoxyacetyl chloride for guanosine.

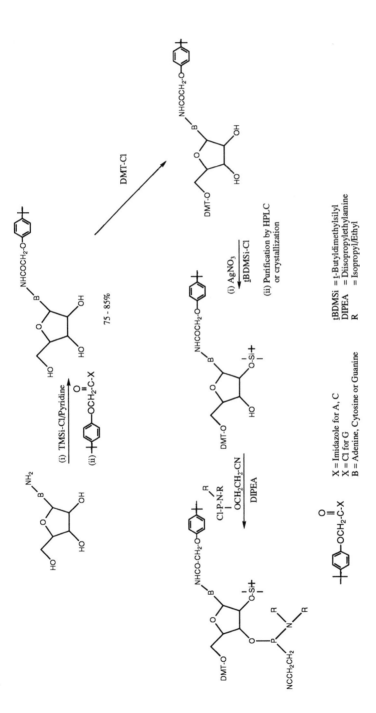

Scheme 1   Synthesis Steps Involved in the Preparation of Suitably Protected Ribonucleoside Phosphoramidites

tBDMSi = t-Butyldimethylsilyl
DIPEA = Diisopropylethylamine
R = Isopropyl/Ethyl

X = Imidazole for A, C
X = Cl for G
B = Adenine, Cytosine or Guanine

(b)   N-ţ-BPA-5′-*O*-DMT-ribonucleosides: 5′-Hydroxyl protection of ribonucleosides was performed with dimethoxytrityl chloride in pyridine following
the standard method. Purification of these was either carried out by HPLC or
crystallization.

(c)   N-ţ-BPA-2′-*O*-ţBDMSi-5′-*O*-DMT-ribonucleosides: Silylation reaction was
carried out following the method of Ogilvie et al. (*23*), i.e., treatment of N-ţ-
BPA-5′-*O*-DMT-ribonucleoside with t-butyldimethylsilyl chloride in the
presence of anhydrous silver nitrate in pyridine/THF. With the use of N-t-
butylphenoxyacetyl nucleoside, formation of the desired 2′-*O*-ţ-BDM silyl
ether was found to be selective (70% or greater). In addition to this selectivity,
the 2′-*O*-ţ-BDM silyl derivatives of C and G were isolated as a crystalline
material with greater than 99% purity by HPLC, NMR. The desired 2′-*O*-ţ-
BDM silyl derivatives of A and U with similar purity were obtained by silica
gel chromatography using a mixture of hexane-ethylacetate.

(d)   N-ţ-BPA-2′-*O*-ţ-BDMSi-5′-*O*-DMT-ribonucleoside-3′-*O*-(N,N-dialkylamino-
ß-cyanoethyl)-phosphoramidites: 2′-*O*-Silyl protection is prone to migration
under basic conditions especially in the presence of moisture. It is essential to
prevent this migration since the separation of two regioisomeric
phosphoramidites becomes very difficult. In order to prevent this migration of
silyl group, a combination of bases has been suggested instead of N,N-
diisopropylethyl amine (*24*). When the phosphitylation of the 3-hydroxyl
ribonucleoside containing the 2′-*O*-ţ-BDMSi group with monochloro-ß-
cyanoethoxy-N,N-diisopropylamino/N,N- diethylamino phosphine (1.2 to 1.5
eq.) was carried out in the presence of anhydrous N,N-diisopropylethyl amine
(2.5 to 3.0 eq.) in anhydrous tetrahydrofuran, formation of undesired 2′-*O*-
phosphoramidite was not detected.   The presence of an undesired
phosphoramidite isomer (2′-*O*-phosphoramidite) formed due to the migration
of silyl group can be detected by ^{31}P-NMR. The chemical shifts of 2′-*O*-
phosphoramidite generally lie between the chemical shifts of two
diastereoisomers of 3′-*O*-phosphoramidites, e.g., ^{31}P chemicals shifts for "C"-
2′-phosphoramidite and "C"-3′-phosphoramidites are 150.24 and 150.3 ppm
and 149.43 and 151.51 ppm respectively.

Oligoribonucleotide Synthesis

Synthesis of oligoribonucleotides was carried out at two different scales using a
nucleoside loaded onto CPG supports (30-35 µmol/g). The 1.0 µmol scale syntheses
were carried out on the 8900 Expedite Nucleic Acid synthesizer system, and 30-100
µmol large scale syntheses were performed on the 8800 DNA synthesizer. The
protocol for 1.0 µmol synthesis follows the DNA synthesis cycle with the exception
that an extended coupling time (9-10 min.) was used. Longer oligoribonucleotide
sequences were synthesized with N,N-diethylaminophosphoramidites (100 mg/ml)
and ethylthio tetrazole solution in acetontrile (0.4 M) to give slightly better coupling
(98.0 to 99% based on DMT cation measurement). The protocol used for larger scale
synthesis (up to 100 µmol) is detailed in Scheme 2. Using this protocol, several
oligonucleotide chains up to 28 bases long were synthesized. The average coupling
efficiency on large scale syntheses using either methylthio or ethylthio tetrazole was
found to be 96-97%. The coupling times for these syntheses were 25-30 minutes.
Longer coupling time or higher concentration of amidites did not improve the coupling
efficiency.   Furthermore, use of p-nitrophenyl tetrazole to increase the coupling
efficiency did not improve the quality of oligonucleotide, rather the presence of n+1,

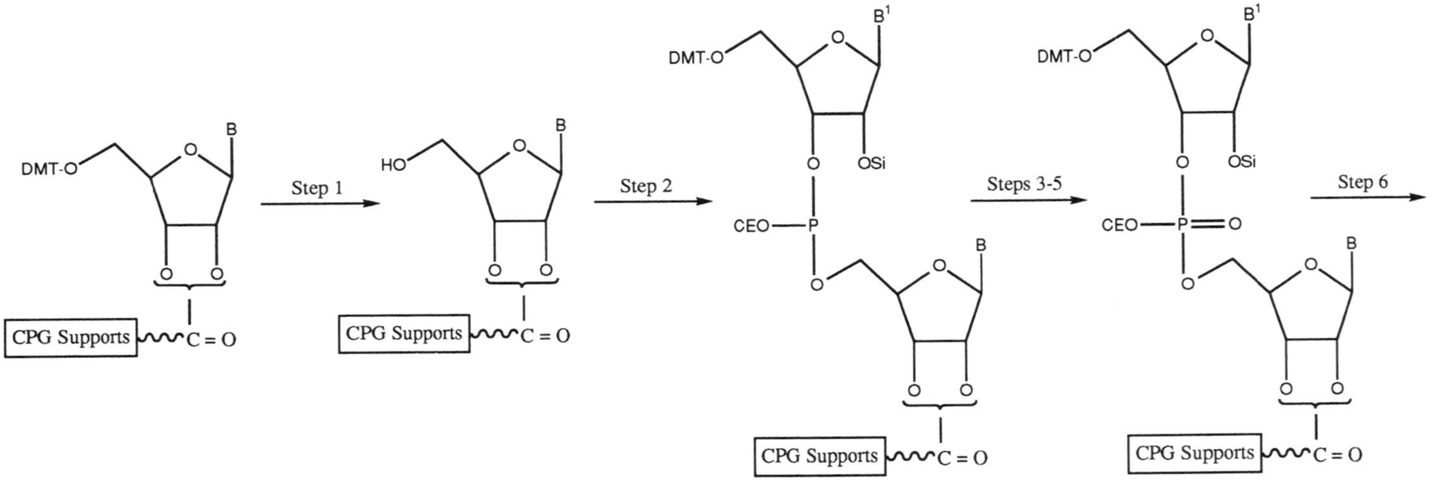

Steps

1. Detritylation with DCA/TCA solution
2. Activation and coupling with tetrazole/ethyl-thio-tetrazole and phosphoramidite monomer
3. Capping with t-butylphenoxyacetic anhydride, N-methylimidazole and pyridine in THF
4. Oxidation with iodine-pyridine-water mixture in THF
5. Capping with t-butylphenoxyacetic anhydride
6. Repeat of steps 1-5

Scheme 2   Steps Involved in Automated Oligoribonucleotide Synthesis on Solid
            Supports

n+2, n+3... oligomers were observed. The 5'-terminal DMT was left on the oligonucleotide chain for facilitating purification.

Removal of Exocyclic Amine Protection. After synthesis, the solid support was dried by flowing argon through the column and then transferred into a screw cap Eppendorf tube. For 1.0 µmol scale synthesis, the support was treated with a mixture of conc. $NH_4OH/EtOH$ (3:1, 1 ml) either at 55° C for 2 hr or at R.T. for 16 hr. Complete removal of exocyclic amine protecting group (tBPA) from oligonucleotide chains can be achieved with this mixture within 2 hr at room temperature. Surprisingly, however, treatment of CPG beads with this reagent for 2 hr at room temperature does not release oligonucleotide chains completely from solid supports. This solution requires longer incubation at room temperature or incubation at higher temperature (2 hr, 55° C) unlike concentrated NH_4OH (1 hr, R.T.) for complete chain release. In the case of large scale synthesis (60 to 100 µmol), the supports were incubated with 60-100 ml of ammonia solution for 16 hr at R.T. After this treatment, the solution was filtered through a Millex filter and the CPG beads were washed thoroughly with a mixture of ethanol and water. The combined solution was dried in a centrifugal speed vac. The material so obtained was redissolved in ethanol-water, and, if necessary, the solution was filtered again to remove traces of CPG particles. The filtered solution was finally dried in the speed vac.

Desilylation. Various desilylating agents, as a source of fluoride ion, have been suggested for complete removal of the 2'-O-t-BDM silyl protecting group; tetrabutylammonium fluoride (TBAF) and triethylamine-hydrofluoric acid (Et_3NHF) (26) are generally recommended. Tetrabutylammonium fluoride is the preferred desilylating agent in the case of an oligoribonucleotide with 5'-terminal DMT-group as a purification handle. In order to achieve satisfactory deprotection of the t-BDM silyl group with TBAF solution, we found it is essential to have anhydrous material. As a general method, 1.0 ml TBAF solution (1.0 M) is used for 1.0 µmol scale synthesis. In order to evaluate complete desilylation, we monitored this step by anion exchange HPLC using Gen Pac Fax column at different intervals (4 to 24 hr). Complete removal of silyl groups can be achieved within 24 hr at room temperature as shown in Figure 1. Completely desilylated material results in a single major peak, whereas incomplete desilylated material results in several peaks next to the desired peak. Thus, incubation of the silylated RNA chain with TBAF solution for 20-24 hr at R.T. removes the silyl group completely. Commercially available colorless solution or freshly prepared solution can be used without a problem. However, yellow colored TBAF solution tends to cause problems or does not desilylate completely. For our 60 to 100 µmol scale synthesis, the material obtained after removal of exocyclic amine protection was divided into 3 to 5 equal portions, and each was treated with 20 ml of TBAF solution (1.0 M) for 24 hr. After complete desilylation, the excess TBAF solution was quenched with ammonium bicarbonate solution (20 mM, pH = 7 to 7.5), which allowed us to retain the 5'-terminal DMT-group on the oligomer chain. The reaction mixture was diluted to ten times its volume with ammonium bicarbonate solution (pH = 7.5) made from sterilized water.

Removal of Excess TBAF. Purification of synthetic RNA is complicated due to the presence of the tetrabutylammonium ion in the crude mixture. Several techniques have been recommended for this purpose: (a) gel exclusion chromatography, (b) dialysis, (c) precipitation or cartridge-based RP, (d) ion exchange, or (e) extraction. Most of these methods work satisfactorily for small scale synthesis. However, we found adapting one of these techniques for large scale synthesis to be either inconvenient or inefficient and time consuming. Presently, we are using a weak ion exchange membrane cartridge (QMA-MemSep) for the removal of TBAF.

Protocol for Large Scale RNA Synthesis on the 8800

Synthesis Steps (mode)	Reagents/Solvents	Repeat/Duration (Scale Dependent)[a]
1. Wash (flow through)	Acetonitrile	2 x 6 sec (5.0 ml)
2. Deblock (batch)	3% DCA in methylene chloride	6 x 6 sec (8.5 ml) with 30 sec mix
3. Wash (flow through)	Acetonitrile	4 x 6 sec (6.0 ml)
4. Precouple wash	Dry acetonitrile	1 x 6 sec (6.0 ml) with 30 sec mix and 60 sec argon flow
5. Couple (batch)		
(a) Tetrazole wash	Tetrazole[b]/alkyl thio tetrazole[c]	1.0 sec (1.5 ml) 15 sec mix and 15 sec argon flow
(b) 1st coupling	Tetrazole/amidite[d]/ tetrazole	1.7 ml/1.0 ml/1.7 ml mix for 30 sec, pause for 4.5 min. and repeat the process twice.
(c) 2nd coupling	Tetrazole/amidite[d]/ tetrazole	1.7 ml/1.0 ml/1.7 ml mix for 30 sec, pause for 4.5 min. and repeat the process twice.
	Tetrazole	1.2 ml mix for 2 min.
6. Wash (flow through)		3 x 6.0 sec (6.0 ml)
7. Capping (batch)	Cap A (anhydride in THF) Cap B (pyridine, N-methyl imidazole in THF)	Cap A = 2.7 ml Cap B = 5.6 ml mix for 2 min.
8. Wash (flow through)	Acetontrile	4 x 6.0 sec (6.0 ml)
9. Oxidation	I₂ in pyridine, water and THF	6.4 ml mix for 4.0 sec
10. Wash (flow through)	Acetontrile	4. x 6.0 sec (6.0 ml)
11. Capping	Cap A mixture Cap B mixture	Cap A = 2.7 mll Cap B = 5.6 ml mix for 2 min.
12. Wash (flow through)	Acetonitrile	4 x 6.0 sec (6.0 ml)

a = Scale depends on loading of CPG and amount used
b = Tetrazole solution (0.46 M in acetonitrile)
c = Ethylthiotetrazole (0.4 M in acetonitrile)
d = Amidite (100 mg/ml in acetontrile)

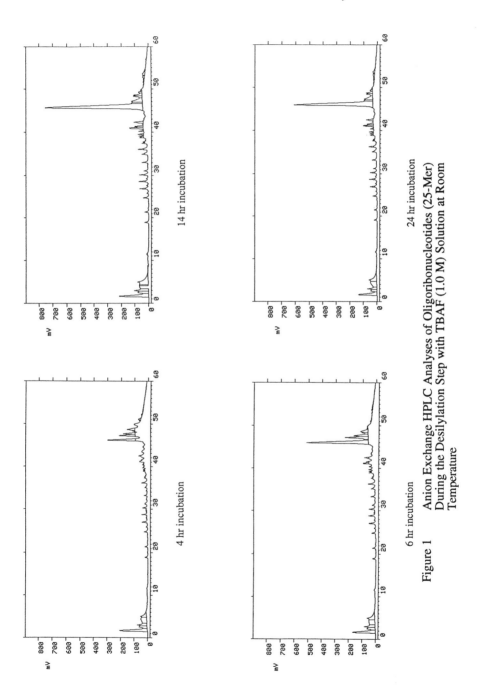

Figure 1   Anion Exchange HPLC Analyses of Oligoribonucleotides (25-Mer) During the Desilylation Step with TBAF (1.0 M) Solution at Room Temperature

The cartridge is equilibrated with dilute ammonium bicarbonate (20 mM) solution containing 10% acetonitrile. The crude synthetic RNA solution obtained after desilylation and quenching with ammonium bicarbonate is loaded onto the QMA-MemSep cartridge at a flow rate of 5 to 15 ml/min. (depending on cartridge capacity). The cartridge is then washed with ammonium bicarbonate (20 mM) solution for 5 minutes followed by a step gradient, 2% at a time, with ammonium bicarbonate (1.5 M) solution for 3-5 minutes, and finally oligonucleotide sequences were eluted with 20 to 100% ammonium bicarbonate (1.5 M) solution. The fractions collected were analyzed for the desired material and pooled for further purification. We like to mention that during loading and washing only short failure and protecting groups are removed, and there is no loss of full length product.

Purification by HPLC. In the case of synthetic DNA, one step purification by HPLC results in a higher quality product (90-95%). However, due to the secondary structure of synthetic RNA, it is not possible to achieve higher purity by one step chromatography. In our method, two step purifications were used to achieve purity greater than 96%. The first step chromatography is based on RP column, and the second step is strong anion exchange chromatography using Protein PAC Q-15.

(a) RP-HPLC: Preparative PRP-1 250 x 21.5 mm (Hamilton, Reno, Nevada) was equilibrated with triethylammonium acetate solution (TEAA, 0.1 M, pH = 7.0). The pooled solution (after MemSep desalting ~ 2000 O.D. at 260 nm) was loaded directly onto the column at a flow rate of 5 ml/min. After loading, the column was washed with 0.1 M TEAA solution for 5 min. The full length DMT-on product was eluted using the following gradient Table I.

Table I. Gradient Table for Large Scale RP-Purification (2000 O.D. Crude Material)

Time	Buffer A	Buffer B	Flow Rate
0	100%	0%	5.0 ml
10.0	75%	25%	5.0 ml
11.0	75%	25%	3.0 ml
40.0	65%	35%	3.0 ml
42.0	65%	35%	3.0 ml
50.0	50%	50%	3.0 ml
55.0	0%	100%	10.0 ml
65.0	0%	100%	10.0 ml

Buffer A =   TEAA solution (0.1 M, pH = 7.0)
Buffer B =   Acetonitrile

The oligomer with 5′-terminal DMT-group eluted between 30 to 35% acetonitrile. The pooled materials (1000 O.D.) thus obtained in most cases were found to contain about 85 to 90% of the desired full length product. The pooled and lyophilized material was detritylated with 50% acetic acid at R.T. and concentrated to dryness.

(b) Ion exchange purification: The ion exchange chromatographic purification was carried out to achieve purity greater than 96%. Several buffers and different anion exchange columns have been recommended for this purification. Anion exchange chromatography based on Protein-Pac Q-15 HR™ column 100 X 20 mm from Waters Chromatography gave the most satisfactory result using sodium phosphate and sodium chloride solution. The oligomer obtained from RP-HPLC step was dissolved in minimum volume of phosphate buffer (20 mM Na_2HPO_4) and loaded onto the column at a flow rate of 2.0 ml/min. The desired product is eluted following the gradient Table II.

Table II. Gradient Table for Large Scale Anion Exchange Purification up to 2000 O.D.

Time	Buffer A	Buffer B	Flow Rate
0	80%	20%	2.0 m
10 min.	70%	30%	2.0 ml
90 min.	40%	60%	2.0 ml
100 min.		100%	2.0 ml
120 min.		100%	3.0 ml

Buffer A = 20 mM Sodium phosphate in 10% acetonitrile water
Buffer B = 20 mM Sodium phosphate and 1.0 M sodium chloride in 10% acetontrile water

The desired pooled fractions were desalted by a second reverse phase column. The pooled solution was loaded onto the column at a flow rate of 4.0 ml/min. Salt was removed by passing triethylammonium acetate (0.1 M, pH = 7.0) for 5 min. at a flow rate of 4.0 ml/min. The desired material (800 O.D.) was eluted as a single peak by increasing the percentage of acetonitrile to 100% in 45 min. The analysis of the desalted material is shown in Figures 2 and 3. The results obtained from various scales of syntheses and purification are given in Table III.

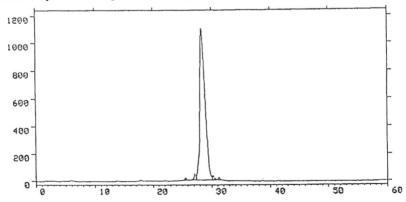

HPLC Condition: Column: GenPak FAX, Eluent A: 20 mM Ammonium phosphate + 10% CH_3CN, Eluent B: Eluent A + 1M NaCl, Gradient: 20-100% B in 60 minutes.

Figure 2   Anion-Exchange HPLC Chromatogram of Purified AUG-UAC-AAA-UAU After Preparative Reverse-Phase HPLC

Table III.   Synthesis and Purification Results

Sequences	Scale µmol	Average Coupling (%)	Material Crude	(O.D.) Purified	Final Purity (%)
GCCGCGAGC	32	97	1756	850	98
CUGCGCGAG	32	97	1930	925	98
AAUGAGGAAAUU	66	96.5	4150	2025	97
CAUCAAUGCUUGC-ACCGAUG	30	96.3	2550	1150	97
UAUGUGGAAGAC-AGCGGGUGGUUC	a. 1.0 b. 32.0	98 96	206 5450	88 1810	97
UGCCAGCUAUGA-GGUAAAGUGUCA-UAGC	64	96.5	10970	3550	96.5
Fl-GCUCUCGUCUG-AUGAGGCUUCGG-CCGAAAGACCGU	1.0	98	250	100	95

Evaluation of Synthetic RNA

One common test of synthetic oligoribonucleotides is the determination of enzymatic activity, i.e., cleavage oligoribonucleotides by another oligoribonucleotide strand in the presence of magnesium or manganese ion. To evaluate the quality of synthetic and purification methods, we synthesized two RNA sequences of an S24-R19 Hammerhead ribozyme shown in Figure 4; S24 and R19 sequences are enzyme and substrate strands respectively. These sequences were purified in good yield following deprotection and purification methods described above. Small aliquots from these sequences were radiolabelled with ^{32}P, and ribozyme activity studies were carried out according to the literature method (24). The result obtained from this experiment is shown in Figure 5, indicating that synthetic oligoribonucleotides are also effective in cleaving other synthetic RNA strands. Another important way of evaluating the quality of synthetic oligoribonucleotides is crystal formation. Several short sequences (12- to 14-mers) synthesized and purified by the above methods have resulted in a nice single crystal (27) as shown in Figure 6.

Discussion. Although several reports dealing with the synthesis of oligoribonucleotides are available in the literature, none of these reports (28-30) deals with scales beyond 10.0 µmol. Small scale synthesis of RNA and materials obtained from these syntheses may be sufficient to carry out initial biological studies but may not be sufficient to explore RNA molecules for therapeutic agents. It is, therefore, essential to have the ability to synthesize and purify these molecules in scales larger than previously described. This report describes some advancements in synthesizing RNA molecules up to 100 µmol scale. By introducing t-butylphenoxyacetyl as exocyclic amine protection, we have been able to remove this protecting group at a

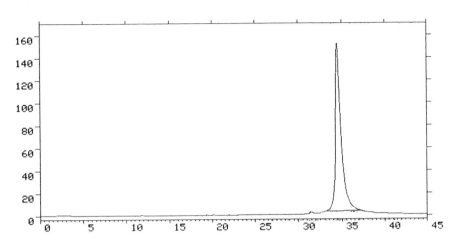

HPLC Condition: Column: GenPak FAX, Eluent A: 20 mM Ammonium phosphate + 10% CH3CN, Eluent B: Eluent A + 1M NaCl, Gradient: 20-100% B in 60 minutes.

Figure 3   Anion-Exchange HPLC Chromatogram of Purified 5'-UGCCAG-CUAUGA-GGUAAA-GUG-UCAUAGC After Preparative Reverse-Phase HPLC

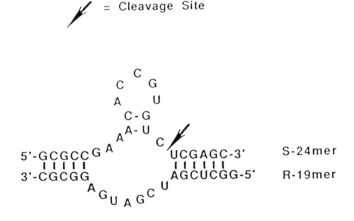

Figure 4   S24-R19 Hammer Head Ribozyme (Reproduced with permission from *Biochimie* 1993, *75*, 13-23. Copyright 1993 Elsevier, Paris)

ORGc

XC

BPB

END

Lanes (L-R) 1 = 24 mer, 2 = 19 mer, 3 = 24 & 19-mer mixture, 4 = 24-mer and 5 = 19-mer with incubation buffer 50 mM Tris pH = 7.5, 0.5 mM EDTA and 20 mM Mg^{2+} and 6 & 7 = 24 & 19-mer mixture with incubation buffer and 20 mM Mg^{2+} at 35 and 50 °C; 8 & 9 = 24 & 19-me mixture with 20 mM Mg^{2+} where 19-mer is not 31P-labelled.

Figure 5   S-24-R19 Hammer Head Ribozyme Cleavage Reaction (Reproduced with permission from *Biochimie* 1993, *75*, 13-23. Copyright 1993 Elsevier, Paris)

Figure 6   Crystal Structure of Oligoribonucleotide AUG-UAC-AAA-UAU After Purifications

milder condition using conc. NH4OH-ethanol mixture within 2 hr instead of 16 hr at 55° C. This shorter exposure at 55° C minimizes the extent of desilylation, chain cleavage and subsequent loss of full length material. The use of N,N-diethylamino phosphoramidites instead of standard N,N-diisopropylamino phosphoramidite and alkylthiotetrazole allows us to increase the coupling efficiency for longer sequences or larger scales by 2 to 3 percent. Standard N,N-diisopropylaminophosphoramidite and tetrazole mediated synthesis on 30 to 100 μmole scale results in 92-94% stepwise coupling. The desilylation with tetrabutylammonium fluoride solution instead of triethylamine hydrofluoric acid permits us to retain the 5′-terminal DMT-group, which helps in the RP-HPLC purification method. We find that the use of ammonium bicarbonate solution (.1 M, pH = 7.5) instead of triethylammonium acetate (.1 M, pH = 7.0) as a quenching buffer for excess TBAF reagent allows us to retain the 5′-terminal DMT-group quantitatively. The removal of excess TBAF and other species generated from it can be achieved by MemSep Cartridge QMA more efficiently than by other methods. The material obtained from desalting can be directly loaded onto the RP-column. The reverse-phase chromatography using DMT-on strategy helps to remove failure and shorter sequences with no DMT-group. The anion exchange chromatography using Protein-Pac A-15 HR material as a stationary phase eliminates secondary structure and removes any n-1, n-2 sequences carried over from RP-HPLC. Finally, desalting on a reverse-phase column removes inorganic salts present in buffers used. We have been able to synthesize RNA molecules in several hundred milligram quantities with purity greater than 95%. Synthetic oligoribonucleotides obtained by these methods also possess biological activity as demonstrated by ribozyme activity. Various other laboratories are able to synthesize RNA up to 60 bases long using our labile base-protected ribonucleotide phosphoramidite successfully (*31*).

Conclusion

It is certain that synthesis of oligoribonucleotide (RNA) has yet to reach the status of synthesis of oligodeoxynucleotide (DNA). However, continuing improvements are being carried out to synthesize longer and larger scales of oligoribonucleotides. We have been able to synthesize RNA in larger scales than previously described with suitable protecting strategies. Our goal is to optimize the scale of synthesis, deprotection, and purification so it becomes possible to achieve results similar to DNA synthesis.

Acknowledgments

We would like to thank Margaret Simpson for her help in preparing this manuscript and Jack Johansen for his support and encouragement.

Literature Cited

1. Khorana, H. G. *Pure Appl. Chem.* **1968**, *17*, 349-381.
2. Lohrmann, R.; Söll, D.; Hagatsu, H.; Ohtsuka, E.; Khorana, H. G. *J. Am. Chem. Soc.* **1966**, *88*, 819-829.
3. (a) Letsinger, R. L.; Makadevan, V. *J. Am. Chem. Soc.* **1965**, *87*, 3526.
    (b) Ikehona, K.; Babl, C. P.; Katagiri, N.; Michniewicz, J.; Wightman, R. H.; Narang, S. A. *Can. J. Chem.* **1973**, *51*, 3469.
    (c) Reese, C. B. *Tetrahedron* **1978**, *36*, 3075.

4. (a) Letsinger, R. L.; Finnan, J. L.; Heavner, G. A.; Lunsford, W. B. *J. Am. Chem. Soc.* **1975**, *97*, 3278-3279.
    (b) Letsinger, R. L.; Lunsford, W. B. *J. Amer. Chem. Soc.* **1976**, *98*, 3655-6570.
5. (a) Beaucage, S. L.; Caruthers, M. H. *Tetrahedron* **1981**, *22*, 1859-1862.
    (b) McBride, L. J.; Caruthers, M. H. *Tetrahedron* **1983**, *24*, 245-248.
    (c) Sinha, N. D.; Biernat, J.; Köster, H. *Tetrahedron Lett.* **1983**, *24*, 5843-5846.
6. Froehler, B. C.; Ng, P. G.; Matteucci, M. D. *Nucl. Acids Res.* **1986**, *14*, 5399-5407.
7. (a) Caruthers, M. H. In *Synthesis and Applications of DNA and RNA*; Narang, S. A., Ed.; Academic Press, Inc., Harcourt Brace Jovanovich: Orlando, FL, 1987; pp 47-94.
    (b) Ohtsuka, E.; Iwai, S. In *Synthesis and Applications of DNA and RNA*; Narang, S. A., Ed.; Academic Press, Inc., Harcourt Brace Jovanovich: Orlando, FL, 1987; pp 115-136.
8. Griffin, B. E.; Reese, C. B. *Tetrahedron Lett.* **1964**, 2925.
9. Takaku, H.; Kamaike, K.; Kasuga, K. *J. Org. Chem.* **1982**, *47*, 4937-4940.
10. Ogilvie, K. K.; Sadana, K. L.; Thompson, E. A.; Quillian, M. A.; Westmore, J. B. *Tetrahedron Lett.* **1974**, 2861.
11. Reese, C. B.; Saffhill, R.; Sulston, J. *J. Am. Chem. Soc.* **1967**, *89*, 3366.
12. Reese, C. B.; Thompson, E. A. *J. Chem. Soc. Perkin I* **1988**, 2881.
13. Usman, N.; Ogilvie, K. K.; Jiang, M.-Y.; Cedergren, R. J. *J. Am. Chem. Soc.* **1987**, *109*, 7845.
14. Wu, T.; Ogilvie, K. K. *J. Org. Chem.* **1990**, *55*, 4717.
15. Iwai, S.; Ohtuska, E. *Nucl. Acids Res.* **1988**, *16*, 9443.
16. Lehmann, C.; Xu, Y.-Z.; Christodoulou, C.; Tan, Z.-K.; Gait, M. J. *Nucl. Acids. Res.* **1989**, *17*, 2379.
17. Beijer, B.; Sulston, I.; Sproat, B. S.; Rider, P.; Lamond, A. I.; Neurer, P. *Nucl. Acids Res.* **1990**, *18*, 5143.
18. Lyttle, M. H.; Wright, P. B.; Sinha, N. D.; Bain, J. D.; Chamberlin, A. R. *J. Org. Chem.* **1991**, *56*, 4608.
19. Gasparutto, D.; Teoule, R. *Nucl. Acids Res.* **1992**, *20*, 5159.
20. Chaix, C.; Molko, D.; Teoule, R. *Tetrahedron Lett.* **1989**, *30*, 71.
21. Sinha, N. D.; Davis, P.; Usman, N.; Perez, J.; Hodge, R.; Kremsky, J.; Casale, R. *Biochimie* **1993**, *75*, 13.
22. Ti, G. S.; Gaffney, B. L.; Jones, R. A. *J. Am. Chem. Soc.* **1982**, *104*, 1316.
23. Hakimelahi, G. H.; Proba, Z. A.; Ogilvie, K. K. *Can. J. Chem.* **1982**, *60*, 1106.
24. Scaringe, S.; Francklyn, C.; Usman, N. *Nucl. Acids Res.* **1990**, *18*, 5433.
25. Paulsin, N. N.; Morocho, A. M.; Chen, B.-C.; Cohen, J. S. *Nucl. Acids Res.* **1994**, *22*, 639.
26. Gasparutto, D.; Livache, T.; Bazin, H.; Duplaa, A.-M.; Guy, A.; Khorlin, A.; Molko, D.; Roget, A.; Teoule, R. *Nucl. Acids Res.* **1992**, *20*, 5159.
27. Dr. ChuHee Keng, MIT, Cambridge, personal communication.
28. Gait, M. J.; Pritchard, C.; Slim, G. In *Oligoribonucleotides and Analogues, A Practical Approach*; Eckstein, F., Ed.; IRL Press at Oxford University Press: Oxford, 1991; pp 25.
29. Damha, M. J.; Ogilvie, K. K. In *Protocols for Oligonucleotides and Analogs*; Agrawal, S., Ed.; Humana Press: Totwa, NJ, 1993; pp 81.
30. Vinayak, R. A. *Companion to Methods in Enzymology* **1993**, *5*, 7.
31. Eckstein, F. Max Planck Institute, Göttingen, Germany, personal communication.

RECEIVED July 19, 1994

Chapter 13

Anti-Human Immunodeficiency Virus Activity of a Novel Class of Thiopurine-Based Oligonucleotides

Rich B. Meyer, Jr., Alexander A. Gall, and Vladimir V. Gorn

**Microprobe Corporation, 1725 220th Street, S.E.,
Bothell, WA 98021**

A novel class of oligonucleotides with broad-spectrum antiviral activity is described. These oligonucleotides contain 1-methyl-6-thiopurine heterocyclic bases, which are essential for activity. They also contain the 2'-O-methyl ribose modification, and the most active members of the class have normal phosphodiester linkages. Members of this class of oligonucleotides are potent inhibitors of HIV replication in cultured fresh peripheral mononuclear blood cells at concentrations in the range of 30 nM.

The search for oligonucleotide-based agents which inhibit virus replication by a sequence-directed mechanism, i.e. antisense, has often resulted in antiviral activity without sequence specificity. Phosphorothioate oligonucleotides most often show this effect. (*1*) It was initially observed in the anti-HIV activity of a series of phosphorothioates in which an oligomer containing only deoxycytidine was the most active. (*2*) Phosphorothioate oligonucleotides have also been shown to have non-sequence specific effect against a wide variety of other viruses, such as Herpes Simplex, (*3*) flu, cytomegalovirus, and others. Oligo-nucleotides bearing lipophilic endcapping modifications (*4,5*) and guanine-rich oligonucleotides (*6,7*) also have non-sequence specific antiviral effects.

An ideal antiviral drug should have good potency in inhibition of virus replication and the minimum toxicity possible to the host. Few compounds currently under investigation have such properties. Although the potential for non-toxic, sequence-specific antisense agents as antiviral drugs has not yet been realized, the possibilities of exploiting the antiviral properties of non-sequence specific oligonucleotides still remains of interest, *if there is antiviral specificity*.

0097–6156/94/0580–0199$08.00/0
© 1994 American Chemical Society

In 1981, Chan and co-workers described the inhibition of a number of retroviral reverse transcriptases by an unusual class of high molecular weight polynucleotides. (*8*) These nucleotides, synthesized over several years time by Broom and co-workers, (*9*) contained few or none of the conventional Watson-Crick hydrogen bonding sites of the normal nucleic acids. Instead the 1-position of the purines was methylated and the 6-position of the purines, instead of an oxygen or an amino group, contained a non-hydrogen bonding sulfur atom. If the 1-methyl group were not on these thiopurine polynucleotides, they were quite toxic to cells in culture, since their breakdown products were the potent anti-metabolites 6-mercaptopurine or 6-thioguanine. Furthermore, without the sulfur the compounds were much

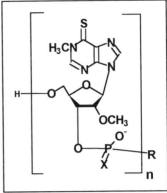

Figure 1. Structure of *N*-1, 2'-*O*-dimethyl-6-thioinosine based oligonucleotides. X = O or S, R = 3'-endcapping groups.

less active. These observations led us to propose that similarly modified oligo-nucleotides may have antiviral specificity by virtue of their inhibition of viral enzymes that make DNA. In this paper, we report the synthesis of a series of oligonucleotides containing the 1-methyl-6-mercaptopurine moiety, as shown in Figure 1, and their antiviral specificity.

Materials & Methods

1-*N*-Methyladenosine Hydroiodide (1) In a 1 L, one-necked, round-bottom flask, with a magnetic stirrer and septa, was suspended 267 g (1.0 mole) of adenosine in 600 mL of N,N-dimethylacetamide (DMA). To the mixture was added with stirring 500 g (219 mL, 3.5 mole) iodomethane. The mixture was covered with argon gas and stirred at room temperature in the dark for 3 days to give a clear solution. HPLC showed one major peak of the product and traces of starting adenosine. The mixture was poured into 2 L of acetone with stirring, filtered, washed with acetone, dried *in vacuo* to give 369 g (91%) of white powder, identical to that synthesized by Jones and Robins. (*10*)

1-Methyl-6-thioinosine (2). In a 2 L, one-necked, round-bottom flask, with a magnetic stirrer and septa, was suspended 121.8 g (0.3 mole) of 1-*N*-methyladenosine hydroiodide (**1**) and sodium hydrogen sulfide monohydrate (45 g, 0.6 mole) in 180 mL of *N,N*-dimethylformamide (DMF). The mixture is stirred under argon at 40-50°C for 15-20 hr to give a greenish mixture. The mixture was allowed to cool to room temperature and neutralized by addition of pyridinium hydrochloride (45 g, 0.4 mol). DMF was removed *in vacuo* (40° water bath), then the residue was dissolved in water and placed on an Amberlite XAD-4 column. The column was washed with water followed by elution with 50%

(v/v) methanol. Fractions with 320 nm absorption were analyzed by TLC and appropriate fractions combined and evaporated giving a white precipitate characterized as 1-methyl-6-thioinosine, m.p. 196-200°. R_f 0.27 (Kieselgel TLC plates, ethylacetate-acetone-methanol-water, 12:1:1:1). λ_{max} (0.1 M triethylamine-acetate, pH 7.5): 230, 320 nm. This is identical to the compound previously prepared by Ratsep, et al. (*11*)

1,2'-N,O-Dimethyl-3',5'-di(p-toluoyl)-6-thioinosine (6). A mixture of 53.5 g (0.20 mol) of 1-methyl-6-thioinosine (2) and 12.0 g (0.30 mol) of 60% sodium hydride (suspension in oil) in 500 mL of dry N,N-dimethylformamide was stirred for 30 min under argon at room temperature and treated with 100 g (2.0 mol) of liquid chloromethane (Aldrich). The mixture was stirred for 3 hr at - 10°, neutralized with 16.0 g (0.13 mol) of pyridinium hydrochloride, and partially evaporated at 40°/1 mmHg to 100 mL volume. The residue was dissolved in 300 mL of dry pyridine, cooled in ice bath and treated with *p*-toluoyl chloride (90 ml, 0.82 mol) dropwise. The mixture was stirred for 3 hr at room temperature, treated with methanol (50 mL), and evaporated *in vacuo*. The oily residue was washed with water (3x300 mL). The water layers were decanted leaving a sticky oil in the flask. The oil was dissolved in 0.4 L of boiling methanol and 20 mL of water was added to start crystallization. After cooling, a crystalline solid was filtered, washed with cold 90% methanol and dried *in vacuo* to give 33 g (30%) of white powder, m.p. 188-193°. The product was recrystallized from xylenes to give 30 g of pure **6**, m.p. 195-196°. Anal. Calc. for $C_{28}H_{28}N_4O_6S$: C, 61.30; H, 5.14; N, 10.21; S, 5.84. Found: C, 59.97; H, 5.00; N, 9.91; S, 5.65.

1,2'-N,O-Dimethyl-6-thioinosine (4). To a suspension of 50.0 g (91 mmol) of 1,2'-N,O-dimethyl-3',5'-di(*p*-toluoyl)-6-thioinosine (3) in 400 ml of anhydrous methanol at 60° was added 20 ml (20 mmol) of 1 M NaOMe in methanol, and the reaction mixture was stirred for 15 min at 60°. The starting material completely dissolved in 10 min after addition of NaOMe. The reaction mixture was neutralized with 1.2 ml (20 mmol) of acetic acid and methanol was evaporated *in vacuo*. The residue was washed with hexanes (3x50 ml), evaporated twice with water, stirred in 100 ml of water, and filtered. 1,2'-N,O-Dimethyl-6-thioinosine (4) was isolated as a colorless crystalline solid, 27.6 g (97% yield), m.p. 190-193°. The product was pure enough for use in the next step. Recrystallization from methanol gave a colorless solid with m.p. 193-195°. R_f 0.30 (Kieselgel TLC plates, ethylacetate-acetone-methanol-water, 12:1:1:1). λ_{max} (0.1 M triethylamine-acetate, pH 7.5): 230, 320 nm. Anal. Calc for $C_{12}H_{16}N_4O_4S$: C, 46.15; H, 5.16; N, 17.94; S, 10.26%. Found: C, 46.00; H, 5.32; N, 17.79; S, 10.48%.

5'-O-Dimethoxytrityl-N-1,2'-O-dimethyl-6-thioinosine (7). 27.0 g (86 mmol) of dry 1,2'-O-dimethyl-6-thioinosine (4) was stirred under argon with 31.0 g (90 mmol) of 4,4'-dimethoxytritylchloride in 450 mL of dry pyridine at room temperature for 4 hr. The mixture was treated with 10 mL of methanol, evaporated to 200 ml volume, and diluted with 500 ml of water. The water layer was decanted from the oil, and the residue was washed three times with water (3 x 200 mL) and dried *in vacuo* giving a solid foam. The foam was stirred for 2 days with 300 mL of dry ether and filtered giving a crystalline solid, 44.5 g (84% yield). The product was purified by precipitation from acetone with 30% methanol containing 0.1% NH_4OH. The precipitate was filtered, washed with 30% methanol, and dried *in vacuo* at 60° for 2 days to give 40 g product as a fine crystalline solid, m.p. 100-110° (decomp.), R_f 0.43 (Kieselgel TLC plates, ethyl acetate-methanol-triethylamine, 100:5:1). λ_{max} (0.1 M triethylammonium-acetate, pH 7.5): 228, 320 nm. Anal. Calc for $C_{33}H_{34}N_4O_6S$: C, 64.48; H, 5.58; N, 9.11; S, 5.22. Found: C, 64.64; H, 5.74; N, 8.73; S, 5.23.

5'-*O*-Dimethoxytrityl-1,2'-*N,O*-dimethyl-6-thioinosine-3'-*O*-(2-cyanoethyl-*N,N*-diisopropylphosphoramidite) (8). A solution of 9.7 g (16 mmol) of 5'-O-dimethoxytrityl-1,2'-N,O-dimethyl-6-thioinosine in 270 mL dry dichloromethane and 11 mL of diisopropylethylamine was treated dropwise with 4.7 mL (21 mmol) of 2-cyanoethyl-*N,N*-diisopropylchlorophosphoramidite (Aldrich), and the mixture stirred at room temperature under argon. After 1.5 hour, 5 mL of methanol was added and the mixture poured into 1 L of ethyl acetate containing 30 mL of triethylamine. The resultant solution was washed successively with saturated $NaHCO_3$ (2x250 mL) and saturated NaCl (2x250 mL), then dried over anhydrous Na_2SO_4 and evaporated to dryness. The residue was dissolved in 40 mL $CHCl_3$-NEt_3 and purified by flash chromatography over a silica gel column (5x24 cm, preflushed with hexanes-triethylamine, 10:1). The column was washed with 500 mL of hexanes-ethyl acetate-triethylamine (10:20:1) and the phosphoramidite eluted with additional 600 mL of the same solvent mixture. The product was precipitated by adding a solution in 20 mL of dichloromethane-triethylamine (100:1) into vigorously stirred hexanes, to afford 7.6 g (57%) of phosphoramidite (6) with m.p. 99-115° (decomp.), R_f 0.60 (Kieselgel TLC plates, ethyl acetate-dichloromethane-triethylamine, 5:5:1). λ_{max} (0.1 M triethylammonium acetate, pH 7.5): 234, 320 nm. Anal Calc for $C_{42}H_{51}N_6O_7PS$: C, 61.90; H, 6.31; N, 10.31; P, 3.80; S, 3.93. Found: C, 62.03; H, 6.53; N, 9.92; P, 3.71; S, 3.69.

Synthesis of Oligonucleotides. All of the oligonucleotides reported here contained a custom 3'-modification developed by us. This is a 3'-(6-aminohex-1-yl) phosphate, added by using a custom solid support for oligo synthesis, (*12*) as shown in Figure 3.

Oligonucleotides were synthesized on a Pharmacia Oligopilot automated DNA synthesizer using the standard cyanoethyl-*N,N*-diisopropylamino-phosphoramidite (CED-phosphoramidite) chemistry. ODNs were purified by adaptations of standard methods. (*13*) Detachment from the solid support was accomplished with concentrated ammonia at 25° for 1 day. HPLC for the trityl ODNs was performed with a Hamilton PRP-1 (7.0 x 305 mm) reversed-phase column employing a gradient of 20% to 45% CH_3CN in 0.1 M $Et_3NH^+OAc^-$, pH 7.5. After detritylation with 80% acetic acid for 1 hr at room temperature, the ODNs were precipitated by addition of 3M sodium acetate and 1-butanol. (*14*) The detritylated oligonucleotides were purified on a C-18 reverse phase column using a gradient similar to above. Phosphorothioates were prepared using the Beaucage reagent.

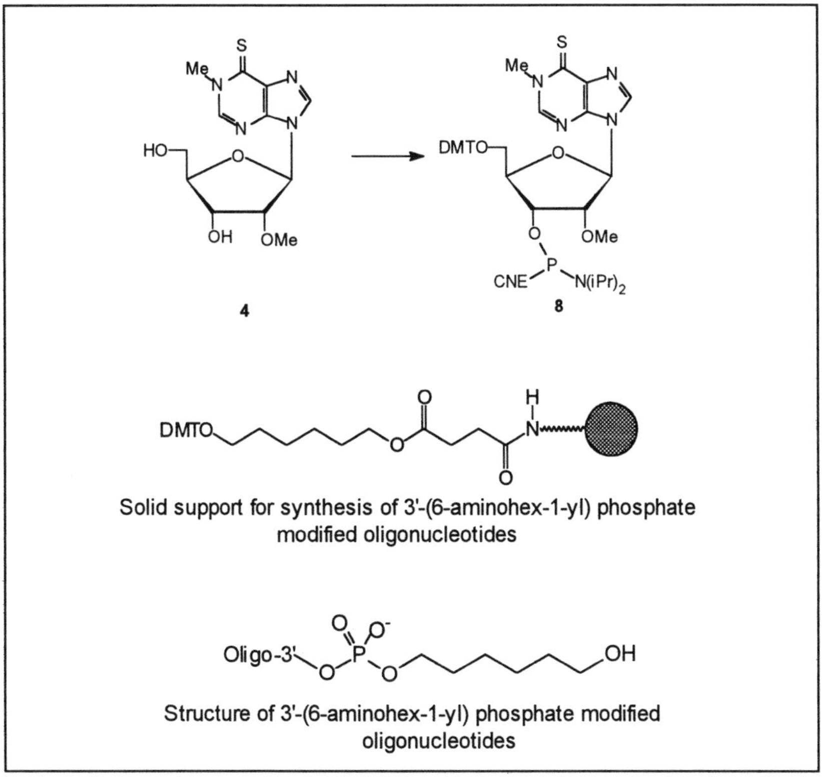

Figure 3. Synthesis of Oligonucleotides.

Inhibition of HIV Replication. Assays to determine the ability of the compounds to inhibit HIV-1 in human peripheral blood mononuclear (PBM) cells were as described by Schinazi *et al.* (*15*) 3'-Azido-3'-deoxythymidine (AZT) was included as a positive control for the antiviral assays.

Cell Culture. Human PBM cells from healthy HIV-1 seronegative and hepatitis B virus seronegative donors were isolated by Ficoll-Hypaque discontinuous gradient centrifugation at 1,000 x g for 30 minutes, washed twice in phosphate-buffered saline (pH 7.2; PBS), and pelleted at 300 x g for 10 min. Before infection, the cells were stimulated by phytohemaglutinin (PHA) at a concentration of 6 μG/ML for 2-3 days in RPMI 1640 medium supplemented with 15% heat-inactivated fetal calf serum, 1.5 mM L-glutamine, penicillin (100 U/ml), streptomycin (100 µg/ml), and 4 mM sodium bicarbonate buffer.

Viruses. HIV-1 (strain LAV-1) was obtained from Dr. P. Feorino (Emory University, Atlanta, GA). The virus was propagated in human PBM cells using RPMI 1640 medium, as described previously (*16*) without PHA or fungizone and supplemented with 26 units/ml of recombinant interleukin-2 (Cetus Corporation, Emeryville, CA), 7 µg/ml DEAE-dextran (Pharmacia, Uppsala, Sweden), and 370 U/ml anti-human leukocyte (alpha) interferon (ICN, Lisle, IL). Virus obtained from cell-free culture supernatant was titrated and stored in aliquots at -70°C until use.

Inhibition of Virus Replication in Human PBM cells. Uninfected PHA-stimulated human PBM cells were infected in bulk with suitable dilutions of virus. The mean reverse transcriptase (RT) activity of the inocula was about 60,000 dpm RT activity/10^6 cells/10 ml. This represents, by a limiting dilution method in PBM cells, a multiplicity of infection of about 0.01. After 1 h, the cells were uniformly distributed among 25 cm^2 flasks to give a 5 ml suspension containing about 2 x 10^6 cells/ml. The drugs at twice their final concentrations in 5 ml of RPMI 1640 medium, supplemented as described above, were added to the cultures. The cultures were maintained in a humidified 5% CO_2-95% air incubator at 37°C for six days after infection at which point all cultures were sampled for supernatant RT activity. Previous studies had indicated that maximum RT levels were obtained at that time.

Cytotoxicity Studies in PBM Cells. The drugs were evaluated for their potential toxic effects on uninfected PHA-stimulated human PBM cells. The cells were cultured with and without drug for 24 hr at which time radiolabeled thymidine (0.5 µCi in 20 µl/well) was added. The assay was performed as described previously (*17*). Alternately, cells are counted on day 6 using a hemacytometer or Coulter counter as described previously (*5*).

Median-Effect Method. EC_{50} and IC_{50} values were obtained by analysis of the data using the median-effect equation (*18*). These values were derived from the computer-generated median effect plot of the dose-effect data using a commercially available program (*19*).

Results

Chemistry. The preparation of the monomeric nucleoside used in these new oligonucleotides is shown in Figure 2. The modified ribonucleoside 1-methyl-6-thioinosine (**2**) was prepared from adenosine by an adaptation of the literature method (*11*). The major modification, which is particularly useful for very large scale preparations, was the substitution of sodium hydrosulfide for liquid H_2S in the sulfhydrolysis of 1-methyladenosine (**1**).

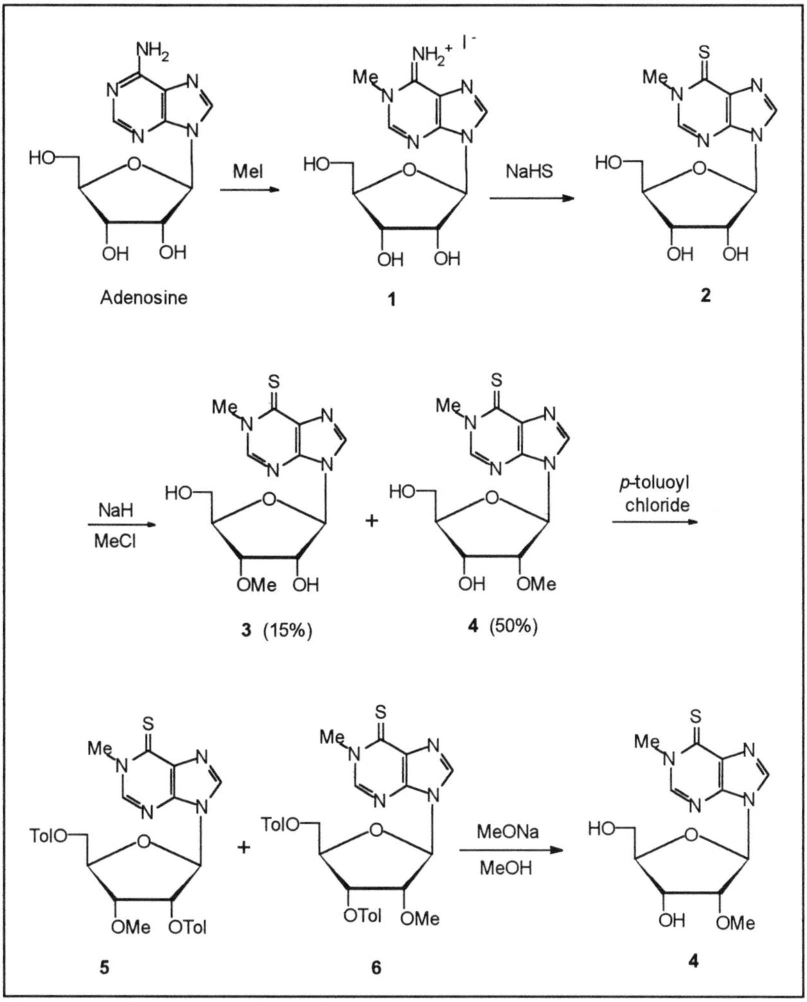

Figure 2. Synthesis of 1,2'-O-dimethyl-6-thioinosine

The introduction of the 2'-*O*-methyl substituent, while straightforward on a small scale, presented challenges on a large scale. We sought a method to circumvent the use of the Markiewicz reagent (*20,21*) to selectively methylate the 2'-hydroxyl group. We have found that side reactions of the sulfur-containing nucleoside was problematic when methyl iodide was used as the methylating reagent. We have found, however, that sodium hydride and methyl *chloride* can be used to selectively methylate the 2'-hydroxyl of nucleosides.

We have purified the 2'-*O*-methyl product **4** without column chromatography. This was accomplished by esterification free hydroxyls of the nucleosides in the reaction mixture with toluyl chloride, followed by direct crystallization of the desired 2'-*O*-methyl isomer (**3**). One recrystallization gave essentially pure product. Base catalyzed removal of the toluyl groups gave the desired 2'-O-methyl nucleoside in high yield, devoid of the 3'-*O*-methyl isomer by HPLC.

The preparation of the 5'-*O*-dimethoxytrityl-3'-*O*-*N*,*N*-diisopropyl-phosphoramidite cyanoethyl ester of **4** was accomplished by standard methods.

Oligonucleotide Synthesis. All of the oligonucleotides described in this report, save one, contained a 3'-endcapping group. These groups have been reported by us earlier *(12)*. Most of the oligonucleotides use the "hexanol tail" which is a 3'-(6-hydroxyhexyl) phosphate ester (see Figure 3).

All oligonucleotides were prepared on Pharmacia Oligo Pilot automated DNA synthesizers, using polystyrene-based solid support (Pharmacia). The solid support was supplied in the "amine-linker" form, and derivatized by the methods previously published by us for CPG supports *(22)*. Reagent ratios of 2 to 2.5 equivalents of amidite per cycle were used. This gave extremely high coupling efficiencies (greater than 98%) while still sparing the amidite reagents. Overall yields of oligonucleotides, based on amidite utilization, were approximately 20%.

There was concern initially about whether the modified heterocyclic base moiety would remain intact during the course of oligonucleotide synthesis. In particular, we were concerned that the 6-mercapto group would be stable to the iodine oxidation cycles in the course of DNA synthesis and that the methylated heterocyclic ring would be stable to the ammonia deprotection step. Iodine oxidation steps differ on the various commercially available DNA synthesizers. We found, however, no evidence of side reactions with the sulfur-containing heterocycle.

Quality control for the integrity of the heterocyclic base is quite straightforward with these oligonucleotides. The base has an absorbance maximum at 318 nm and an absorbance minimum at 260 nm. The ratio of the pure material is approximately 16 to 1. Any changes in the heterocyclic base, either from sulfur oxidation or ring opening, is readily detectable by a change in this ratio.

The phosphorothioates were prepared using the Beaucage reagent for thiolation of the phosphites. This procedure was non-problematic, generally giving higher yields than the preparation of the phosphodiester oligos.

Inhibition of HIV Replication. This new class of homo-oligonucleotides proved to be potent inhibitors of replication of Human Immunodeficiency Virus in vitro. As discussed below, the length of the oligonucleotides was a major factor in determining potency. The addition of a phosphorothioate backbone to the longer oligomers gave no increase in activity. The addition of a lipophilic end group also did not enhance the antiviral activity of the compounds.

Effect of Length of Oligonucleotide. The effect of the length of the oligonucleotides with phosphodiester internucleotide linkages on their ability to inhibit HIV replication was dramatic, as shown in Table 1. The activity in cell culture increased with length about two orders of magnitude from the 16-mer to

Table 1.   Inhibition of replication of HIV-1 by thiopurine oligonucleotides

Oligonucleo-tide length[a]	Backbone[b]	Endcap[c]	IC_{50}, μM[d]
16 - Mer	PO	3'-Hex	2.8
20 - Mer	PO	3'-Hex	1.9
	PS	3'-Hex	0.6
24 - Mer	PO	3'-Hex	0.8
	PS	3'-Hex	0.1
28 - Mer	PO	3'-Hex	0.4
	PS	3'-Hex	0.05
	PO	3'-Chol	0.2
32 - Mer	PO	3'-Hex	0.03
	PS	3'-Hex	0.05
36 - Mer	PO	3'-Hex	0.07
	PS	3'-Hex	0.06
	PO	3'-Chol	0.07
	PO	$3'-C_{16}$	0.08
$(dC)_{28}$[e]	PO	3'-Hex	20
	PS	3'-Hex	0.2

[a] Length of homopolymer of 1,2'-O-dimethyl-6-thioinosinate, as shown in Fig 1.
[b] PO=phosphodiester internucleoside linkages, PS=phosphorothioate linkages.
[c] 3'-Hex = 3'-(6-hydroxyhex-1-yl) phosphate; 3'-Chol = 1-[(cholesteryloxy)carbonyl]-5-hydroxymethyl-(3R-trans)-pyrrolidin-3-yl phosphate (ref. 22); $3'-C_{16}$ = (16-hydroxy-hexadec-1-yl) phosphate.
[d] Conc for 50% inhibition of virus replication, as described in Methods
[e] 28-mer comprised of deoxycytidylate.

the 32-mer, for nucleotides having the same 3'-hexanol endcapping modification.   At about 32 nucleotides long, the change in activity leveled off. The 16-mer possessed quite weak activity ($IC_{50} \sim 2.8 \mu M$), while the 32-mer was approximately 100-fold more potent ($IC_{50} \sim 30$ nM).

Effect of Backbone Modification. The oligomers of varying lengths were prepared with both phosphodiester and phosphorothioate backbones and the activity compared, and the results are shown in Table 1.   In addition, Table 1 shows the activity of a 28-mer of deoxycytidylate for comparison.   Quite strikingly, the phosphorothioates maintain the same level of activity over several differing lengths, and the deoxycytidylate phosphorothioate analog (SdC_{28}) is equally as active as the thiopurine-containing phosphorothioate.   The phosphodiesters containing the thiopurines, however, fall off sharply in activity below 30 nucleotides long, and the phosphodiester 28-mer of deoxycytidylate is virtually devoid of anti-HIV activity. The novel feature of these new phosphodiester oligonucleotides is that they are as or more active than the corresponding phosphorothioates when the optimum length is reached

Effect of Endcapping Group on Activity. We and others have shown that the addition of a 3'-phosphate ester to oligo*deoxy*nucleotides increases the stability to degradation in cell culture by several fold. While it is not known whether the 3'-exonucleases that degrade the 2'-deoxy oligos will also degrade the 2'-*O*-methyl ribonucleotides in this report, we routinely use the 3'-hexanol group for ease of synthesis, in addition to possible enhancement in stability. The 2'-*O*-methyl group, of course, is expected to give these oligos much higher resistance to degradation than the 2'-deoxy or ribosyl oligos, as has been reported for oligonucleotides of mixed sequence. (*23*)

The addition to these oligonucleotides of 3' endcapping groups of various sizes does not substantially change their activity. As shown in Table 1, the addition of a cholesterol or hexadecanol group to the 3' terminus of these oligo-nucleotides, via a phosphodiester linkage, causes very little change, and certainly no improvement in the inhibitory properties. This is in sharp contrast to activity found by others from the addition of cholesterol to oligonucleotides which improves antiviral potency. (*4,5*)

Cytotoxicity. No cytotoxicity was observed for any of the phosphodiester compounds tested in PMBCs at concentrations up to 100 µg/ml (approximately 5 - 15 µM). This indicates a selectivity index of at least 100 for the most active compounds.

Discussion

In reports of antiviral activity of both sequence specific and non-specific oligonucleotides, the best IC_{50} values seen are in the range of tenths of micromolar in cell culture systems. Inevitably, these potencies are seen in the phophorothioate oligonucleotides, and the corresponding phosphodiesters have little, if any, activity. The phosphodiester oligonucleotides described here, however, are among the most potent phosphodiester oligonucleotides for which activity has been reported in cell culture.

The agents described here, while being non-specific with regard to sequence, seem to differ in their mode of action from the non-specific actions of the phosphorothioates. This is most vividly shown by comparing the effect of length on activity for the PO and PS series of oligomers. The phosphoro-thioates show inhibitory activity over a wide range of lengths, while the phosphodiester agents only begin to show good inhibition when they reach a length of 28 nucleotides. In addition, other phosphodiester oligonucleotides do not show this antiviral action regardless of length, further indicating the unique nature of these thiopurine oligos.

The mechanism of action of these agents is under investigation. They inhibit the HIV reverse transcriptase (Y.C. Cheng, unpublished results), but at concentrations no lower than those at which they inhibit HIV replication in vitro. This indicates that while reverse transcriptase inhibition may be a component of their activity, it is not likely to be the major factor. The phosphorothioate $S(dC)_{28}$ is a strong inhibitor of reverse transcriptase, with a K_i = 2.8 nM (*24*), yet is less potent than the thiopurine phosphodiesters in inhibition of virus replication (Table 1). This again emphasizes the differences in the action of these compounds when compared to the known non-sequence-specific anti-HIV activity of PS oligonucleotides (1).

Acknowledgments. We thank Dr. Raymond Schinazi and Dr. Y.C. Cheng for conducting the assays for inhibition of HIV in culture. We are grateful for the preparation of the derivatized solid support by Eugene Lukhtanov, and for the preparation of certain intermediates by A. David Adams.

Literature Cited

1.   Stein, C.A.; Cheng, Y.C. *Science* **1993**, *261*, 1004-12.
2.   Matsukura, M.; Shinozuka, K.; Zon, G.; Mitsuya, H.; Reitz, M.; Cohen, J.S.; Broder, S. *Proc. Natl, Acad. Sci. USA* **1987** *84*, 7706.
3.   Gao,W.Y.; Hanes, R.N.; Vazquez-Padua, M.A.; Stein, C.A.; Cohen, J.S.; Cheng, Y.C. *Antimicrob. Agents Chemother.* **1990**, *34*, 808-12.
4.   Letsinger, R.L.; Zhang, G.R.; Sun, D.K.; Ikeuchi. T.; Sarin, P.S. *Proc. Natl. Acad. Sci. USA* **1989**, *86*, 6553.
5.   Shea, R.G.; Marsters, J.C.; Bischofberger, N. *Nucleic Acids Res.* **1990**, *18*, 3777-83.
6.   Wyatt, J.R.; Vickers, T.A.; Roberson, J.L.; Buckheit, R,W, Jr; Klimkait, T.; DeBaets, E.; Davis, P.W.; Rayner, B.; Imbach, J.L.; Ecker, D.J.. *Proc Natl Acad Sci U S A.* **1994**, *91*, 1356-60.
7.   Ojwang, J.; Elbaggari, A.; Marshall, H.B.; Jayaraman, K.; McGrath, M.S.; Rando, R.F.. *J Acquir Immune Defic Syndr.* **1994**, *7*, 560-70.
8.   Chan, E.W.; Lee, C.K.; Dale, P.J.; Nortridge, K.R.; Hom, S.S.; and Seed, T.M. *J. Gen. Virol.* **1981**, *52*, 291-299.
9.   Broom, A.D.; Amarnath, V. Polyribonucleotides containing Thiopurines: Synthesis and Properties *Biochem. Biophys. Res. Comm.* **1976**, *70*, 1029-1034, and references cited in Project 2 of this application.
10.   Jones, J.W.; Robins, R.K. *J. Am. Chem. Soc.* **1963**, *85*, 193-201.
11.   Ratsep,P.C.; Mishra,N.C.; Broom A.D. *Nucleosides & Nucleotides* **1991**, *10*, 1641-1655.
12.   Gamper, H. B.; Reed, M. W.; Cox, T.; Virasco,J. S.; Adams, A. D.; Gall, A. A.; Scholler, J. K.; Meyer, Jr., R. B. *Nucleic Acids Res.* **1993**, *21*, 145-150.
13.   McLaughlin, L. W. In *Oligonucleotide Synthesis-A Practical Approach*; Gait, M. J. (ed.); IRL Press: Oxford, 1984
14.   Van Ness, J.; Kalbfleisch, S.; Petrie, C. R.; Reed, M. W.; Tabone, J. C.; Vermeulen, N. M. *J. Nucleic Acids Res.* **1991**, *19*, 3345-3350.

15.   Schinazi, R.F.; Sommadossi, J.-P.; Saalmann, V.; Cannon, D.; Xie, M.-
      W.; Hart, G.; Smith, G.; Hahn, E. *Antimicrob. Agents Chemother.* **1990**,
      34, 1061-1067.
16.   McDougal, J.S.; Cort, S.P.; Kennedy, M.S.; Cabridilla, C.D.; Feorino,
      P.M.; Francis, D.P.; Hicks, D.; Kalyanaramen, V.S.; Martin, L.S. *J.
      Immun. Meth.* **1985**, *76*, 171-183.
17.   Bardos, T.J.; Schinazi, R.F.; Ling, K.-H.J.; Heider, A.R. *Antimicrob.
      Agents Chemother.* **1992**, *36*, 108-114.
18.   Chou, T.-C.; Talalay, P. *Adv. Enz. Regul.* **1984**, *22*, 27-55.
19.   Chou, J.; Chou, T.-C. A computer software for Apple II Series and
      IBM-PC and Instruction Manual; Elsevier-Biosoft; Elsevier Science
      Publishers, Cambridge, U.K. 1985.
20.   Markiewicz, W. T.; Wiewiorowski, M. *Nucleic Acids Res. Spec. Publ.*
      1978, S158.
21.   Robins, M. J.; Wilson, J. S.; Hansske, F. *J. Am. Chem. Soc.*, **1983,**
      4059.
22.   Reed, M.W.; Adams, A. D; Nelson, J. S.; Meyer, Jr., R. B. *Bioconj.
      Chem.* **1991**, *2*, 217-225.
23.   Sproat, B.S.; Lamond, A.I.; Beijer, B.; Neuner, P.; Ryder, U. *Nucleic
      Acids Res.* **1989**, *17*, 3373-3386.
24.   Majumbar, C.; Stein, C.Y.; Cohen,J.S.; Broder, S.; Wilson, S.H.
      Biochemistry **1989**, *28*, 1340-1346.

RECEIVED September 27, 1994

STRUCTURAL STUDY

Chapter 14

New Twists on Nucleic Acids

Structural Properties of Modified Nucleosides Incorporated into Oligonucleotides

Richard H. Griffey, Elena Lesnik, Susan Freier, Yogesh S. Sanghvi, Kelly Teng, Andrew Kawasaki, Charles Guinosso, Patrick Wheeler, Ventrankaman Mohan, and P. Dan Cook

Isis Pharmaceuticals, 2292 Faraday Avenue, Carlsbad, CA 92008

The structures and physical properties of oligonucleotide duplexes depend strongly on the limited conformational flexibility of (deoxy)ribose-phosphodiester linkages and the sugar. Molecular mechanics and NMR spectroscopy have been employed to understand and predict alterations in sugar pucker and conformation resulting from replacement of oxygen, hydrogen, and carbon in the sugar with other heteroatoms. Studies on model monomer, dimer, and longer single stranded nucleosides are correlated with data on duplex stability obtained from oligomers incorporating 2′, 3′, and 4′-modified sugars to establish a set of guidelines for the types of alterations that can be tolerated within oligonucleotide duplexes. The results suggest that increased affinity of an antisense oligomer for a complementary RNA target can be achieved by decreasing the entropic motion of the sugar while maintaining a preorganized structure with an RNA-like conformation.

Utilization of antisense oligonucleotides in the treatment of disease has emerged as an exciting new therapeutic paradigm (*1*). Pharmaceutical "antisense" encompasses strategies where gene expression is inhibited through hybridization of an exogenous oligomer to a specific intracellular messenger RNA target (*2*). This provides a very high potential for specificity of action, since theoretically a small oligomer of ~15-20 nucleotides has the sequence specificity required to inhibit the expression of a target gene. To be successful, antisense therapeutics have to fulfill additional criteria, including sufficient nuclease resistance, biodistribution, and ease of synthesis (*3*). These issues have spurred a quest for oligonucleotide replacements with favorable specificity, affinity, and stability (*4*). Since many types of nucleotide modifications which provide nuclease resistance reduce the stability of the resulting duplex with the RNA target, increasing affinity for the target RNA is a paramount concern (*5*).

Conformational Properties of Oligonucleotides

The solution conformations of nucleic acids vary dynamically on the picosecond time scale. The structure of an oligonucleotide is governed by the interaction of dipolar, torsional, electrostatic, steric, and London forces from the bases, sugar, and backbone (*6*). Figure 1 demonstrates the result of dynamic motion on the structure of an 8-mer

0097–6156/94/0580–0212$08.00/0

DNA duplex. Four conformations of the duplex obtained from a molecular dynamics study at 25 psec intervals have been superimposed. The structures all fall within the classic 'B' family of helices, but illustrate the perturbations induced by the entropy of thermal motion. This random dynamic motion is balanced by the hydrogen bonds among the complementary bases, which provide the Watson-Crick sequence specificity, and by the limited range of conformations which can generate a helical structure. The small free energy of the duplex results from a cancellation of the large favorable enthalpy of the single strands, whose conformations are defined by seven backbone and glycosidic dihedral angles and their large entropy, or the energetic cost of organizing the strands into a low energy conformation. The residual thermal motion in the strands has been proposed to have an "entropic benefit", which allows duplex structures to be melted for recognition processes (7).

The importance of preorganization and rigidity in the formation of stable duplexes is supported by the work of Benner, who incorporated glycerol units into a DNA backbone (8). The trisubstituted glycerol is equivalent to a ribose sugar without the C2′ carbon. While the glycerol moieties can adopt conformations suitable for formation of a double helix, each incorporation produced a 9-15° C decrease in the T_m for the duplex. Constrained cyclic nucleotides should decrease the entropy of the duplex, and two families have been synthesized (9, 10). However, their incorporation into oligomers reduces the affinity of the oligo for target DNA or RNA by 1.5°-2.0° C/modification. Hence, reducing the entropy of the system is insufficient to improve the affinity unless the rigid analog adopts a conformation that is compatible with the structure of the target for duplex formation.

The backbone and the sugar of an oligomer can be modified to change the entropy and enthalpy of interaction and improve the affinity. Atomic substitutions along the phosphodiester backbone generally are not well tolerated, since a limited number of dihedral solutions exist for the DNA or RNA duplexes. A variety of alternate backbones have been prepared, including substitution of sulfonates, sulfonamides, hydrazines, hydroxylamines, formacetals, alkylphosphonates, phosphorothioate, phosphorodithioates, and amide linkages for the phosphate diester (11, 12, 13). The affinity of these oligomers for an RNA target generally is reduced, except where substitutions with limited rotational freedom, e.g. hydrazines and hydroxylamines, are incorporated (14, 15).

The conformation of the sugar might be used to alter the entropy and geometry of the duplex. There are twenty possible conformations of the ribo- and deoxyribofuranose ring. These can be plotted on a "pseudorotation" cycle, which relates each "twist" or "envelope" conformation to a specific value of the five internal dihedral angles (16). Crystal structures obtained to date suggest that two low-energy states predominate, centered about the C3′-exo and C2′-exo conformations. A two-state exchange model among these forms has been proposed for RNA and DNA and is shown in Figure 2 for RNA. The two forms are separated by a relatively small barrier of 1-2 Kcal, which molecular mechanics studies suggest is traversed on the picosecond time scale (17). The same conformations also are observed for fragments of oligomers down to the nucleoside level using X-ray crystallography, NMR, circular dichroism, laser Raman spectroscopy, and other physical techniques. The furanose conformation is not "fixed", but is governed by anomeric and gauche interactions of functional groups and the ring oxygen, which stabilize certain geometries (18).

In this chapter, modified nucleosides are examined using NMR as monomers and when incorporated into oligonucleotides. The conformational properties are correlated with biophysical data including duplex stability. The conformational preferences of nucleosides are retained from monomers into single stranded oligonucleotides. The biophysical properties of oligomers and their duplexes can be predicted from analysis of monomers. The discovery of alternate structural motifs, such as peptide nucleic acids (PNAs) which have increased affinity for complimentary RNA and DNA targets, suggests that other oligonucleotide analogs have yet to be discovered (19).

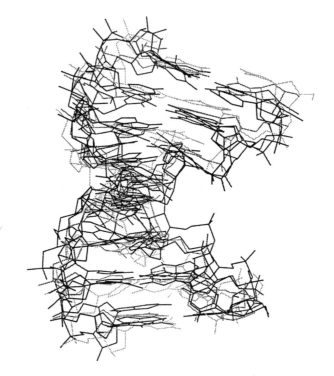

Figure 1. Superposition of 4 Conformations for an 8-Mer DNA Duplex Obtained at 10 Psec Intervals. The AMBER force field has been employed using Biosym Insight and Discover software.

Figure 2. Conformational exchange between 'N' and 'S' forms of RNA.

Results and Discussion

The ribofuranose ring has been substituted at C2′, C3′, and C4′ with substituents varying in electronegativity, polarizability, and steric bulk. The pseudorotation parameters have been determined from crystal structures or in solution from an analysis of the three-bond proton-proton (HH) coupling constants observed in high-resolution proton NMR spectra. The coupling constants are fit to a periodic Karplus-type equation, which predicts the values of the internal dihedral angles of the sugar unit (20, 21). The calculated pseudorotation properties can be correlated with the changes in affinity for the RNA and DNA complements that the modification produces. This correlation suggests that an increased %N conformation is associated with increased T_m values against RNA complements. This observation suggests that the affinity of antisense therapeutics may be improved via selective incorporation of specific modified nucleosides into oligomers.

2′-F-Deoxynucleotides. The predominant conformation of 2′-F nucleosides shown in Figure 3a has been determined by X-ray crystallography to be C3′-endo, although 2′-F uridine crystallizes in an unusual form P=74° (22). Theoretical calculations on model systems suggest that the barrier for conversion to a C2′-endo form is increased along both the "eastern" and "western" paths (23). The P_s conformer is destabilized by a larger gauche effect to O4′ for the more electronegative F compared to OH. This differs from RNA where the 2′-O and 3′-O have equal gauche effects to O4, and the conformation is governed by the anomeric effects of the C1′ and C4′ substituents with the ring oxygen.

The conformation of oligonucleotides containing 2′-F units has been examined using ^{1}H NMR analysis of coupling constants and a parameterized version of the Karplus equation for 2′-F derivatives. This represents an improvement over older studies, where a simple analysis failed to account for the alpha and beta electronic effects of the F atom (24, 25). The original NMR data on mono- and dinucleotides published by Ikehara and co-workers have been used to fit new pseudorotation parameters which are listed in Table I.(25) These calculations show that the best amplitude of the sugar pucker is 40-43°, slightly greater than observed in crystal structures. For three 2′-F purines, the %N is calculated to be 86-92% with P_n=-10° to -16°. The pseudorotation of the residual P_s could be fit to 93° or 172°. These values are shifted toward C2′-exo by 20-30° from the previous calculations (25). Conformational analyses of fluoropiperidines also show that fluorine prefers such an axial orientation (26). The compound 3′-5′-TIPS-2′-F-uridine (Figure 3b) has been prepared as a rigid 'N' derivative to identify the predominant 'N' pseudorotation. The 3′-5′-TIPS protecting group is known to shift the pseudorotation to entirely 'N' (27). Calculations reveal that the derivative adopts a conformation with >99% P_n=-19.0° in chloroform solution at 20° C. A pentamer G*A*U*C*dT containing four 2′-F nucleotides also has been prepared and studied. Analysis of the three-bond H-H coupling constants shows that the G, C, and U 2′-F nucleotides exist in a conformation with 91-97% P_n= -6±1°, while the 2′-F A is 84% P_n=-27±5°. The ground state P_n must be favored by 2-2.5 Kcal/mol over the P_s conformer from the ratio of P_n to P_s determined from NMR. This compares to a theoretical value of 1.5 Kcal/mol determined for model systems. The %N at 20° C would be expected to be even higher in a 2′-F-modified oligomer.

The stability of oligomers containing 2′-F substitutions for RNA targets is increased by ~2°/modification in an RNA strand (28). This effect may result from increased preorganization of 2′-F strand for the RNA target due to a reduced rate of interconversion between 'N' and 'S' forms, since the barrier to pseudorotation through 'E' quadrant is increased (23). The 2′-F derivatives offer no increase in resistance to endo- or exonuclease degradation (29).

Table I. Pseudorotation Parameters for 2′-F Nucleosides and
Oligomers

Compound	Temperature (° C)	%N	P_n	P_s
fGpU[a]	20	87	-16±1	137±5
fIpC	20	91	-11	94
fApU	20	93	-10	103
fGfAfUfCdT	50			
fG		94	-6±1	170±5
fA		84	-27±5	149
fU		97	-7±1	168
fC		91	-5	165
3′,5′-TIPS fU[b]	20	100	-15±1	161

[a] Coupling constants taken from ref. 25. [b] NMR experiments performed on sample dissolved in CDCl₃.

2′-O-Methyl Ribosides. This family of nucleosdes (Figure 4a) commonly occurs in tRNAs and have been proposed to stabilize RNA duplexes at specific locations in the anticodon loop (*30*). The conformational properties of 2′-O-Me uridine and cytidine as nucleosides and in dinucleotides have been studied in an attempt to rationalize the prevalence of this modification (*31*). Not all coupling constants could be determined from proton NMR spectra, and the conformations have been estimated only from the H1′H2′ coupling constant. The proportion of the C3′-endo conformation is increased from 55% to 60% in the 2′-O-Me derivative compared to uridine. In a UpU dinucleotide, no increase in the 63% population of the C3′-endo conformer is observed upon methylation. The stabilization of the C3′-endo form has been attributed to an unfavorable steric interaction between the 2′-O-Me and the 3′-phosphate, which orients the 2′-O-Me toward the base, and a second steric interaction of the O-Me group with the C2-carbonyl of the base. Substitution of 2′-O-Me nucleotides into DNA oligomers increases the T_m of duplexes against RNA complement by 1.5-2° C per modification on average, but local effects can be greater (up to 3° C per point modification) (*32*). In a fully modified 12-mer, 2′-O-Me RNA has a greater affinity for RNA target by +1.5-1.8° C compared to a phosphorothioate DNA oligomer, and by +0.4° C compared to the same 2′-O-allyl sequence (*33, 34*).

2′-O-Ethyl Ribosides. In contrast to 2′-O-Me nucleosides, the pseudorotation equilibrium for 2′-O-Et nucleosides (Figure 4b) at 20° C is shifted toward P_s. Analysis of NMR data for 2′-O-ethyl adenosine provides pseudorotation parameters of 23% P_n= -2°, P_s= 160°. These results are similar to adenosine where calculations reveal 28% P_n=-24°, P_s=165°. Incorporation of an alkyl group may result in a balance between the gauche effect from the 2′-O-alkyl group and an unfavorable steric interaction between the O-CH₂ and the heterocycle. The proton chemical shifts of the two O-CH₂ protons differ by 0.2 ppm, which suggests that rotation about O-CH₂ bond is hindered and subject to a local ring current from the base. Incorporation of 2′-O-ethyl A as a point modification provides a +0.4° C/mod. stabilization in a DNA background against an RNA target (*32*).

2′-O-Propyl Ribosides. This modification (Figure 4c) has been studied by NMR in more detail as a monomer and in SS oligonucleotides. Pseudorotation parameters calculated from NMR studies of 2′-O-propyl G and C presented in Table II show that P_s/P_n ratio is ~3 in the G nucleoside, but the values are similar to those observed for the 2′-O-ethyl derivatives. In a single-stranded (SS) 8-mer containing four 2′-O-

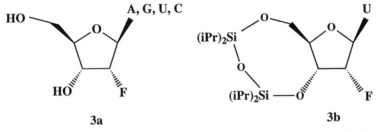

Figure 3. A) Structure of 2'-F-Uridine. B) Structure of 3',5'-TIPS-s'-F-Uridine.

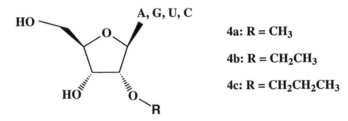

4a: R = CH₃

4b: R = CH₂CH₃

4c: R = CH₂CH₂CH₃

Figure 4. Structures of 2'-*O*-alkyl nucleosides. A) 2'-O-Methyl. B) 2'-*O*-Ethyl. C) 2'-*O*-Propyl.

propyl nucleosides, the P_s/P_n ratio is shifted to ~1, and the value of P_s shifts from ~160° to 140°. This may result from a destabilization of the "ground" state observed for the nucleoside as the stacking interactions in the SS force the propyl group into contact with the 3-flanking base, and as relief from unfavorable steric interactions of the propyl chain with the sugar protons. The shift of P_n and P_s toward C4'-exo/O4'-endo would lead to higher energy ground states, which would be expected to interconvert more readily, since the barrier to interconversion has been reduced. Unfavorable steric interactions of the propyl groups in the minor groove may also shift some backbone dihedrals in the SS.

Table II.   Pseudorotation Parameters for 2'-O-Propyl Nucleosides and Oligomers

Compound	Temperature (° C)	%N	P_n (°)	P_s (°)
2'-O-PrU[a]	20	22	4	156
2'-O-PrC	20	23	14	160
2'-O-PrG[a]	30	24	-3	156
2'-O-PrG	70	26	-17	160
pCpGpApU[b]	30			
C		50	41	160
A		30	25	147
G		42	41	176
U		45	34	147

[a] A 1.0 mg sample was dissolved in 0.6 mL D_2O and coupling constants measured from the 400 MHz proton NMR spectrum. [b] A 100 OD sample of a pCpGpApUdGdTdGdC 8-mer was synthesized using conventional solid phase chemistry. Coupling constants were measured from the 400 MHz proton NMR spectrum using conventional 2D TOCSY methods.

The 2'-O-propyl nucleotide is stabilizing by +1.2° C (P=O) and +0.9° C (P-S)/modification compared to DNA in three fully modified sequences against an RNA target (*34*). The propyl is destabilizing by -0.9° C/modification compared to the 2'-F and by -0.4° C/modification relative to the 2'-O-Me modification. As a point modification, the propylA is a neutral substitution in a DNA background (-0.1° C/mod.; 6 sequences) (*32*). The 2'-O-propyl nucleotides are stabilizing in a duplex with RNA, possibly because the d dihedral in the sugar is the same in C1'-exo pucker as in the C3'-endo pucker for RNA. This allows the 2'-O-propyl nucleotide to maintain A-form stacking while the equilibrium sugar pucker is shifted toward a P_s value. This idea is supported by CD data from 2'-O-propyl duplexes with RNA, which generate an A-form pattern (*32*).

2'-O-Butylimidazoyl Adenosine. A much larger 2'-O-alkyl substituent, 2'-O-butylimidazoylA (Figure 5) has been synthesized and incorporated into an SS oligomer (*35*). As determined from proton coupling constants, the 2'-O-butylimidazoylA nucleoside at 20° C adopts a conformation with 20% 'N', P_n=31°, P_s=165°. The modification has been incorporated into a dCdGdCA*dCdGdC 7-mer and at 37° C, the A* adopts a conformation with 20% 'N', P_n=6° and P_s= 140°. These pseudorotation values for A* in the 7-mer are comparable to those observed by Altona for the adenosine unit in a dA*r*A dA trimer having a conformation with 23% 'N', $P_n = -1°$, $P_s = 140°$ (*36*).

These studies suggest that all 2'-*O*-alkyl groups larger than Me have roughly the same effect on the pseudorotation of the ribofuranose ring. In the nucleoside, the unfavorable steric interactions among the OCH_2 protons, the heterocycle, H1', and O3' can be minimized in a C2'-endo conformation. However, in an oligomer with 2'-*O*-alkyls the value of P_s is shifted by 20-30° toward a C1'-exo pucker. This conformation is required to avoid unfavorable contacts with the heterocycle on the 3'-nucleotide, and places additional constraints on the orientation of the alkyl chain. The C1'-exo form must be associated with a higher energy in the single strand, since the P_n/P_s ratio is near equilibrium. This increase in the energy of the ground state would facilitate the pseudorotation process and increase the entropy introduced by the 2'-*O*-alkyl modification.

2'-*S*-Me Nucleosides. A complete series of 2'-*S*-Me nucleosides (Figure 6a) has been prepared by Fraser et al. (*37*). Coupling constants have been extracted from proton NMR data and fit to a modified Karplus equation using a sugar pucker amplitude of 38.5°. The conformational properties of 2'-*S*-Me G, C, A, and U are presented in Table III. The 2'-*S*-Me sugar modification produces a shift to the C2'-endo conformation ranging from 80% 'S' with $P_s = 156°$ for 2'-*S*-Me C to 100% 'S' with $P_s = 174°$ for 2'-*S*-Me A. This shift toward an 'S' pucker may result from a combination of reduced electronegativity at C2', which stabilizes the C2'-endo geometry and an unfavorable steric interaction between the sulfur atom, the heterocycle, and H1' in the C3'-endo conformation. The increased %S pucker also may reflect an increase in the barrier to interconversion through the O4-exo transition state caused by a bad steric interaction between the sulfur atom and the 3'-oxygen atom. The lack of a large shift in the %S as the temperature is reduced from 20° C to -40° C supports this hypothesis.

The 2'-*S*-Me U and C have been incorporated into sequences as 3-10 point modifications in 15-20-mer sequences of DNA and 2'-*O*-Me RNA. In a DNA background, the 2'-*S*-Me modifications destabilize a DNA: RNA duplex by -1.5° C/modification (5 sequences, 26 substitutions). When incorporated into a 2'-*O*-Me RNA background, the 2'-*S*-Me U and C destabilize a 2'-*O*-MeRNA: RNA duplex by -1.9° C/modification (3 sequences, 25 substitutions). Given the propensity of the 2'-*S*-Me nucleosides for a C2'-endo pucker, their destabilization of both DNA and 2'-*O*-Me RNA oligomers for their RNA complement is not surprising.

Table III. Pseudorotation Parameters for 2'-*S*-Me Nucleosides as a Function of Temperature

Compound	Temperature (° C)	%N	P_n	P_s
2'-*S*-Me A[a]	20	1.0±0.25	25±1	174±1
	-40	0.0	24	174
2'-*S*-Me G	20	10.75	28	163
	-20	4.5	7	174
2'-*S*-Me C	20	24.75	25	154
	-40	21.0	20	163
2'-*S*-Me U	20	17.75	-5	156
	-40	15.75	7	163

[a] All studies have been performed in methanol. Proton coupling constants have been measured directly from spectra obtained at 400 MHz.

2′-NH2. The nucleoside (Figure 6b) has been proposed to exist predominantly in the 2′-endo conformation. Eckstein and co-workers have shown that the incorporation of 2′-NH₂-Cytidine nucleotides into a DNA/DNA, RNA/DNA, or RNA/RNA strand is destabilizing by ≥4° C (*38*). They propose that the steric bulk of the NH₂ group prevents adoption of a 2′-endo conformation in the duplex. A pK of 6.2 has been measured for the amino group in a dimer.

4′-Oxofuran. Teng et al. have reported the synthesis of a novel furan derivative (Figure 7a), where the C5′ carbon atom has been interchanged with an oxygen (*39*). This modification is expected to have a dramatic effect on the pseudorotation of the sugar. The C2′-endo conformation of deoxyribonucleosides is stabilized by the gauche effect of the 3′-*O* with ring oxygen (*40*). This effect is offset by the anomeric effect from the ring N of the base and O4′, and the deoxyribose ring interconverts between C2′-endo and C3′-endo conformers through a low-energy barrier. Introduction of a counterbalancing anomeric effect at C4′ would be expected to "rigidify" the sugar by decreasing the free energy of the 'S' conformation of the deoxyribose ring and by increasing the energetic barrier for interconversion to the 'N' conformer. The free energy of the 'S' conformation may be increased by a bad "trans" interaction of the new O4′ with O3′, but this may not be so critical. A similar "bad" gauche interaction exists between the 2′-F and the ring N of the base.

The conformation of the 4′-oxofuran has been studied via molecular dynamics performed in a solvent box at 300° C using the AMBER force field. A rapid conversion of classical 'N' and 'S' starting conformations to puckers in the 'S' hemisphere is observed. Variations in pseudorotation are observed, but these alterations reflect changes in the internal v0 and v1 dihedrals, while the v3 (d) dihedral, which orients the phosphate backbone, is less flexible.

The proton NMR spectra of a 4′-oxofuran monomer and a T*T dimer containing a 4′-oxofuran substitution have been analyzed. Very small H3′H4′ and large H1′H2′,2″ HH coupling constants are observed over a range of temperatures. The pseudorotation parameters have been fit in the monomer and dimer using a parameterized Karplus equation with value of 34° for the amplitude of ring pucker. Larger values of the pucker amplitude produced poor fits to the NMR data. The nucleoside exhibits a strong preference for an 'S' pucker with 11% N, P_s=203.5°, and P_n=15.5°. This corresponds to a predominate C3′-exo state, as might be expected when the anomeric effect at C4′ forces the new O4≤ to adopt an axial orientation.

3′-CH2 Substituted Sugars. This modification would be expected to shift the pseudorotation of the sugar toward the 'N' quadrant, since removal of the 3′-*O* will remove the gauche effect with O4′, which provides the driving force for the C2′-endo conformation. Molecular dynamics on 3′-CH₃-dT (Figure 8a) in a water box suggest that this hypothesis is correct, with a 30-50% increase in the %N conformation observed over a 100 psec time period (V. Mohan, unpublished observations).

The pseudorotation parameters for a T*T dimer, where T* is a 3′-methylenedimethylhydrazino linkage (Figure 7b) have been determined at 60° C and are listed in Table IV (*41*). The study has been performed at elevated temperature to provide adequate resolution of the overlapping signals from the 5′-CH₂, 3′CH₂, and 2′,2≤ protons. The calculated pseudorotation properties of the 3′-CH₂ nucleoside are shifted strongly toward the 'N' conformation compared to a standard deoxynucleoside.

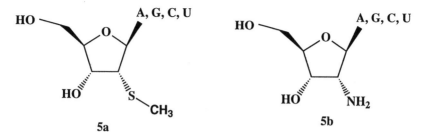

Figure 5. Structure of 2'-*O*-Butylimidaozyladenosine.

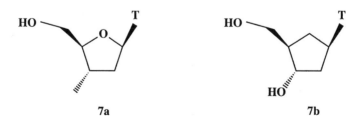

Figure 6. Structure of 2'-substituted nucleosides. A) 2'-*S*-Methyl. B) 2'-amino.

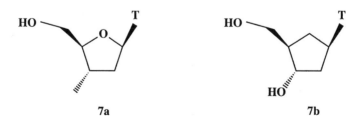

Figure 7. Structures of nucleoside dimers with modified backbones. A) 4'-Oxo-furan thymidine dimer. B) 3'-methylenedimethylhydrazino thymidine dimer.

Table IV. Pseudorotation Properties for Methylenedimethylhydrazino
T*T in Water

Compound[a]	Temperature (° C)	%N	P_n	P_s
3´-CH₂	60	86	6	178
3´-OH	60	24	15	164

[a] All studies have been performed in deuterium oxide. Proton coupling constants
have been measured directly from spectra obtained at 400 MHz.

Carbocyclic Sugars. Substitution at O4´ with carbon (Figure 8b) would be
expected to reduce the conformational preference of the ribose ring and possibly lead
to a decrease in T_m against RNA complement. Sagi et al. have prepared c-dT12 and c-
dT20 (42). They have solved the crystal structure for the cT monomer and shown the
carbocyclic sugar adopts a C1´-exo conformation with P=119° (43). The CD spectrum
of a cT₂₀/dA₂₀ duplex is nearly identical to a dT₂₀/dA₂₀ duplex. In their hands, the
carbocycle provides a +0.3° C increase in T_m against DNA complement. Perbost et al.
have synthesized T4 and T12 oligomers containing carbocyclic oligodeoxynucleotides
(44). The c-dT12 oligomer has an increased affinity of +0.8° C/modification for a
DNA complement. The slight improvement in T_m for the carbocyclic oligomers is
difficult to rationalize if the C1´-exo conformation is the predominant form in solution,
since the orientation of the base pairs relative to the backbone would be altered from
the canonical B-form.

Conclusions

Incorporation of modifications at the C2´, C3´, and C4´ positions can alter
dramatically the conformational properties of the furanose ring. Introduction of
electronegative substituents at C2´ produces a large gauche effect, which stabilizes the
ground state conformation and increases the barrier to interconversion among the 'N'
and 'S' conformers. This reduction in 'unproductive' motion may explain the
favorable binding properties of 2´-F oligos for RNA complement. Introduction of less
electronegative substituents, NH₂ or S-Me, enhances the gauche effect from the 3´-O
to the ring oxygen and increases the percentage of the 'S' conformer. The decrease in
affinity for RNA probably results from a diminished enthalpy, which offsets the
decreased entropy of the system. For 2´-O-alkyl substitutions, the Me group shifts the
equilibrium toward the 'N' conformer and may increase the barrier to pseudorotation.
Larger alkyl groups may introduce unfavorable steric interactions with the base and
sugar, which offset the favorable loss in the entropy of the system. In all cases, the
higher %N conformer correlates with an increase in binding affinity for RNA
complement. No 2´-sugar modification can produce a larger increase in the T_m than
the 2´-F nucleotide, since no other substituent will have as large a gauche effect or as
strong an orienting drive. Other heteroatomic substitutions, such as CH₂ for O4´ or
2´-O-alkyl chains, may alter the hydrophobic character of the oligomer in a manner that
improves affinity for a target molecule. These changes in hydrodynamic character may
offset unfavorable conformational properties. An increase in the hydrophobicity of the
SS relative to the duplex has been proposed to account for the improved affinity of 5-
MeC and 5-MeU relative to the standard nucleotide and probably explains part of the
effect of C5-propyne substitutions in pyrimidines (45, 46). The same effect may
account for the dramatic increase in affinity of a PNA for an RNA target (47). If so,
identification and placement of hydrophobic groups into the major and minor grooves
may provide additional improvements in affinity of modified oligomers for RNA
compliment.

References

1. Crooke, S. T. *Ann. Rev. Pharmacol. Toxicol.* **1992**, *32*, 329-376.
2. *Oligodeoxynucleotides as Antisense Inhibitors of Gene Expression*; Cohen, J. S., Ed.; CRC Press: Boca Raton, FL, 1987.
3. *Antisense Research and Applications*; Crooke, S. T.; Lebleu, B., Eds.; CRC Press: Boca Raton, FL, 1993.
4. Uhlman, E.; Peyman, A. *Chem. Rev.* **1990**, *90*, 543-583.
5. Milligan, J. F.; Matteucci, M. D.; Martin, J. C. *J. Med. Chem.* **1993**, *36*, 1923-1937.
6. Saenger, W. *Principles of Nucleic Acid Structure*; Springer-Verlag: New York, 1988.
7. Searle, M. S.; Williams, D. H. *Nucl. Acids Res.* **1993**, *21*, 2051-2056.
8. Schneider, K. C.; Benner, S. A. *J. Am. Chem. Soc.* **1990**, *112*, 453-455.
9. Egli, M.; Lubini, P.; Bolli, M.; Dobler, M.; Leuman, C. *J. Am. Chem. Soc.* **1993**, *115*, 5855-5856.
10. Jones, R. J.; Swaminathan, S.; Milligan, J. F.; Wadwani, S.; Forehler, J. F.; Matteucci, M. D. *J. Am. Chem. Soc.* **1993**, *115*, 9816-9817.
11. Sanghvi, Y. S.; Cook, P. D. In *Nucleosides and Nucleotides as Antitumor and Antiviral Agents*; Chu, C. K.; Baker, D.; Eds.; Plenum Press, 1993; pp. 311-324.
12. Kiely, J. S. *Annu. Rep. Med. Chem.* **1994**, *29*, in press.
13. Cook, P. D. *Anticancer Drug Des.* **1991**, *6*, 585-607.
14. Vasseur, J. J.; Debart, F.; Sanghvi, Y. S.; Cook, P. D. *J. Am. Chem. Soc.* **1992**, *114*, 4006-4007.
15. DeMesmaeker, A.; Waldner, A.; Lebereton, J.;Hoffman, P.; Fritsch, V.; Wolf, R.M.; Freir, S.M.; *Ang. Chem. Int. Ed.* **1994**, *33*, 226-229.
16. Altona, C.; Sundaralingham, M. *J. Am. Chem. Soc.* **1973**, *95*, 2332-2344.
17. McCammon, J. A.; Harvey, S. C. *Dynamics of Proteins and Nucleic Acids*; Cambridge University Press: Cambridge, 1987.
18. Ellervik, U.; Magnusson, G. *J. Am. Chem. Soc.* **1994**, *116*, 2340-2347.
19. Egholm, M.; Buchardt, O.; Christensen, L.; Behrens, C.; Freier, S.M.; Driver, D.A.; Berg, R.H.; Kim, S.K.; Norden, B.; Nielsen, P.E.; *Nature* **1993**, *365*, 566-568.
20. deLeeuw, F. A. A.; Altona, C. *J. Comp. Chem.* **1983**, *4*, 428-437.
21. Haasnoot, C. A. G.; deLeeuw, F. A. A.; Altona, C. *Tetrahedron* **1980**, *36*, 2783-2792.
22. Marck, C.; Lesyng, B.; Saenger, W. *J. Mol. Str.* **1982**, *82*, 77-94.
23. Olson, W. K. *J. Am Chem. Soc.* **1982**, *104*, 278-286.
24. Uesugi, S.; Kaneyasu, T.; Imura, J.; Ikehara, M.; Iwahashi, H.; Kyogoku, Y. *Tet. Letters* **1979**, *42*, 4073-4076.
25. Ikehara, M. *Heterocycles* **1984**, *21*, 75-90.
26. Lankin, D. C.; Chandrakuman, N. S.; Rao, S. N.; Spangler, D.P.; Snyder, J.P. *J. Am. Chem. Soc.* **1993**, *115*, 3356-3357.
27. Robbins, M.; Wilson, J. S.; Sawyer, L.; James, M. N. G. *Can J. Chem.* **1983**, *61*, 1911-1920.
28. Kawasaki, A. M.; Casper, M. D.; Frier, S. M.; Lesnik, E. A.; Zounes, M. C.; Cummins, L. L.; Gonzales, C.; Cook, P. D. *J. Med. Chem.* **1993**, *36*, 831-841.
29. Cummins, L., personal communication.
30. Kawai, G.; Yamamoto, Y.; Kamimura, T.; Masegi, T.; Sekine, M.; Hata, T.; Iimori, T.; Watanabe, T.; Miyazawa, T.; Yokoyama, S.; *Biochemistry* **1992**, *31*, 1040-1046.

31. Kawai, G.; Hashizume, T.; Yasuda, M.; Miyazawa, T.; McCloskey, J.A.; Yokoyama, S.; *Nucleosides Nucleotides* **1992**, *11*, 759-771.
32. Lesnik, E. A.; Guinosso, C. J.; Kawasaki, A. M.; Sasmor, H.; Zounes, M.; Cummins, L.; Ecker, D.; Cook, P.D.; Freier, S.M. *Biochemistry* **1993**, *32*, 7832-7838.
33. Morvan, F.; Porumb, H.; Degols, G.; Lefebre, I.; Pompon A.; Sproat, B.S.; Malvy, C.; Lebleu, B.; Imbach, J-M *J. Med. Chem.* **1993**, *36*, 280-287.
34. Cummins, L.; Owens, S.R.; Risen, L.M.; McGee, D.; Guisnosso, C.; Zounes, M.; Greig, M.; Lesnik, E.; Freir, S.M.; Sasmor, H.; Griffey, R.; Cook, P.D.; *Nucl. Acids Res.* **1994**, submitted for publication.
35. Guinosso, C., unpublished observation.
36. Altona, C. *Recl. Trav. Chim. Pay-Bas.* **1982**, *101*, 413-433.
37. Fraser, A.; Wheeler, P.; Cook, P. D.; Sanghvi, Y.S. *J. Heter. Chem.* **1993**, *30*, 1277-1287.
38. Aurup, H.; Tuschl, T.; Benseler, F.; Ludwig, J.; Eckstein, F. *Nucl. Acids Res.* **1994**, *22*, 20-24.
39. Teng, K.; Cook, P. D. *J. Org. Chem.* **1994**, *59*, 278-280.
40. Plavec, J.; Tong, W.; Chattopadhyaya, J. *J. Am. Chem. Soc.* **1993**, *115*, 9734-9746.
41. Sanghvi, Y. S.; Vasseur, J-J.; Debart, F.; Cook, P. D. *Coll. Czech. Chem. Comm.* **1993**, *58*, 158-162.
42. Sagi, J.; Szemzo, A.; Szecsi, J.; Otvos, L. *Nucl. Acids Res.* **1990**, *18*, 2133-2140.
43. Kalman, A.; Koritsanszky, T.; Beres, J.; Sagi, G. *Nucleosides. and Nucleot.des* **1990**, *9*, 235-243.
44. Perbost, M.; Lucas, M.; Chavis, C.; Imbach, J.-L. *Biochem. Biophys. Res. Comm.* **1989**, *165*, 742-747.
45. Plaxco, K. W.; Goddard, W. A. *Biochemistry* **1994**, *33*, 3050-3054.
46. Froehler, B.C.; Jones, R.J.; Cao, X.; Terhorst, T.J. *Tet. Letters* **1993**, *34*, 1003-1006.
47. Mohan, V.; Griffey, R.H. *Bioorg. Med. Chem. Lett.* **1994**, in press.
48. Weiner, S.J.; Kollman, P.A.; Case, D.A.; Singh, U.C.; Ghio, C.; Alagona, G.; Profeta Jr., S.; Weiner, P. *J. Am. Chem. Soc.* **1984**, *106*, 765-784.

RECEIVED August 23, 1994

Author Index

Augustyns, Koen, 80
Barascut, Jean-Louis, 68
Bellon, Laurent, 68
Bolli, M., 100
Cook, P. Dan, 1,212
Damha, M. J., 133
De Winter, Hans, 80
Doboszewski, Bogdan, 80
Dong, Beihua, 118
Freier, Susan, 212
Fritsch, Valérie, 24
Fry, Stephen, 184
Fulcrand-El Kattan, Géraldine, 169
Gall, Alexander A., 199
Ganeshan, K., 133
Gorn, Vladimir V., 199
Griffey, Richard H., 212
Guinosso, Charles, 212
Hatta, T., 154
Hendrix, Chris, 80
Herdewijn, Piet, 80
Hudson, R. H. E., 133
Imbach, Jean-Louis, 68
Just, George, 52
Kawai, Stephen H., 52
Kawasaki, Andrew, 212
Khamnei, Shahrzad, 118
Kim, S.-G., 154
Lebreton, Jacques, 24
Lesiak, Krystyna, 118
Lesnik, Elena, 212
Lesnikowski, Zbigniew J., 169
Leumann, C., 100

Leydier, Claudine, 68
Li, Guiying, 118
Lubini, P., 100
Maddry, Joseph A., 40
Maitra, Ratan, 118
Maran, Avudaiappan, 118
Maury, Georges, 68
Mesmaeker, Alain De, 24
Meyer, Rich B., Jr., 199
Mohan, Ventrankaman, 212
Montgomery, John A., 40
Ranter, Camiel De, 80
Reynolds, Robert C., 40
Saison-Behmoaras, Tula, 80
Sanghvi, Yogesh S., 1,212
Schinazi, Raymond F., 169
Secrist, John A., III, 40
Silverman, Robert H., 118
Sinha, Nanda D., 184
Suzuki, S., 154
Takaki, K., 154
Takaku, H., 154
Tarköy, M., 100
Teng, Kelly, 212
Torrence, Paul F., 118
Van Aerschot, Arthur, 80
Verheggen, Ilse, 80
Waldner, Adrian, 24
Wheeler, Patrick, 212
Wilson, David W., 169
Wolf, Romain M., 24
Xiao, Wei, 118

Affiliation Index

Atlanta Veterans Affairs Medical
    Center, 169
Chiba Institute of Technology, 154
Ciba-Geigy Ltd., Switzerland, 24

Cleveland Clinic Foundation, 118
Emory University School of Medicine, 169
Federal Institute of Technology,
    Switzerland, 100

Georgia State University, 169
Isis Pharmaceuticals, 1,212
Katholieke Universiteit Leuven, 80
McGill University, 52
MicroProbe Corporation, 199
Millipore Corporation, 184
Muséum National D'Histoire Naturelle, 80

National Institutes of Health, 118
Rega Institute, 80
Southern Research Institute, 40
Université Sciences et Techniques
    du Languedoc, 68
University of Bern, 100
University of Toronto, 133

Subject Index

A

2'-Amino nucleosides
  properties, 220
  structure, 220,221f
Avian myeloblastosis virus (AMV)
    transcriptase, complementary DNA
    synthesis, 155–159
Antihuman immunodeficiency virus
    activity, thiopurine-based
    oligonucleotides, 199–209
Antisense concept of drug discovery, 1
Antisense oligodeoxynucleotides
  function, 24
  need for chemical modifications, 24
  novel backbone replacements, 24–37
Antisense oligodeoxyribonucleotides
  DNA sequences of reverse transcription,
    155,156f
  function, 154
  inhibition of reverse transcription
    antiviral effect, 161–167
    complementary DNA synthesis
      by AMV reverse transcriptase, 155–159
      by HIV reverse transcriptase,
        157,159–160
    previous studies, 154–155
    procedure, 155
Antisense oligonucleotides
  analogues, 40
  potential problems, 1–2
  use in treatment of disease, 212
Antisense oligonucleotide therapeutics
  examples, 119

Antisense oligonucleotide therapeutics—
    Continued
  potential applications, 118
  use as gene knock-out reagents, 118–119
Antisense strategy
  development, 52
  branched nucleic acids, 143
Antisense technology, carboranyl
    oligonucleotides, 169–180
Antisense therapeutics
  applications of carbohydrates, 1–17
  criteria, 212
Antiviral action of interferon, role in
    2',5'-oligoadenylate system and
    2',5'-oligoadenylate-dependent
    endonuclease, 123–124
Antiviral drug, properties, 199
Antiviral effect of oligonucleotide
    phosphorothioates complementary to
    HIV
  nonsequence-specific inhibition,
    163,166–167
  procedure, 161
  sequence-specific inhibition, 161–165
Applications
  branched nucleic acids, 140–149
  carbohydrates, 1–17
Automated chemical synthesis of
    oligoribonucleotides
  desilylation, 189,191f
  evaluation of synthetic RNA, 194–197
  excess tetrabutylammonium fluoride,
    189,192
  exocyclic amine protection removal, 189

Automated chemical synthesis of
oligoribonucleotides—*Continued*
experimental description, 185
purification by high-performance LC,
192–194,195*f*
synthesis of monomeric building blocks,
185–187
synthetic procedure, 187–189

B

Backbone replacements for oligonucleotides, novel, *See* Novel backbone
replacements for oligonucleotides
Bicyclic sugar modifications,
carbohydrate-based modifications in
antisense research, 15,17*f*
α-Bicyclo-DNA
bicyclo-sugar unit synthesis, 101–103
complexes of oligo-α-bicyclodeoxy-
nucleotides
with complementary RNA, 108–112
with complementary DNA,
110,113*f*,114*t*,115
duplex formation between oligo-α- and
oligo-β-bicyclonucleotides, 115–116
stability of oligo-α-bicyclonucleotide
sequences toward phosphodiesterase
action, 116
synthesis
bicyclonucleotides, 103–104
building blocks for oligonucleotide
synthesis, 106–107
oligo-α-bicyclodeoxynucleotides, 108
X-ray analysis of bicyclonucleotides,
105–106, 109*f*
Bicyclonucleosides
structure, 101,102*f*
synthesis of bicyclonucleotides,
105–106,109*f*
Bicyclonucleotides, X-ray analysis,
105–106,109*f*
Biological applications, branched nucleic
acids, 140–149

Boron-10, low energy neutron
absorption, 169
Boron-containing compounds, use in
malignancies, 169
Boron neutron capture therapy
description, 169
molecular basis, 170–171
use of carboranyl oligonucleotides,
169–180
Boronated compounds, rationalization of
synthesis, 171–172
Branched and dendriatic oligo-RNA and
oligo-DNA, carbohydrate-based
modifications in antisense research, 14
Branched nucleic acids
antisense strategy development, 143
applications, 140–149
branched poly(dT) prototype system,
143–146
chemical synthesis, 137–140
conformation, 140–141
discovery, 136
importance, 136–137
second-generation acid, 148–150
significance of branched poly(dT)
hybridization patterns, 146–148
small nuclear RNA-branched–RNA
interaction probing, 141–142
Branched oligoribonucleotides, solid-phase
synthesis, 137–139
Building blocks for oligonucleotide
synthesis, 106–107
2′-*O*-Butylimidazoyl adenosine,
218–219,221*f*

C

C-1′ to C-5′ modifications, carbohydrate-
based modifications in antisense
research, 8–11,12*f*,13,16*f*
C-3′-substituted upper monomeric unit
synthesis from xylose
acetolysis of branched-chain sugars,
54–56

C-3'-substituted upper monomeric unit synthesis from xylose—*Continued*
construction of branched-chain sugars, 53–54
from nucleoside, 57–58
nucleoside formation and activation, 56–58
total synthesis, 58–59
Carbocyclic sugars, structural properties, 222
Carbohydrates, synthesis, 1–17
Carbohydrate-based modifications in antisense research
bicyclic sugar modifications, 15,17*f*
branched and dendriatic oligo-RNA and oligo-DNA, 14
C-1' to C-5' modifications, 8–11,12*f*,13,16*f*
dephosphono linkages, 4–8
future applications, 17
hexose modifications, 14
L and L/D modifications, 14–15,16*f*
O-4' sugar modifications, 15,16*f*
Carboranyl oligonucleotides for boron neutron capture therapy of cancers
advantages, 169
biological properties, 177–178
carboranyl cage structure, 174–175,176*t*
5-*O*-carboranyl-2'-deoxyuridine-containing oligonucleotides, 173–174
di(thymidine *O*-carboranylmethyl-phosphonate), 172–173
molecular basis, 170–171
physicochemical properties, 175–177
rationalization of synthesis, 171–172
stereochemistry vs. properties, 179–180
synthesis, 172–174
5-*O*-Carboranyl-2'-deoxyuridine-containing oligonucleotides, properties, 175–179
Carboxyl-based oligonucleotide analogues, synthesis, 44–47
Characterization, α-bicyclo-DNA, 110–116
Chemical syntheses of DNA and RNA, 184–185

Complementary DNA
complexes with oligo-α-bicyclodeoxy-nucleotides, 110,113*f*,114*t*,115
synthesis
by AMV reverse transcriptase, 155–159
by HIV reverse transcriptase, 157,159–160
Complementary RNA, complexes with oligo-α-bicyclodeoxynucleotides, 108,109*f*,110–112*f*
Covalently reactive antisense reagents, targeted ablation of RNA, 119–120

D

Dendriatic and branched oligo-RNA and oligo-DNA, *See* Branched and dendriatic oligo-RNA and oligo-DNA
5'-Deoxy-5'-thionucleoside lower monomeric units, synthesis, 59
Dephosphono linkages, carbohydrate-based modifications in antisense research, 4–8
Destruction of specific messenger RNAs, catalytic approaches, 120–121
2,3-, 3,4-, and 2,4-Dideoxy-β-D-*erythro*-hexopyranosyl nucleoside building blocks, synthesis and properties, 81–87
Dimer synthesis
amide 1, 25–26
amide 2, 26
amide 3, 28
amide 4, 28
amide 5, 28
carbamate 7, 28
carbon chain internucleoside 9, 33
urea, 29
5'-*O*-Dimethoxytrityl-1,2'-*N*,*O*-dimethyl-6-thioinosine, synthesis, 201–202
5'-*O*-Dimethoxytrityl-1,2'-*N*,*O*-dimethyl-6-thioinosine–3'-*O*-(2-cyanoethyl-*N*,*N*-diisopropylphosphoramidate), synthesis, 202

1,2′-*N,O*-Dimethyl-3′,5′-di(*p*-toluoyl)-6-
thioinosine, synthesis, 201
1,2′-*N,O*-Dimethyl-6-thioinosine,
synthesis, 200*f*,201
Di(thymidine *O*-carboranylmethyl-
phosphonate), properties, 175–180
DNA
analogues, synthesis, 100
complementary, *See* Complementary
DNA
α-DNA, base pairing, 101
oligomers containing sulfide-linked
dinucleosides
3′-C-substituted upper monomer unit
synthesis from xylose, 53–59
5′-deoxy-5′-thionucleoside lower
monomeric unit synthesis, 59
hybridization properties, 60–62
reasons for study, 52–53
sulfone synthesis, 62–63
synthesis, 60–62

E

2′-*O*-Ethyl ribosides, 216,217*f*

F

2′-Fluorodeoxynucleosides, 215,217*f*

G

Gene knock-out reagents, use of antisense
oligonucleotide therapeutics, 118–119

H

Hexopyranosyl-like oligonucleotides
importance of pairing selectivity, 80
modeling of pyranosyl-like
oligonucleotides, 87–93

Hexopyranosyl-like oligonucleotides—
Continued
physicochemical consideration for
synthesis, 80
properties of 1,3,4-substituted
pyranosyl-like oligonucleotides, 92–98
reasons for interest, 80–81
synthesis
general, 81–87
1,3,4-substituted pyranosyl-like
oligonucleotides, 92–98
Hexose modifications, carbohydrate-
based modifications in antisense
research, 14
Human immunodeficiency virus (HIV)
antiviral effect of complementary
oligonucleotide phosphorothioates,
161–167
replication inhibition, thiopurine-
based oligonucleotides, 206–208
reverse transcriptase
complementary DNA synthesis, 155–159
interaction with 4′-thio-RNA, 76–77
Hybridization properties, DNA oligomers
containing sulfide-linked
dinucleosides, 60–62

L

L and L/D modifications,
carbohydrate-based modifications in
antisense research, 14–15,16*f*

M

Messenger RNA, specific, destruction
using catalytic approaches, 120–121
2′-*S*-Methyl nucleosides, 219–220,221*f*
2′-*O*-Methyl ribosides, 216,217*f*
1-*N*-Methyladenosine hydroiodide, 200
3′-Methylene-substituted sugars,
structural properties, 220,222
1-Methyl-6-thioinosine, 200–201

Modified nucleosides, incorporation into oligonucleotides, 212–222

Modified oligonucleotides, use in antisense drug discovery approach, 2

N

Nonionic oligonucleotide analogue synthesis
carboxyl-based analogues, 44–47
silicon-based analogues, 41–44
sulfonyl-based analogues, 46–50

Novel backbone replacements for oligonucleotides
dimer synthesis
amide 1, 25–26
amide 2, 26
amide 3, 28
amide 4, 28
amide 5, 28
carbamate 7, 28
carbon chain internucleoside 9, 33
urea, 29
structures, 24,25f
thermal denaturation of duplexes formed, 33–37

Nuclear intron lariat species, description, 136

Nuclease resistance, 4'-thio-β-D-oligoribonucleotide, 72–74

Nucleic acids, structural properties of modified nucleosides incorporated into oligonucleotides, 212–222

Nucleic acid analogues, development for use as specific inhibitors of gene expression, 52

Nucleosides, modified, incorporation into oligonucleotides, 212–222

O

O-4' sugar modifications, carbohydrate-based modifications in antisense research, 15,16f

2',5'-Oligoadenylate antisense chimeras for targeted ablation of RNA components, 121,122f
2',5'-phosphodiesterase activity, 123
PKR, 126–128
postulated mechanism, 124–125
role in antiviral action of interferon, 123–124
structure, 121
synthetase reactions, 121–123

2',5'-Oligoadenylate-dependent RNase
purification, 123
RNA substrate preference, 123

2',5'-Oligoadenylate synthetases, reactions, 121–123

Oligo-α-bicyclodeoxynucleotide complexes
with complementary DNA, 110,113f, 114t,115
with complementary RNA, 108–112

Oligo-α-, oligo-β-bicyclonucleotides, duplex formation, 115–116

Oligo-α-bicyclonucleotide sequences, stability toward phosphodiesterase action, 116

Oligodeoxynucleotides, antisense, *See* Antisense oligodeoxynucleotides

Oligonucleotides
conformational properties, 212–214
hexapyranosyl like, *See* Hexopyranosyl-like oligonucleotides
modifiable sites, 2
modifications, 2–17
structural properties of incorporated modified nucleosides, 212–222
types of modifications, 2,3f

Oligonucleotide analogue synthesis, nonionic, *See* Nonionic oligonucleotide analogue synthesis

Oligonucleotide-based agents, antiviral activity without sequence specificity, 199

Oligonucleotide phosphorothioates complementary to HIV, antiviral effect, 161–167

Oligoribonucleotides
automated chemical synthesis, 184–197
branched, solid-phase synthesis, 137–139
Oligo-RNA and oligo-DNA, carbohydrate-based modifications, 14
4′-Oxofuran, 220,221f

P

Pairing properties, α-bicyclo-DNA, 110–116
Passive hybridization, description, 119
2′,5′-Phosphodiesterase activity, 2′,5′-oligoadenylate antisense chimeras for targeted ablation of RNA, 123
Phosphorothioate oligonucleotides
antisense drug candidates, 2
antiviral activity without sequence specificity, 199
antiviral effect, 161–167
PKR
function, 126
2′,5′-oligoadenylate antisense chimeras for targeted ablation of RNA, 126–128
2′-Propyl ribosides, 216,217f,218
Pyranosyl-like oligonucleotides
modeling, 87–93
1,3,4 substituted, See 1,3,4-Substituted pyranosyl-like oligonucleotides

R

Regioisomeric oligoribonucleotides, solid-phase synthesis, 139–140
RNA
complementary, See Complementary RNA
occurrence of branches, 133–134
pharmacological receptor, 1
splicing, 134–135
sugar modified β-RNA, 68–77

RNA—Continued
synthesis, quantitative internucleotide bond formation, 185
4′-thio, See 4′-Thio-RNA

S

Silicon-based oligonucleotide analogues, synthesis, 41–44
Small nuclear RNA–branched RNA, interactions, 141–142
Solid-phase synthesis
branched oligoribonucleotides, 137–139
regioisomeric oligoribonucleotides, 139–140
Specific messenger RNA, destruction using catalytic approaches, 120–121
Stability, oligo-α-bicyclonucleotide sequences toward phosphodiesterase action, 116
Structural properties of modified nucleosides incorporated into oligonucleotides
2′-amino nucleosides, 220,221f
2′-O-butylimidazoyladenosine, 218–219
carbocyclic sugars, 222
2′-O-ethyl ribosides, 216,217f
2′-fluorodeoxynucleosides, 215,217f
2′-S-methyl nucleosides, 219–220,221f
2′-O-methyl ribosides, 216,217f
3′-methylene-substituted sugars, 220,222
4′-oxofuran, 220,221f
2′-O-propyl ribosides, 216,218
1,3,4-Substituted pyranosyl-like oligonucleotides, 92–98
Sugars
carbocyclic, 222
3′-methylene substituted, 220,222
Sugar modifications
bicyclic, 15,17f
O-4′, 15,16f
Sugar-modified β-RNA, 4′-thio-RNAs, 68–77

Sulfone, synthesis, 62–63
Sulfonyl-based oligonucleotide analogues,
   synthesis, 46–50
Synthesis
α-bicyclo-DNA, 100–116
branched nucleic acids, 137–140
carbohydrates, 1–17
DNA oligomers containing sulfide-
   linked dinucleosides, 60–62
hexopyranosyl-like oligonucleotides,
   81–87
nonionic oligonucleotide analogues,
   40–50
solid-phase
   branched oligoribonucleotides, 137–139
   regioisomeric oligoribonucleotides,
     139–140
Synthetase reactions, 2′,5′-oligoadenylate
   antisense chimeras for targeted
   ablation of RNA, 121–123
Synthetic RNA
   evaluation, 194–197
   use for therapeutic purposes based on
   ribozyme technology, 185

T

Thermal denaturation, duplexes formed
   between modified oligodeoxynucleotides
   and complementary RNA strands, 33–37
4′-Thio-β-D-oligoribonucleotide
   chemistry of synthesis, 69–72
   hydrogen-bonding properties, 74–76

4′-Thio-β-D-oligoribonucleotide—
   Continued
   interaction with HIV–1 reverse
     transcriptase, 76–77
   nuclease resistance, 72–74
   reason for synthesis, 68
Thiopurine-based oligonucleotides
   action, 208–209
   backbone modification vs. HIV
     replication inhibition, 207–208
   chemistry, 205–206
   cytotoxicity, 208
   development, 200
   end-capping group vs. HIV replication
     inhibition, 208
   length vs. HIV replication inhibition,
     206–207
   mode of action, 208–209
   procedure, 204–205
   synthesis, 206
   synthetic procedure, 203,204*f*
4′-Thio-RNA
   chemistry of synthesis, 69–72
   hydrogen-bonding properties, 74–76
   interaction with HIV–1 reverse
     transcriptase, 76–77
   nuclease resistance, 72–74
   reason for synthesis, 68

X

Xylose, synthesis of 3′-C-substituted
   upper monomeric units, 53–59

Production: Susan Antigone
Indexing: Deborah H. Steiner
Acquisition: Barbara Pralle
Cover design: Peggy Corrigan

Printed and bound by Maple Press, York, PA